AUTOPHAGY IN HEALTH AND DISEASE

ELSEVIER

ELSEVIER *science & technology books*

Companion Web Site:

http://www.booksite.elsevier.com/9780123851017

Autophagy in Health and Disease
Roberta A. Gottlieb, Editor

Resources:

- All figures and tabels from the book are available
- Links to additional web resources.

ELSEVIER

ACADEMIC PRESS

AUTOPHAGY IN HEALTH AND DISEASE

Edited by

ROBERTA A. GOTTLIEB

ELSEVIER

AMSTERDAM • BOSTON • HEIDELBERG • LONDON
NEW YORK • OXFORD • PARIS • SAN DIEGO
SAN FRANCISCO • SINGAPORE • SYDNEY • TOKYO
Academic Press is an imprint of Elsevier

Academic Press is an imprint of Elsevier
32 Jamestown Road, London NW1 7BY, UK
225 Wyman Street, Waltham, MA 02451, USA
525 B Street, Suite 1800, San Diego, CA 92101-4495, USA

First edition 2013

Cover Image: Visualization of autophagosomes (green) and mitochondria (red) in an adult cardiac myocyte overexpressing the pro-apoptotic protein Bnip3. Image obtained by Youngil Lee and Åsa B. Gustafsson Ph.D. , 2008, SDSU BioScience Center.

British Library Cataloguing-in-Publication Data
A catalogue record for this book is available from the British Library

Library of Congress Cataloging-in-Publication Data
A catalog record for this book is available from the Library of Congress

ISBN: 978-0-12-385101-7

For information on all Academic Press publications
visit our website at www.store.elsevier.com

Typeset by TNQ Books and Journals

Printed and bound by CPI Group (UK) Ltd, Croydon, CR0 4YY

Transferred to digital print 2013

This book is dedicated to all of the special people who have encouraged my curiosity: my parents Robert and Naomi Gottlieb; my teachers and mentors, particularly Rex Robinson, Doug Murphy, Bill Lennarz and Genie Kleinerman; my co-inspirators over the years, Robert Engler, Bernie Babior, and Robert Mentzer; and my endlessly curious children, Arielle, Austin, and Sideara. Facts are merely the springboard for developing new ideas: jump in and swim!

Contents

Contributors

E. Dale Abel M.D. Ph.D. Division of Endocrinology Metabolism and Diabetes and Program in Molecular Medicine, University of Utah School of Medicine, Salt Lake City, UT, USA.

Leticia A.M. Carneiro Ph.D. Instituto de Microbiologia Professor Paulo de Góes and Instituto de Biofísica Carlos Chagas Filho, Universidade Federal do Rio de Janeiro, Rio de Janeiro, Brazil.

Francesco Cecconi Ph.D. Dulbecco Telethon Institute at the Department of Biology, University of Tor Vergata, Rome, Italy and Laboratory of Molecular Neuroembryology, IRCCS Fondazione Santa Lucia, 00143 Rome, Italy.

Yan Cheng Ph.D. Department of Pharmacology and The Penn State Cancer Institute, The Pennsylvania State University College of Medicine, and Milton S. Hershey Medical Center, Hershey, PA, USA.

Valentina Cianfanelli Ph.D. Dulbecco Telethon Institute at the Department of Biology, University of Tor Vergata, Rome, Italy and Laboratory of Molecular Neuroembryology, IRCCS Fondazione Santa Lucia, 00143 Rome, Italy.

Kim D. Finley Ph.D. Expression Drug Designs, LLC and Donald P. Shiley BioScience Center, San Diego State University, San Diego, CA, USA.

Roberta A. Gottlieb M.D. Director, Donald P. Shiley BioScience Center, San Diego State University, San Diego, CA, USA.

Timothy E. Graham M.D. Division of Endocrinology Metabolism and Diabetes and Program in Molecular Medicine, University of Utah School of Medicine, Salt Lake City, UT, USA.

Michael Gurney M.S. Donald P. Shiley BioScience Center, San Diego State University, San Diego, CA, USA.

Åsa B. Gustafsson Ph.D. Skaggs School of Pharmacy and Pharmaceutical Sciences, University of California, San Diego, CA, USA.

Deron Herr Ph.D. Expression Drug Designs, LLC, San Diego, CA, USA.

Guy Las Ph.D. Department of Medicine, Section of Molecular Medicine, Boston University School of Medicine, Boston, MA, USA.

Phyllis-Jean Linton Ph.D. Donald P. Shiley BioScience Center, San Diego State University, San Diego, CA, USA.

Bo Liu Ph.D. State Key Laboratory of Biotherapy and Cancer Center, West China Hospital, West China Medical School, Sichuan University, Chengdu 610064, China.

Robert M. Mentzer, Jr. M.D. Donald P. Shiley BioScience Center, San Diego State University, San Diego, CA., WSU Cardiovascular Research Institute, Departments of Physiology, Surgery Wayne State University School of Medicine, Detroit, MI, USA.

Girija Muralidhar Ph.D. Donald P. Shiley BioScience Center, San Diego State University, San Diego, CA, USA.

Kazuhiko Nishida Ph.D. Department of Cardiovascular Medicine, Osaka University Graduate School of Medicine, Suita, Osaka, Japan.

Kinya Otsu M.D. Ph.D. Department of Cardiovascular Medicine, Osaka University Graduate School of Medicine, Osaka, Japan.

Orian S. Shirihai M.D. Ph.D. Department of Medicine, Section of Molecular Medicine, Boston University School of Medicine, Boston, MA, USA.

Leonardo H. Travassos Ph.D. Instituto de Microbiologia Professor Paulo de Góes and Instituto de Biofísica Carlos Chagas Filho, Universidade Federal do Rio de Janeiro, Rio de Janeiro, Brazil.

Gilad Twig M.D. Ph.D. Department of Medicine, Section of Molecular Medicine, Boston University School of Medicine, Boston; Department of Medicine and Talpiot Medical Leadership Program, The Chaim Sheba Medical Center, Tel-Hashomer, Israel.

Stephanie Wohlgemuth Ph.D. Assistant Professor, Department of Animal Sciences, University of Florida, Gainesville, FL, USA.

Osamu Yamaguchi M.D. Ph.D. Department of Cardio-
vascular Medicine, Osaka University Graduate School of Medicine, Suita, Osaka, Japan.

Jin-Ming Yang M.D. Ph.D. Department of Pharmacology and The Penn State Cancer Institute, The Pennsylvania State University College of Medicine, and Milton S. Hershey Medical Center, Hershey, PA, USA.

Foreword

Lysosomes containing cytoplasmic contents were first described in electron microscopy studies by Thomas Ashford and Keith Porter in 1962. For many years, the morphologic phenomenon was noted, but the field was transformed by the work of Ohsumi's group and others, to clone the autophagy genes in yeast and later humans, in the 1990's. Since then, the role of autophagy has been elucidated in an increasing number of diseases. Given the tremendous advances made in recent years, it is appropriate to offer a compilation of our current understanding of the roles of autophagy in health and disease. We have focused on organ systems and diseases where important advances have been made. We have not covered the role of autophagy in the central nervous system, as this complex and rapidly-evolving field, encompassing stroke, seizures, and neurodegenerative diseases, deserves a book of its own.

The book is organized to provide a background on the process of autophagy in Section I. Section II addresses autophagy in tissues and Section III covers autophagy in disease processes. Section IV addresses the development of tools to measure and interventions to modulate autophagy. I would like to thank each of the chapter authors for their thorough coverage of each topic. Their work represents the state of the art in their respective fields and should serve as a catalyst for future investigations.

It is hoped that "Autophagy in Health and Disease" will provide a milestone reflecting how far we have come and forecasting the challenges ahead. Future writers will be able to discuss methods to image autophagy in vivo and the use of small molecule therapeutics to regulate autophagy with tissue-specific and temporal control for treatment of specific diseases. Its successes are a credit to the patience, enthusiasm, and steady encouragement of my editors Mary Preap and Janice Audet at Elsevier. Its shortcomings are mine alone. I hope you will find this book as exciting to read as I have in assembling the contributions of each chapter author.

Sincerely,

Roberta A. Gottlieb M.D.

SECTION I

OVERVIEW

CHAPTER

1

Overview: Selective Removal of Aggregates and Organelles

Roberta A. Gottlieb

Director, Donald P. Shiley BioScience Center, San Diego State University, 5500 Campanile Drive, San Diego, CA 92182-4650

AUTOPHAGY REGULATION AND MACHINERY

The next chapter will describe the machinery of autophagy in great detail; the goal here is to provide a framework. Autophagy is invoked as a cellular response to stress; stressors may include nutrient limitation (carbohydrates, amino acids, or growth factors), misfolded protein and endoplasmic reticulum (ER) stress, intracellular pathogens (bacteria, viruses, or parasites), reactive oxygen species, or organelle dysfunction. A kinase signaling network regulates autophagy initiation, but Ulk1/Atg1 is a key signaling kinase responsible for initiating autophagy. Phosphorylation of ULK1 on Ser 317 and Ser 777 by AMPK, the chief energy-sensing kinase, activates the kinase, whereas phosphorylation by mTOR on Ser 757 dissociates it from AMPK and suppresses autophagy[1]. Autophagic machinery is regulated by additional posttranslational modifications including (but not limited to) acetylation, ubiquitination, and oxidation of sulfhydryls.

Macroautophagy involves protein nucleation and formation of a cup-shaped lipid double membrane structure which elongates to engulf the cargo. This requires the participation of enzymes homologous to E2 and E3 ubiquitin ligases, which transfer Atg12 onto an acceptor lysine of Atg5, and phosphatidylethanolamine onto LC3 after Gly is exposed at the C-terminus by the cysteine protease and delipidating enzyme Atg4 (See Figure 1.1). Deletion of the critical factors Atg5, Atg7, and Atg3 result in inactivation of the canonical pathway of autophagy. Though mice are born and appear developmentally normal, they are unable to survive fasting and die in the newborn period, demonstrating a requirement for canonical autophagy in starvation[2]. LC3 lipidation with phosphatidylethanolamine attaches it to the autophagosomal membrane on both the concave and convex faces, with the result being that LC3 on the concave face is retained all the way through

Autophagy in Health and Disease
http://dx.doi.org/10.1016/B978-0-12-385101-7.00001-2

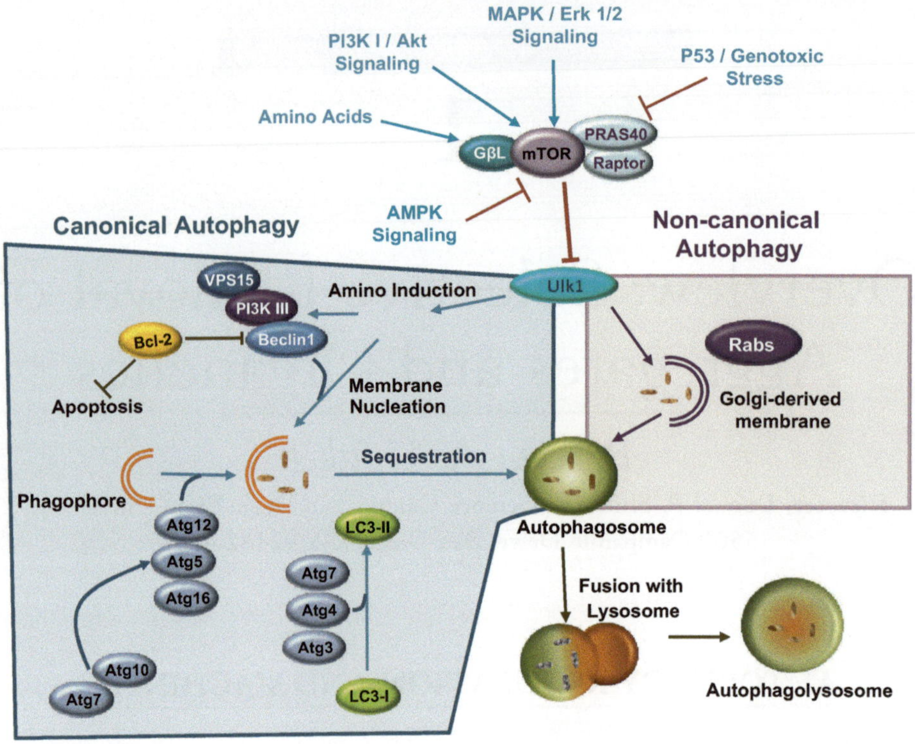

FIGURE 1.1 **Overview of macroautophagy.** Nutrient status regulates mTOR, the major negative regulator of autophagy through Ulk1. Signaling activates the Class III phosphatidylinositol-3-kinase complex that mediates nucleation of the phagophore. Ubiquitin-like ligases Atg7/Atg10 mediate conjugation of Atg12 onto Atg5, and Atg7/Atg3 mediate conjugation of LC3-I onto phosphatidylethanolamine in the phagophore membrane. Atg4 processes LC3-I to expose a glycine residue at the C-terminus. The growing phagophore is recruited to the cargo, often involving specific adaptor proteins, eventually enclosing to form a double-membrane structure, the autophagosome. Following lysosomal fusion and acidification, the contents are degraded and the constituents are exported back to the cytosol for utilization. Less is known about non-canonical autophagy other than the participation of Rabs and membrane derived from the Golgi network. Ulk1 can also activate the non-canonical pathway of autophagy.

lysosomal fusion. Atg4 is important for delipidating LC3 that has become associated with inappropriate membranes[3]. LC3 has multiple homologs which may be expressed in different tissues and may play nuanced roles in selective autophagy. A non-canonical form of macroautophagy has been recognized which involves membrane derived from the Golgi and requires the participation of multiple Rab proteins (Figure 1.1). Ulk1 also activates this pathway, but whether the same signals activate both canonical and non-canonical pathways to the same extent is not clear. Since Atg5-null or Atg7-null mice cannot survive starvation, it suggests that non-canonical autophagy cannot fully replace canonical autophagy, or that starvation doesn't activate this alternative pathway. More work will be required to delineate this pathway, its regulation, and structural and enzymatic components.

MITOCHONDRIAL AUTOPHAGY (MITOPHAGY)

Once the cup-shaped phagophore has formed, it is recruited to the target cargo. In the case of mitochondria, loss of mitochondrial membrane potential results in the accumulation of PTEN-inducible kinase 1 (PINK1), which in turn recruits the ubiquitin ligase Parkin (Figure 1.2). Parkin ubiquitinates multiple proteins on the mitochondrial outer membrane, resulting in recruitment of p62/sequestosome1 (SQSTM1), which has binding sites for both ubiquitin and LC3. SQSTM1 in turn binds LC3 on the concave face of the phagophore, resulting in a zipper-like engulfment of the mitochondrion. It is noteworthy that only depolarized mitochondria are targeted for removal, allowing retention of mitochondria with high membrane potential and functionality for continued ATP production; PINK1 is essential for conferring selectivity to this process. This example illustrates the interplay of general autophagy machinery and selective targeting. Nix and Bnip3, members of the BH3-only family of pro-apoptotic Bcl-2 proteins, also trigger mitophagy in a Parkin-dependent fashion[4]. Nix is particularly important for triggering mitochondrial depolarization to accomplish mitochondrial clearance in reticulocytes[5]. Excessive mitochondrial ROS trigger mitophagy through recruitment of Nix or Bnip3[6, 7]. Mitochondrial ROS may also inactivate Atg4 in the vicinity, preventing the delipidation and release of LC3 from nearby membranes, thus facilitating local formation of phagophores[8]. Mitophagy is induced by starvation, and it appears that this also requires mitochondrial depolarization through the mitochondrial permeability transition pore, as deletion of cyclophilin D reduces starvation-induced mitophagy[9]. The mechanism of mitophagy has been elucidated in elegant work by Youle and others[10]. The importance of this particular form of organellophagy is revealed by mutations in Parkin, which give rise to the familial

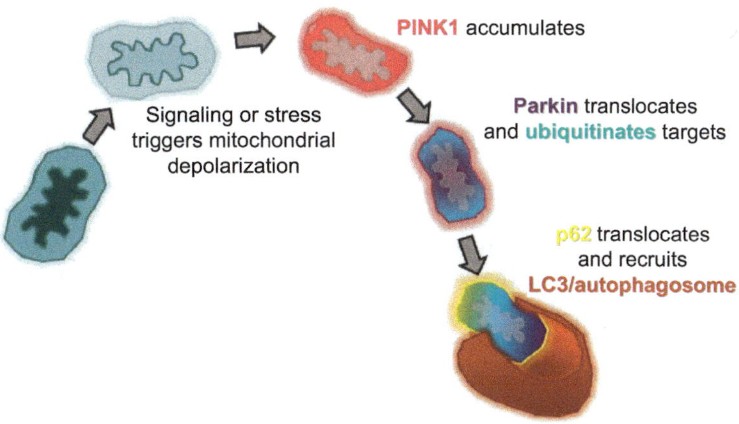

FIGURE 1.2 **Selective mitophagy involves multiple adaptor proteins.** Mitochondria have a high membrane potential unless signaling or cellular stressors trigger mitochondrial depolarization, which allows PINK1 to accumulate on the mitochondrial outer membrane. PINK1 accumulation signals Parkin translocation, resulting in ubiquitination of various proteins on the mitochondrial outer membrane. The ubiquitin modification recruits p62/SQSTM1, which serves as an adaptor to bring the forming autophagosome in proximity to the decorated mitochondrion and facilitate its engulfment.

form of the neurodegenerative disease, Parkinson's Disease, and by work in our lab, showing that Parkin-mediated mitophagy is essential for protection against ischemia/reperfusion injury in the heart[11, 12].

AUTOPHAGIC ELIMINATION OF GRANULES (CRINOPHAGY)

Crinophagy has been described in the growth hormone-secreting cells of the pituitary, after administration of acrylonitrile[13]. Activation of autophagy and crinophagy was recently described in Paneth cells of patients with Crohn's disease[14]. Variants of autophagy-regulating genes Atg16L, IRGM, and NOD2 are associated with Crohn's disease, suggesting a link between the abnormal crinophagy and the disease. Crinophagy also occurs when granule secretion is impaired, as in the secretion of albumin from liver when vinblastine is administered[15], or insulin granules in pancreatic beta islets exposed to interleukin-1β[16]. Interestingly, abnormalities in insulin secretion have been observed in Atg7-deficient beta cells and in Rab3A-null mice; it appears that autophagy is important for maintaining appropriate stores of insulin granules and turning them over to achieve a half-life of 3-5d[17]. Although the phenomenon of crinophagy was widely described in the late 1980s, only recently have efforts been undertaken to elucidate the molecular mechanisms, particularly the specific adaptor proteins that might be involved.

AUTOPHAGY OF AGGREGATED PROTEINS (AGGREPHAGY)

Isolated misfolded proteins are commonly marked with ubiquitin for proteasomal destruction or chaperone-mediated autophagy; large ubiquitinated protein aggregates can only be eliminated by macroautophagy. This is of particular importance in neurodegenerative diseases such as Huntington's disease, in which polyglutamine expansion leads to accumulation of aggregates of misfolded proteins. In the case of aggrephagy, p62/SQSTM1 and another adaptor protein, Alfy (autophagy linked FYVE protein), are associated with the protein aggregates. Alfy acts as a scaffold, bridging the ubiquitin and p62-decorated protein aggregate with Atg5, followed by complex formation with Atg12, Atg16L, and LC3, to assemble the autophagosome at the site of the protein aggregate. Deletion of Alfy results in impaired aggregate clearance, although starvation-induced autophagy remains intact, thus illustrating its role in selective autophagy[18]. There is growing interest in developing small molecule enhancers of autophagy to treat neurodegenerative diseases[19].

SELECTIVE AUTOPHAGY OF THE ENDOPLASMIC RETICULUM (ER-PHAGY)

The endoplasmic reticulum (ER) is responsible for production of secreted proteins; however, when this process is disrupted, the unfolded protein response (UPR) is invoked which regulates transcriptional and translational pathways governing protein synthesis. An integral aspect of the UPR also involves transcriptional upregulation of autophagy proteins including LC3 and Beclin 1[20, 21]. ER stress has been widely studied, but the connection to autophagy is still being explored. HspB8 and Bag3 have been implicated in ER stress-related autophagy[22]. Obesity results in impaired hepatic autophagy, which in turn results in ER stress and insulin resistance[23]. This finding implicates autophagy as an aspect of cellular homeostasis that prevents or mitigates ER stress. Given the importance of ER stress in a wide variety of cell types, it is likely that autophagy will be recognized to participate in a variety of disease processes associated with the UPR.

PIECEMEAL MICROAUTOPHAGY OF THE NUCLEUS (PMN/MICRONUCLEOPHAGY)

This process, best described in yeast, involves sequestration of small pieces of the nucleus by direct invagination of the vacuolar membrane in proximity to a portion of the nuclear membrane[24]. Specific adaptor proteins have been identified in yeast, and although this process is distinct from macroautophagy, many of the core macroautophagy genes (Atg7, Atg8/LC3, and Atg9) are essential[25]. Mutation of the nuclear lamin A gene results in a syndrome of muscular dystrophy and cardiomyopathy. In a mouse model, autophagosomes were observed in proximity with the nuclear envelope, and inhibition of autophagy resulted in nuclear abnormalities and increased cell death[26].

AUTOPHAGY OF PEROXISOMES (PEXOPHAGY)

This process is best characterized in yeast but also takes place in mammalian cells. Like mitophagy, specific adaptor proteins are required for selective removal of peroxisomes. In yeast it is regulated by the type of nutrient available (e.g., oleate vs. glucose), and is regulated by a complex signaling network tied to nutrient sensing. In mammals, pexophagy occurs during the circadian cycle and in response to withdrawal of a peroxisomal proliferating agent (e.g., phthalate esters). Turnover of peroxisomes is necessary for maintaining their functionality, including import of antioxidant enzymes such as catalase. Peroxisomes accumulate in aged cells concomitant with a decrease in functional autophagy, suggesting that pexophagy is important for organelle and cellular homeostasis[27].

AUTOPHAGY OF INTRACELLULAR PATHOGENS AND PHAGOSOMES (XENOPHAGY)

This important process provides the cell with a means to eliminate intracellular pathogens, whether bacterial, viral, fungal or protozoan. However, because many of these organisms have co-evolved with their hosts, complex strategies and counter-strategies have developed. For instance, the protozoan *Trypanosoma cruzi* triggers recruitment of LC3 to the plasma membrane, which facilitates its entry into the cell. The parasite enters the host cell and remains within an autophagosome until it enters the next phase of its life cycle. In contrast, organisms such as *Listeria monocytogenes* secrete factors to escape from the phagosome before it can be engulfed and delivered to the lysosome[28]. Viruses variously utilize the autophagosome, as a scaffold for assembly to escape detection by innate immunity, or suppress autophagosome-lysosome fusion[29].

CONCLUSION

The preceding discussion, focused primarily on selective autophagy, illustrates the diversity of the process and its broad utility for a variety of cellular functions. Further work will be needed to characterize the molecular basis for specificity. However, it can be seen how mutation of a particular gene, such as Parkin, might have effects that are manifest in a tissue-specific, or stimulus-restricted fashion. It is expected that further study will link autophagy to more diseases arising from genetic and environmental triggers.

References

1. Kim J, Kundu M, Viollet B and Guan K-L. AMPK and mTOR regulate autophagy through direct phosphorylation of Ulk1. *Nat Cell Biol* **13**:132–41.
2. Sou Y-s, Waguri S, Iwata J-i, Ueno T, Fujimura T, Hara T, et al. The Atg8 Conjugation System Is Indispensable for Proper Development of Autophagic Isolation Membranes in Mice. *Molecular Biology of the Cell* 2008;**19**:4762–75.
3. Nakatogawa H, Ishii J, Asai E and Ohsumi Y. Atg4 recycles inappropriately lipidated Atg8 to promote autophagosome biogenesis. *Autophagy* **8**.
4. Lee Y, Lee H-Y, Hanna RA and Gustafsson ÃsB. Mitochondrial autophagy by Bnip3 involves Drp1-mediated mitochondrial fission and recruitment of Parkin in cardiac myocytes. *American Journal of Physiology—Heart and Circulatory Physiology* **301**:H1924–31.
5. Schweers RL, Zhang J, Randall MS, Loyd MR, Li W, Dorsey FC, et al. NIX is required for programmed mitochondrial clearance during reticulocyte maturation. *Proc Natl Acad Sci USA* 2007;**104**:19500–5.
6. Ding WX, Ni HM, Li M, Liao Y, Chen X, Stolz DB, et-al and. Nix is critical to two distinct phases of mitophagy, reactive oxygen species-mediated autophagy induction and Parkin-ubiquitin-p62-mediated mitochondrial priming. *J Biol Chem* **285**:27879–90.
7. Kubli DA, Quinsay MN, Huang C, Lee Y, Gustafsson AB. Bnip3 functions as a mitochondrial sensor of oxidative stress during myocardial ischemia and reperfusion. *Am J Physiol Heart Circ Physiol* 2008;**295**:H2025–31.
8. Scherz-Shouval R. Oxidation as a Post-Translational Modification that Regulates Autophagy. *Autophagy* 2007;**3**:371–3.
9. Carreira RS, Lee Y, Ghochani M, Gustafsson AB, Gottlieb RA. Cyclophilin D is required for mitochondrial removal by autophagy in cardiac cells. *Autophagy* 2010;**6**.
10. Youle RJ and Narendra DP. Mechanisms of mitophagy. *Nat Rev Mol Cell Biol* **12**:9–14.
11. Jones R. The Roles of PINK1 and Parkin in Parkinson's Disease. *PLoS Biol* **8**:e1000299.
12. Huang C, Andres AM, Ratliff EP, Hernandez G, Lee P, Gottlieb RA. Preconditioning Involves Selective Mitophagy Mediated by Parkin and p62/SQSTM1. *PLoS One* 2011;**6**:e20975.
13. Kamijo K, Kovacs K, Szabo S, Bollinger-Gruber JN, Reichlin S. Effect of acrylonitrile on the rat pituitary: enlargement of Golgi region in prolactin cells, crinophagy in prolactin cells and growth hormone cells. *Br J Exp Pathol* 1986;**67**:439–51.
14. Thachil E, Hugot J-P, Arbeille B, Paris R, Grodet A, Peuchmaur M, et al. Abnormal Activation of Autophagy-Induced Crinophagy in Paneth Cells from Patients with Crohn's Disease. Short title: Crinophagy in Crohn's disease. *Gastroenterology* 2012. In press.
15. Glaumann H. Crinophagy as a means for degrading excess secretory proteins in rat liver. *Revis Biol Celular* 1989;**20**:97–110.
16. Sandberg M, Borg LA. Intracellular degradation of insulin and crinophagy are maintained by nitric oxide and cyclo-oxygenase 2 activity in isolated pancreatic islets. *Biol Cell* 2006;**98**:307–15.

17. Chen ZF, Li YB, Han JY, Wang J, Yin JJ, Li JB, et al. The double-edged effect of autophagy in pancreatic beta cells and diabetes. *Autophagy* **7**:12–16.
18. Filimonenko M, Isakson P, Finley KD, Anderson M, Jeong H, Melia TJ, et al. The selective macroautophagic degradation of aggregated proteins requires the PI3P-binding protein Alfy. *Mol Cell* **38**:265–79.
19. Sarkar S, Rubinsztein DC. Small molecule enhancers of autophagy for neurodegenerative diseases. *Mol Biosyst* 2008;**4**:895–901.
20. Copetti T, Bertoli C, Dalla E, Demarchi F, Schneider C. p65/RelA Modulates BECN1 Transcription and Autophagy. *Molecular and Cellular Biology* 2009;**29**:2594–608.
21. Bernales S, Schuck S, Walter P. ER-Phagy: Selective Autophagy of the Endoplasmic Reticulum. *Autophagy* 2007;**3**:285–7.
22. Carra S, Brunsting JF, Lambert H, Landry J, Kampinga HH. HspB8 participates in protein quality control by a non-chaperone-like mechanism that requires eIF2{alpha} phosphorylation. *J Biol Chem* 2009;**284**:5523–32.
23. Yang L, Li P, Fu S. Calay ES and Hotamisligil GS. Defective hepatic autophagy in obesity promotes ER stress and causes insulin resistance. *Cell Metab* **11**:467–78.
24. Mijaljica D, Prescott M and Devenish RJ. The intricacy of nuclear membrane dynamics during nucleophagy. *Nucleus* **1**:213–23
25. Krick R, Muehe Y, Prick T, Bremer S, Schlotterhose P, Eskelinen EL, et al. Piecemeal microautophagy of the nucleus requires the core macroautophagy genes. *Mol Biol Cell* 2008;**19**:4492–505.
26. Park YE, Hayashi YK, Bonne G, Arimura T, Noguchi S, Nonaka I, et al. Autophagic degradation of nuclear components in mammalian cells. *Autophagy* 2009;**5**:795–804.
27. Manjithaya R, Nazarko TY, Farre JC and Subramani S. Molecular mechanism and physiological role of pexophagy. *FEBS Lett* **584**:1367–73.
28. Birmingham CL, Higgins DE, Brumell JH. Avoiding death by autophagy: interactions of Listeria monocytogenes with the macrophage autophagy system. *Autophagy* 2008;**4**:368–71.
29. Sir D, Ou JH. Autophagy in viral replication and pathogenesis. *Mol Cells* **29**:1–7.

CHAPTER

2

Molecular Machinery and Genetics of the Autophagy Pathway

Deron Herr and Kim D. Finley

Expression Drug Designs, LLC, 5500 Campanile Drive, San Diego, CA 92182
Expression Drug Designs, LLC and Donald P. Shiley BioScience Center,
San Diego State University, San Diego, CA, 92182-4650

INTRODUCTION

The *de novo* formation of an autophagosome, a key aspect of the pathway's activation, requires the expansion and sealing of a small membrane cistern known as the phagophore or isolation membrane. Morphologically, the size of the completed autophagosome depends on the target that is being sequestered and in part reflects Atg8/LC3 levels. Typically, AVs cross-sectional radii range from 100-300nm but can be much larger. Once complete, the AVs deliver their cargo along microtubules via a dynein-mediated process to the lysosomes[15, 16]. Following fusion of the two vesicle types, the cargo is degraded in the hydrolytic lumen of lysosomes and the component parts recycle back to the cell. A diverse set of proteins and complexes are involved in the biogenesis of autophagosomes and include the autophagy-related proteins (Atg)[17]. Most Atg genes were initially identified and characterized in yeast. Subsequent studies in higher eukaryotes have revealed that these key factors are highly conserved. In humans, 36 Atg proteins have been identified to date, with 16 genes making up the core Atg machinery and these appear to be essential for all autophagy-related pathways.

Upon the induction of the pathway, the Atg proteins associate and interact following a hierarchical order, initially to start the formation of the phagophore and then to expand it into an autophagosome[17-20]. The molecular functions of individual human Atg genes and their precise contribution during the biogenesis of double-membrane vesicles require additional study. However, based on other model systems they can be classified into five primary functional groups:

1. the Atg1/ULK initiation complex
2. phosphatidylinositol 3-kinase (PI3K) signaling complex

http://dx.doi.org/10.1016/B978-0-12-385101-7.00002-4

3. Atg9 trafficking and lipid transfer systems
4. dual ubiquitin-like conjugation systems
5. nucleation and lysosomal/endosomal fusion.

All of these component parts are necessary but not sufficient for the full initiation and completion of AVs, and there is considerable integration and interdependence between the different functional groups. We now understand that depending on cellular requirements, autophagy can also be a highly selective process. This has led to the identification of additional pathway components that facilitate a specific type of macroautophagic process that includes xenophagy (clearance of pathogens), aggrephagy (clearance of aggregates), mitophagy (mitochondria) and pexophagy (peroxisomes).

THE ATG1/ULK INITIATION COMPLEX

Most tissues appear to have basal rates of autophagosome formation and pathway activity, which can be greatly enhanced or up-regulated following an appropriate physiological signal. Studies have shown that feeding and insulin pathway activation suppress autophagy via the activity of the downstream effector kinase mTOR (Target of Rapamycin)[21, 22]. The mTOR protein is the active kinase in two distinct protein complexes, mTORC1 (regulates autophagy) and mTORC2. Nutrient deprivation, fasting and/or suppressing mTORC1 activity with drugs (rapamycin) can go on to stimulate the initiation complex and facilitate the activation of autophagy[23-26]. The yeast Atg1 protein and its human homologs, ULK1 and ULK2, are members of the serine/threonine class of kinases[12-14, 27, 28]. They function together with other proteins to form the initiation complex that is essential for full activation of the pathway. In yeast, the complex consists primarily of the Atg1:Atg13:Atg17 proteins and appears to be important in the formation of the initial pre-autophagic structure (PAS)[28, 29]. The mammalian homologs ULK1 and ULK2 (Uncoordinated-51 Like Kinase) form a complex with the mAtg13, Atg101 and FIP200 proteins (Figure 2.1)[21, 30, 31]. The ULK/Atg1-interacting

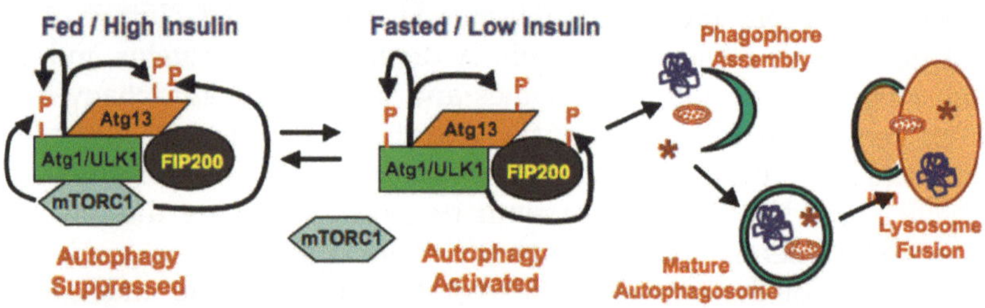

FIGURE 2.1 **Phosphorylation Profile of the ULK1/Atg1 and Atg13 Initiation Complex.** The Atg1/ULK1 proteins are highly conserved kinases that are essential for the activation of the autophagy pathway and the early steps of autophagosome formation. These proteins form higher order complexes with several other autophagy proteins, including Atg13 and FIP200. Both hULK1 and hAtg13 are phosphorylated in a TORC1-dependent manner. This in turn reduces Atg1's kinase activity and suppresses autophagic induction and the clearance of substrates by the pathway. Conversely, both fasting and rapamycin treatment suppress TORC1 activity, causing the rapid dephosphorylation of Atg13, the activation of Atg1, and the initiation of autophagosome formation. Recently, other signaling kinases that are involved with metabolic homeostasis (PKA and AMPK) have been shown to directly interact and regulate these critical pathway components.

proteins serve to regulate Atg1 kinase activity by controlling its phosphorylation state and association with Atg13. In addition to its regulation by TORC1 activity, ULK1 can also associate with and be directly phosphorylated by AMPK kinase (5′ AMP-activated protein kinase)[32-34]. With nutrient withdrawal the ULK1-AMPK complex quickly dissociates resulting in the formation of the ULK1-mAtg13-Atg101-FIP200 complex and the rapid formation of autophagosomes[34, 35]. The downstream phosphorylation targets of ULK1/Atg1 require additional clarification, but loss-of-function mutations or dsRNAi inactivation of the ULK1 gene substantially reduces the ability of cells to activate the pathway and form autophagosomes. Other kinases involved with the regulation of metabolic signaling have recently been discovered to regulate autophagy. They include the AMPK and PKA pathways and appear to further modulate the phosphorylation status of Atg1/ULK1 and Atg13 and their downstream targets[28-36].

PHOSPHATIDYLINOSITOL 3-KINASE (PI3K) SIGNALING COMPLEX

The PI 3-kinases (**PI3K**) are responsible for phosphorylating phosphatidylinositol (**PtdIns**) to form phosphoinositides (PIs, phosphorylated derivatives of PtdIns). Mammals have several PI3K classes that are categorized by their associated regulatory proteins. PI3K class-I is principally implicated in the modulation of signaling cascades and produces $PI^{(3,4,5)}P3$[37-40]. This phospholipid subtype has been shown to inhibit autophagy by upregulating Akt/PKB, which goes on to activate the mTOR (mammalian target of rapamycin) kinase. TOR signaling is the primary regulatory factor that suppresses or inactivates the autophagy pathway. In contrast, the PI3K class-III (PI3KC3) and its product PI3P are essential for enhancing autophagic activity and regulating organelle biogenesis[23, 41, 42]. As in the yeast system, this complex contains three common structural components: hVps34, p150 (Vps15 ortholog) and Beclin 1 (Atg6 ortholog) and is regulated by Atg14 and Vps38 proteins. The mammalian counterparts of Atg14 and Vps38 are Atg14L/Barkor and UVRAG (UV radiation resistance-associated gene), respectively[12, 14, 43]. The Atg14L-containing complex plays a central role in autophagy by directing the PI3K class-III complex to the site of phagophore initiation thereby enhancing PI3P production. This in turn further directs the recruitment of the Atg machinery to the growing phagophore.

Upon starvation, Atg14L localizes to autophagosomal membranes[41]. Importantly, depletion of Atg14L reduces PI3P production, impairs the formation of autophagosomal precursor structures and inhibits autophagy[14, 41, 44, 45]. The UVRAG-containing class III PI3K complex also regulates autophagy, but appears to act at the intersection between autophagy and the endosomal trafficking pathways (see Nucleation and lysosomal/endosomal fusion section). UVRAG initially associates with the BAR-domain (Bin-Amphiphysin-Rvs) of Bif-1, which may regulate mAtg9 trafficking from the trans-Golgi network (TGN)[46, 47]. UVRAG then interacts with the Class C Vps/HOPS complex, promoting the fusion of autophagosomes with late endosomes or lysosomes[48]. Finally, the UVRAG-containing Class III protein complex binds to Rubicon, a late endosomal and lysosomal protein that can suppress autophagosome maturation by reducing hVps34 activity[44, 45]. Importantly, both the Atg14L- and

UVRAG-containing complexes interact through Beclin-1 via Ambra-1, which in turn tethers the protein complexes to the cytoskeleton via an interaction with dynein[49, 50]. Following the induction of autophagy, ULK1 phosphorylates Ambra-1, thus releasing the class III PI3K complexes from dynein; their subsequent re-localization triggers autophagosome formation. Therefore, Ambra-1 constitutes a direct regulatory link between the Atg1/ULK1 and the PI3K complexes[22, 49].

ATG9 TRAFFICKING AND LIPID TRANSFER SYSTEMS

Together with the Atg1/ULK and the PI3K complexes, the Atg9 protein is one of the first factors that localizes to the yeast PAS or to the site of phagophore assembly in mammalian systems (Figure 2.2)[19, 31]. While most autophagy factors are soluble cytoplasmic proteins that are lipidated or contain lipid interacting motifs (i.e. FYVE), Atg9 is the only conserved integral membrane protein that contains multiple transmembrane domains. Atg9 is typically distributed between the site of phagophore development and multiple cytoplasmic tubulo-vesicular compartments derived from the **TGN** and other organelles[29, 51, 52]. Atg9 cycles between these two subcellular locations and is thought to facilitate the transport of lipid building blocks to

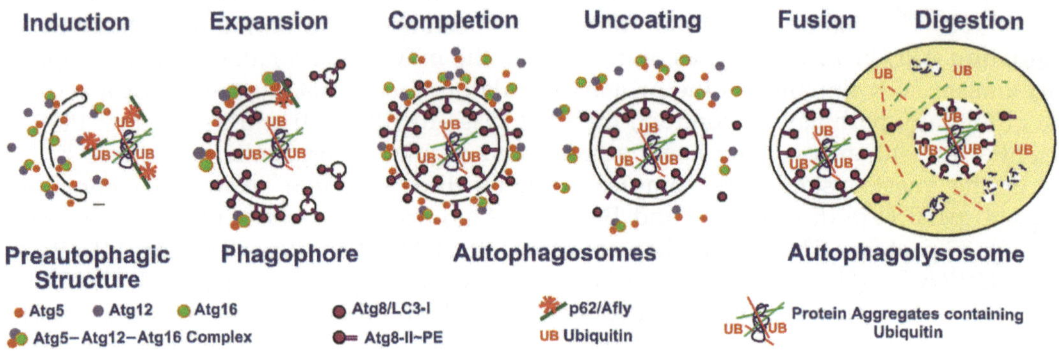

FIGURE 2.2 **Stages of Phagophore Induction, Autophagosome Completion and Lysosomal Fusion.** The generation of autophagic vesicles (AV) requires a complex series of interactions by multiple factors to initiate phagophore formation and to mature the organelles. Following the complete formation of double membrane AV, additional stages are required to remove autophagic proteins (uncoating), as well as trafficking to and fusing with lysosomes. A large complex is formed between the Atg5, Atg12 and Atg16 proteins, which form near the site of phagophore assembly, near targeted substrates. Both the Atg8/LC3 (microtubule-associated protein light chain 3, MAP-LC3) and the Atg12 families are small critical proteins that are ubiquitin-like. For full function the Atg8/LC3 proteins must undergo a carboxyl-terminal cleavage by the cysteine protease Atg4, exposing a reactive glycine (arrow), followed by sequential activation by the Atg7 (E1-like) and Atg3 (E2-like) proteins. Atg8 family members are finally covalently conjugated to phosphatidylethanolamine (PE) via an amide bond between the C-terminal glycine and the amino group on the hydrophilic head of the lipid molecule. Once AVs are fully formed they are trafficked along microtubules via the dynein motor proteins to acidic lysosomes or late multi-vesicular endosomes. Following fusion, intermediate vesicles are formed that include autophagolysosomes (lysosome) and amphisomes (endosome). The contents within autophagolysosomes are then degraded and recycled back to the cell.

the expanding phagophore membrane[29]. Mouse Atg9 (mAtg9) shows characteristics that are similar to its yeast counterpart. Both Atg9 proteins are essential for the recruitment of the PI3K Complex to the phagophore. Atg9 retrieval and recycling from the complete autophagosome is mediated by the Atg2:Atg18 protein complex[53, 54]. This cycling appears to be regulated in part by activation of the Atg1/ULK (protein) and PI3 (lipid) signaling kinases[29]. Mouse Atg9 (mAtg9) also localizes to the TGN and late endosomes, and redistributes to autophagic structures upon the induction of the pathway[11], further promoting the activation of autophagy[11, 55-58]. Cycling of mAtg9 between the different subcellular locations requires both mULK1 (mAtg1) and mVps34 kinase activity[11, 21].

DUAL UBIQUITIN-LIKE CONJUGATION SYSTEMS

Typically, formation of autophagosomes requires the full function of dual ligation systems that include formation of the **Atg5:Atg12:Atg16** complex and the cleavage and lipidation of the **Atg8/MAPLC3** protein family (Figure 2.3)[15, 16, 59]. Both the Atg12 and Atg8/MAP1-LC3 families are comprised of small proteins that have structural and functional similarities to ubiquitin.

Atg12 Pathway

Following activation of the initiation complex, one of the earliest steps in phagophore formation is the generation of the Atg5:Atg12:Atg16 protein complex (Figure 2.3)[9, 60-62]. Initially the ubiquitin-like Atg12 protein interacts with and is conjugated to the Atg7 protein

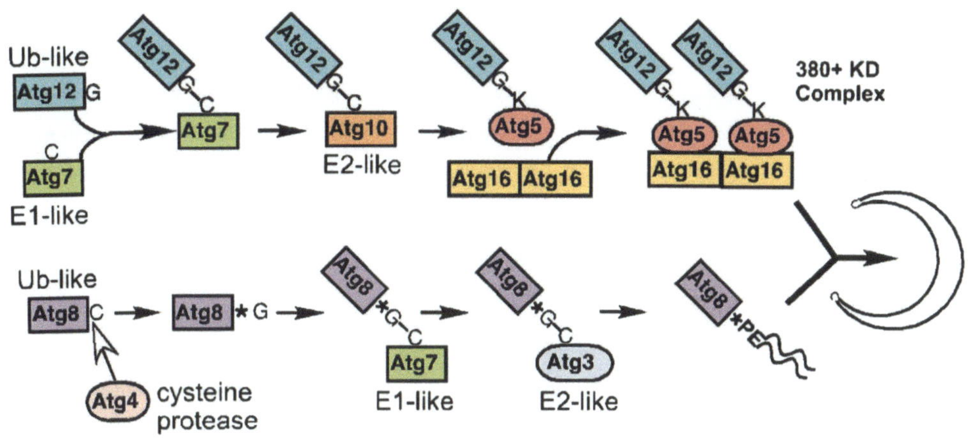

FIGURE 2.3 **Dual Ubiquitin-Like Conjugation Systems.** The key components and interactions of the conserved ubiquitin-like ligation systems that is required for full activation of the autophagy pathway. There is considerable overlap and multiple interactions between the parallel systems. Both require Atg7 activity, which is functionally similar to E1 ubiquitin ligases. Both systems contain E2-like ligases, Atg3 and Atg10, which transfer Atg8/LC3 and Atg12 to their final partners, the PE lipid and Atg5. The full formation of the Atg12:Atg5:Atg16 complex enhances and concentrates the lipidation of Atg8/LC3 at the site of phagophore assembly. Emerging studies are identifying additional other points of convergence and interaction.

TABLE 2.1 Conserved Autophagy Genes Including Duplicate Human Components

Yeast	Human Gene	Loci	Protein Type	Function and Disease Association
INITIATION COMPLEX (KINASE ACTIVITY)				
Atg1	ULK1	*12q24.3*	Serine/threonine kinase	Regulates autophagy, axon, neurite growth
Atg1	ULK2	*17p11.2*	Serine/threonine kinase	Smith-Magenis syndrome region
Atg13	Atg13/ FLJ20698	*11p11.2*	Regulatory subunit	Regulatory and phosphorylated by TOR
Atg17	FIP200/RB1CC1	*8q11.23*	ULK complex	frequently mutated in breast cancer
-	Atg101/C12orf44	*12q13.13*	Atg13 interacting protein	
ATG5:ATG12:ATG16 COMPLEX FORMATION (UBIQUITIN-LIKE)				
Atg7	Atg7	3q25.3	E1-like ligase	Post-partum lethal, neural degeneration
Atg10	Atg10	*5q14.1*	E2-ligase	
Atg5	Atg5	*6q21*	ligated to Atg12	Post-partum lethal, neural degeneration
Atg12	Atg12	*5q21-q22*	Ubiquitin-like	
Atg16	Atg16L1	*2q37.1*	Atg5 complex, WD40	Crohn's disease, Inflammatory bowel-10
ATG8/MAP-LC3 CONJUGATION AND LIPIDATION (UBIQUITIN-LIKE)				
Atg4	Atg4A	*Xq22.3*	C-54 cysteine protease	Cleaves LC3 C-terminus and LC3-PE
Atg4	Atg4B	*2q37.3*	C-54 cysteine protease	
Atg4	Atg4C	*1p31.3*	C-54 cysteine protease	
Atg4	Atg4D	*19p13.2*	C-54 cysteine protease	
Atg7	Atg7	*3q25.3*	E1-like ligase	Post-partum lethal, neural degeneration
Atg3	Atg3	*3q13.2*	E2-like ligase	Atg12 conjugation for mito homeostasis
Atg8	MAP1LC3A	*20q11.22*	Ub-like, PE-lipidated	Lipid-microtubule-cytoskeleton interactions
Atg8	MAP1LC3B	*16q24.2*	Ub-like, PE-lipidated	Lipid-microtubule-cytoskeleton interactions
Atg8	MAP1LC3B2	*12q24.22*	Ub-like, PE-lipidated	Lipid-microtubule-cytoskeleton interactions
Atg8	MAP1LC3C	*1q43*	Ub-like, PE-lipidated	Lipid-microtubule-cytoskeleton interactions
LIPID SIGNALING AND RECRUITMENT				
Atg9	Atg9A/ABCB6	*2q35*	Membrane protein	Golgi-endosome-AV cycling, lipid recruit
Atg9	Atg9B	*7q36.1*	Membrane protein	Golgi-endosome-AV cycling, lipid recruit
Atg2	Atg2A	*11q13.1*		Atg18:Atg9 complex
Atg2	Atg2B	*14q32.2*		Atg18:Atg9 complex
Atg18	WIPI1/Atg18A	*17q24.2*	WD40 repeat protein	Regulate Atg2:Atg9 complex, binds PI3P
Atg21	WIPI2/18B/Atg21	*7p22.1*	WD40 repeat protein	

(*Continued*)

TABLE 2.1 Conserved Autophagy Genes Including Duplicate Human Components—cont'd

Yeast	Human Gene	Loci	Protein Type	Function and Disease Association
Vps15	PI3KR4	*3q22.1*	PI3K subunit, p150	PI-3-K regulatory subunit, Activates Vsp35
Vps34	PI3KC3/MEM3	*16q12*	PtdIns 3-kinase	Produces PtdIns(3)P signaling lipid
NUCLEATION AND LYSOSOMAL FUSION				
Atg6	BECN1/Vps30	*17q21.31*	Beclin-1	Regulation, development, cancer, neurodeg
Atg14	Atg14L/Barkor	*14q22.3*	BECN1 complex	Beclin-1 translocation, enhances PIK3C3
-	RUBICON	*3q29*	BECN1 complex	
-	AMBRA1	*11p11.2*	BECN1 regulator	Autophagy regulation, neural development
-	LAMP1	*13q34*	Lysosomal protein-1	Membrane glycoprotein, tumor metastasis
-	LAMP2	*Xq24*	Lysosomal protein-2	Membrane glycoprotein, tumor metastasis
-	DRAM1	*12q23.2*	Damage-reg auto mod	Autophagy induction by p53
RAB7	RAB7A	*3q21.3*	RAS-GTP binding pro	Endo-Auto trafficking, neuropathies
SELECTIVE AUTOPHAGY				
-	ALFY/ WDFY3	*4q21.23*	Autophagy-linked FYVE	Beach, WD40, FYVE motifs, Aggrephagy
-	SQSTM1/ *p62*	*5q35*	Sequestosome-1	UBA, regulates NF-KB, Paget Disease-3
Pex5	Pex5	*12p13.31*	Peroxisomal factor 5	Biogenesis & autophagy

(E1 ubiquitin-like ligase) via its C-terminal glycine before being transferred to the Atg10 protein (E2 ubiquitin-like ligase)[16]. The Atg12 C-terminal glycine is used to form a final covalent bond with a specific lysine residue in the Atg5 protein. This protein heterodimer promotes the recruitment and tight association of the Atg16 protein. At the site of phagophore formation this facilitates the formation of a larger protein complex (380KD+), via the homo-oligomerization activity of Atg16[61]. Generating the Atg12:Atg5 conjugate is a critical early step needed to trigger the elongation of the double-bilayer and to recruit additional autophagy components. In yeast, the formation of the 380KD+ complex is typically at the PAS site or at the leading edge of the growing phagophore membrane[16]. The formation of this initial protein complex is also important for the enhanced lipidation of Atg8/MAP-LC3 and its recruitment to the growing double membrane structure.

Atg8/MAP1LC3 Pathway

The second ligation component of autophagy involves the cleavage and linkage of Atg8/MAP1LC3 to PE (Figure 2.3)[8, 16, 63]. The Atg8/MAP1-LC3 protein family (known as LC3-I in the unlipidated state) requires Atg4 (cystine protease) to cleave their C-terminus[63, 64]. This initial step exposes a reactive glycine residue that serves as the site for subsequent ligations. A series of ubiquitin-like conjugation steps sequentially ligate Atg8/MAP-LC3 to Atg7 (E1-like ligase) and Atg3 (E2-like ligase). Atg8/MAP-LC3 is then finally conjugated to the **PE** lipid (Figure 2.3)[63, 64]. At this point the hybrid protein-lipid molecule is called

LC3-II or **Atg8-PE** and migrates slightly faster than the original LC3-I molecule on SDS-PAGE Western blots[65]. LC3-II is typically associated with the completion of autophagosome formation. The change in LC3-I/LC3-II ratios is used to assess acute changes in autophagic activity or rates. Although in many studies the absolute levels of LC3-II are thought to be a better reflection of autophagosome abundance than an assessment of I/II ratios. At the organelle level, the LC3-II/Atg8-PE molecules are delivered to the tips of the growing phagophore in small membrane "droplets" in part through their interaction with the initiating Atg5:Atg12:Atg16 complex[15, 16]. The Atg8/MAP-LC3 molecules are located on both sides of the organelle and those on the lumenal side remain with AV until they are degraded in the lysosome. Fluorescent-tagged expression constructs of Atg8/MAP-LC3 (i.e. GFP-LC3) or LC3 antibodies are widely used to mark and characterize AV *in vivo*, to study pathway dynamics and vesicle sub-populations[65]. In addition, the sequestered LC3-II molecules remaining on the lumenal side of the autophagosome are eventually degraded in the lysosome. This indicates that during periods of high autophagic demand or when Atg8a gene expression is suppressed, the Atg8/LC3 family of proteins has the potential to become a rate-limiting component of the pathway[66].

NUCLEATION AND LYSOSOMAL/ENDOSOMAL FUSION

While the function of the Atg1/ULK kinase and the PI3K complex are considered essential components for autophagic nucleation, several additional factors are also required. One component is the mammalian UVRAG protein, which is involved at different points within the autophagy pathway. As described previously, UVRAG interacts with the PI3K class-III complex and regulates both autophagic and endosomal trafficking pathways[14, 41, 43, 44]. The protein also controls the TGN trafficking of the Atg9 protein by interacting with the Class C Vps/HOPS protein complex and promotes fusion events between autophagosome and late endosomes and/or lysosomes[11, 46, 48, 57, 58]. Another autophagy associated protein, Rubicon (RUN domain and cysteine-rich domain containing, Beclin 1-interacting) can also partner with the UVRAG-containing PI3K class-III complex to further suppress Vps34 activity and autophagosome maturation[44, 45]. Further, the Atg14L:UVRAG containing complexes interact with dynein to tether the functional unit to the cytoskeleton through Beclin1:Ambra1 interactions[45, 49, 50]. Following the induction of autophagy, the ULK1 protein kinase phosphorylates Ambra-1, releasing the PI3K-III complex from dynein. The complex is then free to re-locate to sites of phagophore formation and further enhance the development of AV. Therefore, Ambra-1 appears to constitute a direct regulatory link between the Atg1/ULK1 and PI3K complexes[49, 57].

Autophagosome Membranes

The source of the lipids that are essential building blocks for the autophagosome double membrane structure is an intense area of study within the field. A significant problem in understanding the lipid source is the lack of clear protein markers that are common between a specific subcellular compartment and autophagosomes[1, 67]. Several studies have implicated the plasma membrane and several intracellular organelles as the possible source of AV lipids/membranes. The plasma membrane and elements of the trafficking machinery to the cell surface have been linked to the formation of early autophagic intermediates including the phagophore[68-72]. It is possible that early endosomal- and/or TGN-derived membranes are also key factors in the initial steps of autophagy[11, 31, 52]. The TGN also appears important for autophagy by supplying at least part of the extra lipids that are required for the phagophore expansion[57, 73-75].

The endoplasmic reticulum (**ER**) also plays a central role in autophagosome biogenesis and phagophore expansion[1-3, 76, 77]. Recently two electron tomography studies have demonstrated the existence of a physical connection between the ER and newly formed autophagosomes[78, 79]. These studies showed the ER was connected to both the outer and the inner membrane of the growing phagophore through distinct points of contact. This supports the idea that lipid components of AV could arise from the direct transfer of ER membranes at the sites of contact. Consistent with this view, the Atg14L protein was found associating with the ER and the generation of the PI3P signaling lipid has been shown to occur on ER specific sub-domains where AVs emerge subsequent to the start of autophagy-inducing signals[27, 80].

Another significant lipid source during AV formation is the outer mitochondrial membrane. While the experimental evidence shows that mitochondria are essential for membrane expansion, it remains unclear whether these organelles play a key role in early stages of phagophore biogenesis and membrane recruitment[81]. The differences found between these studies have not yet allowed researchers in the field to draw a definitive model of the membrane dynamics during autophagosome biogenesis. However, the supply of lipid from divergent intracellular sources may reflect the use of diverse model systems and experimental conditions used to study the pathway. In addition, it may reveal the inherent flexibility and redundancy of autophagy in a wide range of cell and tissue types and physiological conditions[82, 83]. A series of elegant imaging studies has strengthened the concept that the ER and mitochondria function in tandem during the production of the phosphatidylethanolamine (**PE**) lipid, which is an essential component of the growing phagophore membrane[81, 84, 85]. The ER membrane is a rich source of the phosphatidylserine (**PS**) lipid[86, 87]. Detailed imaging studies of cells under fasting conditions demonstrate the juxtaposition of the mitochondria with the ER membrane, which correlated with the induction of autophagy. At this time PS was translocated from ER to mitochondria, the location of the PS-decarboxylase (**PSD**) enzyme[88]. Here the PS lipid was converted to PE, typically close to the ER-mitochondria association sites. This conversion was enhanced by the presence of the mitochondrial-ER tethering/fusion factor, Mitofusin-2[81]. Cells lacking Mitofusin-2 fail to produce mature AVs. Mitochondria-derived PE was then transferred to the growing phagophore together with the Atg5:Atg12 heterodimeric protein complex. The Atg8/LC3 protein in turn was then localized and conjugated to PE-enriched membranes. Following autophagosome closure the new vesicle was released from the mitochondria and the Atg5:Atg12:Atg16 complex is removed from the AV de-coating, thus limiting the mitochondrial protein component in AVs. These studies also demonstrated that cytochrome b5 (cb5-Mito) promotes the association of the mitochondria with the growing autophagosome[85, 89, 90]. Consistent with previous proteomic analyses it appears that there is remarkably little transfer of mitochondrial membrane proteins to growing phagophore or final AV membranes despite the significant contribution of PE that is generated from mitochondrial membranes.

Cytoskeletal, Motor, Endosomal And Lysosomal Proteins

Once autophagosome formation is complete, the external "coat" proteins and complexes dissociate from the mature vesicle. Lipidated Atg8/LC3-II proteins on the external autophagic surface are removed by Atg4 activity, while Atg8/LC3-II inside the lumen remains with the AV. The autophagosome is now capable of fusing with the lysosome (Figure 2.2). Typically this requires the transport of AVs along microtubules toward the lysosome using the dynein motor complex[91-94]. The dynein complex also facilitates the recruitment of key signaling regulatory factors to the AV[22,49]. In neuronal studies, where the dynein motor was inhibited, there was marked buildup of vesicles within axons and a decreased retrograde transport to the soma (cell body) where the majority of lysosomes are located[49, 95]. Once the two vesicles are in close approximation, additional factors are required to mediate fusion events[93, 95-98].

There appears to be considerable overlap or cross talk between factors known to mediate late endosomal-lysosomal trafficking and fusion events and the autophagy pathway[42, 99]. Endocytosis involves the formation of intracellular vesicles that originate primarily from the plasma membrane, which eventually go on to fuse with lysosomes[100, 101]. The progression and trafficking of endosomes is a complex and highly regulated series of events that produce late-endosomal structures or multivesicular bodies (**MVB**)[99, 102, 103]. MVBs typically fuse with lysosomes but can directly interact with autophagosomes leading to the formation of an intermediate vesicle called an amphisome[99, 104, 105]. This process is regulated by several protein complexes known as ESCRTs (endosomal sorting complex required for transport). Four distinct ESCRT complexes function to concentrate cargo on the endosomal membrane and regulate membrane dynamics and vesicle budding[102, 106, 107]. Recent work suggests that ESCRT proteins are also required downstream of these events to enable fusion of MVBs with lysosomes[99, 102, 108-110].

ESCRT protein complexes have also been shown to have a significant role in AV maturation, trafficking and fusion[101]. Loss of ESCRT function results in the accumulation of autophagosomes and amphisomes (endosomal-autophagic hybrid vesicles). Additional studies have shown that deletion of multiple ESCRT proteins including Hrs (ESCRT-0), Vps28/Tsg101 (ESCRT-I), Vps22, Vps25 (ESCRT-II), Vps2B, Vps24, Vps32 (ESCRT-III), or Vps4 (AAA-ATPase) generally enhanced autophagosome accumulation[99, 102, 108-110]. Other studies have shown that these trafficking defects can exacerbate the accumulation of cytoplasmic protein aggregates containing ubiquitinated proteins and p62[108, 111-114]. Indeed, recent work showed that expression of a mutant Vps2B/CHMP2B-Intron5 protein (ESCRT-III subunit) in mouse cortical neurons or in tissue cells results in the accumulation of ubiquitinated proteins and neural degenerative phenotypes[108, 115]. These autophagic defects are similar to those seen with mutations associated with familial cases of frontotemporal dementia (FTD)[103, 110, 116, 117]. This implies that ESCRT complexes likely have a dual function in controlling both the endosomal and autophagy pathways. Further, ESCRT-dependent fusion events have been implicated in Alzheimer's disease, lysosomal storage disorders, myopathies and bacterial pathogen invasion[115, 118-120]. A detailed mechanistic examination of this pathway may lead to novel treatments for these disorders.

Endocytosis is the primary means by which cells internalize extracellular material such as cell-surface receptors and protein ligands[99]. Modulating these powerful signaling cascades is a fundamental function of endosomes and is essential for proper development, cell growth and metabolic homeostasis. The Rab GTPase protein family is another major group of factors that regulate endosomal trafficking and lysosomal fusion. In the case of endocytosis, small GTPases are key players that facilitate membrane organization and are required for endosome-AV maturation and trafficking. Similarly, activation of the Rab7 GTPase is required for the initiation of the early-to-late endosome transition[121, 122]. Rab GTPases are also required for complete AV formation and flux. Atg5/Atg7-independent or non-canonical autophagy appears to require Rab9 function[99, 121, 123]. Rab1, which regulates ER-TGN trafficking, has a secondary role in the early formation of phagophores. Other preliminary studies of endosomal Rab and GTPase proteins, including Rab5, Rab24, Rab33, Rab7 and RalB, show they also play a role during AV biogenesis and maturation[68, 121, 124, 125]. However, the molecular mechanisms underlying endosomal and autophagosome maturation and lysosomal fusion events are not well characterized and require additional studies.

Selective Autophagy

While autophagosomes can engulf entire regions of the cytoplasm there is a growing understanding that autophagy can be highly selective for the targeting and elimination of defined substrates within cells. These processes include aggrephagy (ubiquitinated protein aggregates)[108, 113, 126-130], mitophagy (old or damaged mitochondria) pexophagy (aged peroxisomes)[131-133] and intracellular pathogens (xenophagy)[134-137]. As with endosomal and proteasomal pathways, it appears from these early studies that ubiquitin is used as the molecular identification signal for substrates that should be cleared by autophagy[23, 108, 138-140]. While some of the critical E3-ligases that mediate the final ubiquitination steps are unknown, several intermediate factors have been identified. The best studied is the role of the p62/Sequestosome-1 (**SQSTM1**) and the Alfy proteins (autophagy-linked **FYVE**) in the clearance of ubiquitinated protein aggregates (Figure 2.2). Alfy is a large (400KD) protein that contains multiple lipid (i.e. FYVE, PH) and protein interacting motifs and binds Atg5 (i.e. Beach, WD40)[112, 127, 130]. The p62 protein contains PB1 (polymerization), ZZ, LIR (LC3 interacting region), and UBA (ubiquitin binding) domains[112, 126, 127]. Therefore, p62 is capable of directly interacting with ubiquitin and autophagy components and both proteins strongly co-localize to nuclear PML bodies and cytoplasmic aggregates[130]. Together p62 and Alfy are selectively directed to these structures and act as scaffolding proteins that recruit individual lipid and Atg pathway components (i.e. LC3, Atg5) to specific sites to initiate phagophore formation (Figure 2.2)[112, 127]. Loss-of-function mutations in the Drosophila *blue cheese* gene (*bchs*, Alfy homolog) result in the rapid accumulation of ubiquitinated protein aggregates in the adult nervous system and reduced life span in the fly[128, 141]. Defects in the Drosophila *Ref(2)P* gene (p62 homolog) block normal formation of ubiquitinated aggregates, which is similar to that observed in p62 KO mice[111, 129, 142]. Studies designed to identify the components that facilitate the other forms of selective autophagy are presently ongoing.

Mouse and Human Genetics of Autophagy

Two landmark studies published in 2004, involving the knockout of the mouse Atg5 and Atg7 genes, began the phenotypic characterization of autophagic function in a mammalian model system[113, 114]. Both studies showed that while autophagy has some effect on development, nearly normal appearing neonatal mice died within 24 hours of birth. All indications are that these pups were unable to withstand the transition between continuous placental nutrition and their first milk feeding[113, 114]. The inability of KO pups to survive this stressful transitional period and access existing amino acid and energy stores is entirely consistent with this conserved pathway playing a central role during times of nutritional deprivation and stress. Subsequently, other full and conditional knockout studies in mice have extended the characterization of autophagic function in a complex mammalian system[113, 114]. It is now clear that autophagy plays a critical role in global metabolic homeostasis as well as a wide range of key developmental programs and is a factor in disorders related to neural degeneration, cardiac function, immune and inflammatory responses.

The genetics of human autophagy recently has focused on mutations or allelic variants in the p62/SQSTM1 (Paget's Disease of the Bone, **PDB**) and Atg16-like (Inflammatory Bowel Disease, **IBD-10**) genes. In the case of PDB, the autosomal dominant disorder is most commonly found in older individuals from Anglo-Saxon lineages[143-145]. In familial cases ~30% have mutations within or near the ubiquitin-binding domain (UBA), with a proline-392 to leucine substitution ($p62^{P392L}$) being the most common. In PDB affected individuals the primary cellular defect appears to be abnormal bone osteoclasts (OCL), which are increased in size and number[143, 146-149]. In general the mutations result in increased osteoclastogenesis and bone turnover. However, environmental factors appear to play a key role in disease development and pathogenesis. The p62 protein interacts with atypical protein kinases (λPKC and ζPKC) and has a key role activating NF-κB following exposure to TNF-1, IL-1 and CD40[149-151]. This suggests that autophagy may have a critical role in the long-term modulation of inflammation.

The intersection of autophagy and inflammation has also been implicated by recent linkage analysis studies done on human inflammatory bowel or Crohn's disease (**CD**)[152-154]. In humans this is a complex disorder and is associated with multiple genetic and non-genetic risk factors. Recently autophagy has been shown to play a role in the development of CD by modulating the adaptive or innate immune response[152, 153, 155]. In some familial CD cases, a polymorphism in the human Atg16-like-1 gene ($Atg16L1^{T300A}$) has been linked to an increased risk for disease development[156, 157]. Studies using Atg16 KO mice have shown that, like other autophagy genes, full deletion of Atg16L1 produced neonatal pups that die within 24 hours of parturition, indicating the gene is essential for survival at this critical period[158]. In addition to autophagy's role in cellular homeostasis, the pathway can also protect cells from invading bacterial pathogens via a selective process called xenophagy[134-137, 159]. *In vitro* studies of epithelial cells with the disease associated $Atg16L1^{T300A}$ variant show that xenophagy and the ability to eliminate intracellular pathogens is clearly impaired[160-162]. This resulted in the reduced sequestration and elimination of intracellular *Salmonella* bacteria by autophagy. It has been reported that components of the autophagy pathway are involved in the antiviral response and phagocytosis mediated by the Toll-like receptor pathway (TLR)[135, 137, 163, 164]. An examination of macrophage function in Atg16L1 deficient cells reveals an elevated profile of inflammatory cytokines, including IL-1β and IL-18[165]. This work indicates that the AtgL1 protein not only mediates the clearance of invading bacteria but also regulates the production of cytokines that are highly implicated in the progression of human CD[152, 160, 165].

As shown previously, genetic or molecular lesions in other vesicle trafficking pathways such as the endosomal or lysosomal systems can have a profound effect on autophagic function. Juvenile neuronal ceroid lipofuscinosis (**JNCL**), or Batten disease, is an inherited neurodegenerative disorder that primarily affects the young nervous system[166]. Mutations in at least six genes have been linked with the disorder and most encode proteins that have a direct role in lysosomal function[166, 167]. Histological characteristics of these disorders are the marked accumulation of ceroid lipofuscin and the mitochondrial ATPase subunit C protein within lysosomes, indicating that lysosomal and autophagic function are disrupted[166-170]. Indeed, in a mouse model of JNCL there is a delay in AVs and lysosomes maturation, along with an inappropriate accumulation of subunit C in AVs[170]. Loss-of-function mutations in a lysosomal aspartyl protease, cathepsin D, also show JNCL like phenotypes[171, 172]. Again AVs show an altered maturation rate and LC3-II turnover, suggesting that AVs and their contents are not completely processed and eliminated in neuronal tissues[118, 172, 173]. In the heart, inherited defects in the LAMP2 gene (Danon disease) are associated with perturbation of autophagic function and long-term viability of cardiac tissues[97, 174, 175]. Mutations in the lysosomal hydrolase acid alpha-glucosidase result in Pompe disease, which alters glycogen storage within cells of the heart, liver, skeletal muscle and CNS[176-178]. This is primarily a lysosomal storage disorder that also shows a marked pathology in autophagic function and a progressive buildup of AV defects[176-178]. At this time there is a rapidly expanding list of mutations and progressive human disorders that show a substantial role for the autophagy pathway in disease prevention and/or progression[135, 137, 154, 159, 175, 179-183]. As our understanding of autophagy is integrated with the endosomal and lysosomal pathways, the importance of the pathway with respect to complex disorders will come into sharper focus as a major cellular process involved in human health.

References

1. Arstila AU, Trump BF. Studies on cellular autophagocytosis. The formation of autophagic vacuoles in the liver after glucagon administration. *Am J Pathol* 2008;**53**(5):687–733.
2. Ericsson JL. Studies on induced cellular autophagy. II. Characterization of the membranes bordering autophagosomes in parenchymal liver cells. *Exp Cell Res* 2008;**56**(2):393–405.
3. Ericsson JL. Studies on induced cellular autophagy. I. Electron microscopy of cells with in vivo labelled lysosomes. *Exp Cell Res* 2008;**55**(1):95–106.
4. Harding TM, Hefner-Gravink A, Thumm M, Klionsky DJ. Genetic and phenotypic overlap between autophagy and the cytoplasm to vacuole protein targeting pathway. *J Biol Chem* 1996;**271**(30):17621–4.
5. Kametaka S, Matsuura A, Wada Y, Ohsumi Y. Structural and functional analyses of APG5, a gene involved in autophagy in yeast. *Gene* 1996;**178**(1-2):139–43.
6. Scott SV, Hefner-Gravink A, Morano KA, Noda T, Ohsumi Y, Klionsky DJ. Cytoplasm-to-vacuole targeting and autophagy employ the same machinery to deliver proteins to the yeast vacuole. *Proc Natl Acad Sci USA* 1996;**93**(22):12304–8.
7. Tsukada M, Ohsumi Y. Isolation and characterization of autophagy-defective mutants of Saccharomyces cerevisiae. *FEBS Lett* 1993;**333**(1-2):169–74.
8. Geng J, Klionsky DJ. The Atg8 and Atg12 ubiquitin-like conjugation systems in macroautophagy. Protein modifications: beyond the usual suspects' review series. *EMBO Rep* 2008;**9**(9):859–64.
9. Hanada T, Ohsumi Y. Structure-function relationship of Atg12, a ubiquitin-like modifier essential for autophagy. *Autophagy* 2005;**1**(2):110–8.
10. Chan EY, Tooze SA. Evolution of Atg1 function and regulation. *Autophagy* 2009;**5**(6):758–65.

11. Young AR, Chan EY, Hu XW, Kochl R, Crawshaw SG, High S, et al. Starvation and ULK1-dependent cycling of mammalian Atg9 between the TGN and endosomes. *J Cell Sci* 2006;**119**(Pt 18):3888–900.
12. Itakura E, Kishi C, Inoue K, Mizushima N. Beclin 1 forms two distinct phosphatidylinositol 3-kinase complexes with mammalian Atg14 and UVRAG. *Mol Biol Cell* 2008;**19**(12):5360–72.
13. Obara K, Sekito T, Ohsumi Y. Assortment of phosphatidylinositol 3-kinase complexes—Atg14p directs association of complex I to the pre-autophagosomal structure in Saccharomyces cerevisiae. *Mol Biol Cell* 2008;**17**(4):1527–39.
14. Sun Q, Fan W, Chen K, Ding X, Chen S, Zhong Q. Identification of Barkor as a mammalian autophagy-specific factor for Beclin 1 and class III phosphatidylinositol 3-kinase. *Proc Natl Acad Sci USA* 2008;**105**(49):19211–6.
15. Reggiori F, Klionsky DJ. Autophagy in the eukaryotic cell. *Eukaryot Cell* 2002;**1**(1):11–21.
16. Reggiori F, Klionsky DJ. Autophagosomes: biogenesis from scratch? *Curr Opin Cell Biol* 2005;**17**(4):415–22.
17. Xie Z, Klionsky DJ. Autophagosome formation: core machinery and adaptations. *Nat Cell Biol* 2007;**9**(10):1102–9.
18. Itakura E, Mizushima N. Characterization of autophagosome formation site by a hierarchical analysis of mammalian Atg proteins. *Autophagy* 2010;**6**(6):764–76.
19. Suzuki K, Ohsumi Y. Molecular machinery of autophagosome formation in yeast, Saccharomyces cerevisiae. *FEBS Lett* 2007;**581**(11):2156–61.
20. Yoshimori T, Noda T. Toward unraveling membrane biogenesis in mammalian autophagy. *Curr Opin Cell Biol* 2008;**20**(4):401–7.
21. Chan EY. mTORC1 phosphorylates the ULK1-mAtg13-FIP200 autophagy regulatory complex. *Sci Signal* 2009;**2**(84):pe51.
22. Fimia GM, Di Bartolomeo S, Piacentini M, Cecconi F. Unleashing the Ambra1-Beclin 1 complex from dynein chains: Ulk1 sets Ambra1 free to induce autophagy. *Autophagy* 2011;**7**(1):115–7.
23. Kim PK, Hailey DW, Mullen RT, Lippincott-Schwartz J. Ubiquitin signals autophagic degradation of cytosolic proteins and peroxisomes. *Proc Natl Acad Sci USA* 2008;**105**(52):20567–74.
24. Liao XH, Majithia A, Huang X, Kimmel AR. Growth control via TOR kinase signaling, an intracellular sensor of amino acid and energy availability, with crosstalk potential to proline metabolism. *Amino Acids* 2008;**35**(4):761–70.
25. Ost A, Svensson K, Ruishalme I, Brannmark C, Franck N, Krook H, et al. Attenuated mTOR signaling and enhanced autophagy in adipocytes from obese patients with type 2 diabetes. *Mol Med J* 2010;**16**(7-8):235–46.
26. Yorimitsu T, He C, Wang K, Klionsky DJ. Tap42-associated protein phosphatase type 2A negatively regulates induction of autophagy. *Autophagy* 2009;**5**(5):616–24.
27. Matsunaga K, Morita E, Saitoh T, Akira S, Ktistakis NT, Izumi T, et al. Autophagy requires endoplasmic reticulum targeting of the PI3-kinase complex via Atg14L. *J Cell Biol* 2010;**190**(4):511–21.
28. Stephan JS, Yeh YY, Ramachandran V, Deminoff SJ, Herman PK. The Tor and PKA signaling pathways independently target the Atg1/Atg13 protein kinase complex to control autophagy. *Proc Natl Acad Sci USA* 2009;**106**(40):17049–54.
29. Reggiori F, Tucker KA, Stromhaug PE, Klionsky DJ. The Atg1-Atg13 complex regulates Atg9 and Atg23 retrieval transport from the pre-autophagosomal structure. *Dev Cell* 2004;**6**(1):79–90.
30. Chang YY, Neufeld TP. An Atg1/Atg13 complex with multiple roles in TOR-mediated autophagy regulation. *Mol Biol Cell* 2009;**20**(7):2004–14.
31. Mari M, Griffith J, Rieter E, Krishnappa L, Klionsky DJ, Reggiori F. An Atg9-containing compartment that functions in the early steps of autophagosome biogenesis. *J Cell Biol* 2010;**190**(6):1005–22.
32. Kim J, Kundu M, Viollet B, Guan KL. AMPK and mTOR regulate autophagy through direct phosphorylation of Ulk1. *Nat Cell Biol* 2012;**13**(2):132–41.
33. Lee JW, Park S, Takahashi Y, Wang HG. The association of AMPK with ULK1 regulates autophagy. *PLoS One* 2010;**5**(11):e15394.
34. Shang L, Wang X. AMPK and mTOR coordinate the regulation of Ulk1 and mammalian autophagy initiation. *Autophagy* 2011;**7**(8):924–6.
35. Mercer CA, Kaliappan A, Dennis PB. A novel, human Atg13 binding protein, Atg101, interacts with ULK1 and is essential for macroautophagy. *Autophagy* 2009;**5**(5):649–62.
36. Cebollero E, Reggiori F. Regulation of autophagy in yeast Saccharomyces cerevisiae. *Biochim Biophys Acta* 2009;**1793**(9):1413–21.

37. Barnett SF, Bilodeau MT, Lindsley CW. The Akt/PKB family of protein kinases: a review of small molecule inhibitors and progress towards target validation. *Curr Top Med Chem* 2005;**5**(2):109–25.
38. Doukas J, Wrasidlo W, Noronha G, Dneprovskaia E, Hood J, Soll R. Isoform-selective PI3K inhibitors as novel therapeutics for the treatment of acute myocardial infarction. *Biochem Soc Trans* 2007;**35**(Pt 2):204–6.
39. Liu P, Cheng H, Roberts TM, Zhao JJ. Targeting the phosphoinositide 3-kinase pathway in cancer. *Nat Rev Drug Discov* 2009;**8**(8):627–44.
40. Wittwer F, Jaquenoud M, Brogiolo W, Zarske M, Wustemann P, Fernandez R, et al. Susi, a negative regulator of Drosophila PI3-kinase. *Dev Cell* 2005;**8**(6):817–27.
41. Itakura E, Mizushima N. Atg14 and UVRAG: mutually exclusive subunits of mammalian Beclin 1-PI3K complexes. *Autophagy* 2009;**5**(4):534–6.
42. Lindmo K, Simonsen A, Brech A, Finley K, Rusten TE, Stenmark H. A dual function for Deep orange in programmed autophagy in the Drosophila melanogaster fat body. *Exp Cell Res* 2006;**312**(11):2018–27.
43. Liang C, Feng P, Ku B, Dotan I, Canaani D, Oh BH, et al. Autophagic and tumour suppressor activity of a novel Beclin1-binding protein UVRAG. *Nat Cell Biol* 2006;**8**(7):688–99.
44. Matsunaga K, Saitoh T, Tabata K, Omori H, Satoh T, Kurotori N, et al. Two Beclin 1-binding proteins, Atg14L and Rubicon, reciprocally regulate autophagy at different stages. *Nat Cell Biol* 2009;**11**(4):385–96.
45. Zhong Y, Wang QJ, Yue Z. Atg14L and Rubicon: yin and yang of Beclin 1-mediated autophagy control. *Autophagy* 2009;**5**(6):890–1.
46. Takahashi Y, Meyerkord CL, Wang HG. Bif-1/endophilin B1: a candidate for crescent driving force in autophagy. *Cell Death Differ* 2009;**16**(7):947–55.
47. Takahashi Y, Coppola D, Matsushita N, Cualing HD, Sun M, Sato Y, et al. Bif-1 interacts with Beclin 1 through UVRAG and regulates autophagy and tumorigenesis. *Nat Cell Biol* 2009;**9**(10):1142–51.
48. Liang C, Lee JS, Inn KS, Gack MU, Li Q, Roberts EA, et al. Beclin1-binding UVRAG targets the class C Vps complex to coordinate autophagosome maturation and endocytic trafficking. *Nat Cell Biol* 2008;**10**(7): 776–87.
49. Di Bartolomeo S, Corazzari M, Nazio F, Oliverio S, Lisi G, Antonioli M, et al. The dynamic interaction of AMBRA1 with the dynein motor complex regulates mammalian autophagy. *J Cell Biol* 2010;**191**(1):155–68.
50. Fimia GM, Stoykova A, Romagnoli A, Giunta L, Di Bartolomeo S, Nardacci R, et al. Ambra1 regulates autophagy and development of the nervous system. *Nature* 2007;**447**(7148):1121–5.
51. Noda T, Kim J, Huang WP, Baba M, Tokunaga C, Ohsumi Y, et al. Apg9p/Cvt7p is an integral membrane protein required for transport vesicle formation in the Cvt and autophagy pathways. *J Cell Biol* 2000;**148**(3):465–80.
52. Ohashi Y, Munro S. Membrane delivery to the yeast autophagosome from the Golgi-endosomal system. *Mol Biol Cell* 2010;**21**(22):3998–4008.
53. Obara K, Sekito T, Niimi K, Ohsumi Y. The Atg18-Atg2 complex is recruited to autophagic membranes via phosphatidylinositol 3-phosphate and exerts an essential function. *J Biol Chem* 2008;**283**(35):23972–80.
54. Suzuki K, Morimoto M, Kondo C, Ohsumi Y. Selective autophagy regulates insertional mutagenesis by the Ty1 retrotransposon in Saccharomyces cerevisiae. *Dev Cell* 2011;**21**(2):358–65.
55. He C, Song H, Yorimitsu T, Monastyrska I, Yen WL, Legakis JE, et al. Recruitment of Atg9 to the pre-autophagosomal structure by Atg11 is essential for selective autophagy in budding yeast. *J Cell Biol* 2006;**175**(6):925–35.
56. Sekito T, Kawamata T, Ichikawa R, Suzuki K, Ohsumi Y. Atg17 recruits Atg9 to organize the pre-autophagosomal structure. *Genes Cells* 2009;**14**(5):525–38.
57. Takahashi Y, Meyerkord CL, Hori T, Runkle K, Fox TE, Kester M, et al. Bif-1 regulates Atg9 trafficking by mediating the fission of Golgi membranes during autophagy. *Autophagy* 2011;**7**(1):61–73.
58. Webber JL, Tooze SA. Coordinated regulation of autophagy by p38alpha MAPK through mAtg9 and p38IP. *EMBO J* 2010;**29**(1):27–40. Jan 6.
59. Phillips AR, Suttangkakul A, Vierstra RD. The ATG12-conjugating enzyme ATG10 Is essential for autophagic vesicle formation in Arabidopsis thaliana. *Genetics* 2008;**178**(3):1339–53.
60. Noda NN, Fujioka Y, Ohsumi Y, Inagaki F. Crystallization of the Atg12-Atg5 conjugate bound to Atg16 by the free-interface diffusion method. *J Synchrotron Radiat* 2008;**15**(Pt 3):266–8.
61. Matsushita M, Suzuki NN, Obara K, Fujioka Y, Ohsumi Y, Inagaki F. Structure of Atg5.Atg16, a complex essential for autophagy. *J Biol Chem* 2007;**282**(9):6763–72.

62. Matsushita M, Suzuki NN, Fujioka Y, Ohsumi Y, Inagaki F. Expression, purification and crystallization of the Atg5-Atg16 complex essential for autophagy. *Acta Crystallogr Sect F Struct Biol Cryst Commun* 2006;**62**(Pt 10):1021–3.
63. Tanida I, Ueno T, Kominami E. LC3 conjugation system in mammalian autophagy. *Int J Biochem Cell Biol* 2004;**36**(12):2503–18.
64. Fujita N, Hayashi-Nishino M, Fukumoto H, Omori H, Yamamoto A, Noda T, et al. An Atg4B mutant hampers the lipidation of LC3 paralogues and causes defects in autophagosome closure. *Mol Biol Cell* 2008;**19**(11):4651–9.
65. Klionsky DJ, Abeliovich H, Agostinis P, Agrawal DK, Aliev G, Askew DS, et al. Guidelines for the use and interpretation of assays for monitoring autophagy in higher eukaryotes. *Autophagy* 2008;**4**(2):151–75.
66. Simonsen A, Cumming RC, Brech A, Isakson P, Schubert DR, Finley KD. Promoting basal levels of autophagy in the nervous system enhances longevity and oxidant resistance in adult Drosophila. *Autophagy* 2008;**4**(2):176–84.
67. Stromhaug PE, Berg TO, Fengsrud M, Seglen PO. Purification and characterization of autophagosomes from rat hepatocytes. *Biochem J* 1998;**335**(Pt 2):217–24.
68. Bodemann BO, Orvedahl A, Cheng T, Ram RR, Ou YH, Formstecher E, et al. RalB and the exocyst mediate the cellular starvation response by direct activation of autophagosome assembly. *Cell* 2011 Jan 21;**144**(2): 253–67.
69. Geng J, Klionsky DJ. Quantitative regulation of vesicle formation in yeast nonspecific autophagy. *Autophagy* 2008;**4**(7):955–7.
70. Moreau K, Ravikumar B, Renna M, Puri C, Rubinsztein DC. Autophagosome precursor maturation requires homotypic fusion. *Cell* 2011;**146**(2):303–17.
71. Nair U, Klionsky DJ. Molecular mechanisms and regulation of specific and nonspecific autophagy pathways in yeast. *J Biol Chem* 2005;**280**(51):41785–8.
72. Ravikumar B, Moreau K, Jahreiss L, Puri C, Rubinsztein DC. Plasma membrane contributes to the formation of pre-autophagosomal structures. *Nat Cell Biol* 2010;**12**(8):747–57.
73. van der Vaart A, Griffith J, Reggiori F. Exit from the Golgi is required for the expansion of the autophagosomal phagophore in yeast Saccharomyces cerevisiae. *Mol Biol Cell* 2010;**21**(13):2270–84.
74. van der Vaart A, Reggiori F. The Golgi complex as a source for yeast autophagosomal membranes. *Autophagy* 2010;**6**(6):800–1.
75. Yen WL, Shintani T, Nair U, Cao Y, Richardson BC, Li Z, et al. The conserved oligomeric Golgi complex is involved in double-membrane vesicle formation during autophagy. *J Cell Biol* 2010;**188**(1):101–14.
76. Dunn Jr WA. Studies on the mechanisms of autophagy: formation of the autophagic vacuole. *J Cell Biol* 1990;**110**(6):1923–33.
77. Eskelinen EL, Reggiori F, Baba M, Kovacs AL, Seglen PO. Seeing is believing: the impact of electron microscopy on autophagy research. *Autophagy* 2011 Sep;**7**(9):935–56.
78. Hayashi-Nishino M, Fujita N, Noda T, Yamaguchi A, Yoshimori T, Yamamoto A. A subdomain of the endoplasmic reticulum forms a cradle for autophagosome formation. *Nat Cell Biol* 2009;**11**(12):1433–7.
79. Yla-Anttila P, Vihinen H, Jokitalo E, Eskelinen EL. 3D tomography reveals connections between the phagophore and endoplasmic reticulum. *Autophagy* 2009;**5**(8):1180–5.
80. Axe EL, Walker SA, Manifava M, Chandra P, Roderick HL, Habermann A, et al. Autophagosome formation from membrane compartments enriched in phosphatidylinositol 3-phosphate and dynamically connected to the endoplasmic reticulum. *J Cell Biol* 2008;**182**(4):685–701.
81. Hailey DW, Rambold AS, Satpute-Krishnan P, Mitra K, Sougrat R, Kim PK, et al. Mitochondria supply membranes for autophagosome biogenesis during starvation. *Cell* 2010;**141**(4):656–67.
82. Reggiori F, Tooze SA. The Emergence of autophagosomes. *Dev Cell* 2009;**17**(6):747–8.
83. Tooze SA, Yoshimori T. The origin of the autophagosomal membrane. *Nat Cell Biol* 2010;**12**(9):831–5.
84. Rambold AS, Lippincott-Schwartz J. Mechanisms of mitochondria and autophagy crosstalk. *Cell Cycle* 2011;**10**(23).
85. Rambold AS, Lippincott-Schwartz J. Starved cells use mitochondria for autophagosome biogenesis. *Cell Cycle* 2010;**9**(18):3633–4.
86. Hunt AN, Postle AD. Mass spectrometry determination of endonuclear phospholipid composition and dynamics. *Methods* 2006;**39**(2):104–11.

87. Vance JE. Molecular and cell biology of phosphatidylserine and phosphatidylethanolamine metabolism. *Prog Nucleic Acid Res Mol Biol* 2003;**75**:69–111.
88. Rontein D, Wu WI, Voelker DR, Hanson AD. Mitochondrial phosphatidylserine decarboxylase from higher plants. Functional complementation in yeast, localization in plants, and overexpression in Arabidopsis. *Plant Physiol* 2003;**132**(3):1678–87.
89. Matsuoka K. Chimeric fluorescent fusion proteins to monitor autophagy in plants. *Methods Enzymol* 2008;**451**:541–55.
90. Toyooka K, Moriyasu Y, Goto Y, Takeuchi M, Fukuda H, Matsuoka K. Protein aggregates are transported to vacuoles by a macroautophagic mechanism in nutrient-starved plant cells. *Autophagy* 2006;**2**(2):96–106.
91. Jahreiss L, Menzies FM, Rubinsztein DC. The itinerary of autophagosomes: from peripheral formation to kiss-and-run fusion with lysosomes. *Traffic* 2008;**9**(4):574–87.
92. Kimura S, Noda T, Yoshimori T. Dynein-dependent movement of autophagosomes mediates efficient encounters with lysosomes. *Cell Struct Funct* 2008;**33**(1):109–22.
93. Rubinsztein DC, DiFiglia M, Heintz N, Nixon RA, Qin ZH, Ravikumar B, et al. Autophagy and its possible roles in nervous system diseases, damage and repair. *Autophagy* 2005;**1**(1):11–22.
94. Webb JL, Ravikumar B, Rubinsztein DC. Microtubule disruption inhibits autophagosome-lysosome fusion: implications for studying the roles of aggresomes in polyglutamine diseases. *Int J Biochem Cell Biol* 2004;**36**(12):2541–50.
95. Rubinsztein DC, Ravikumar B, Acevedo-Arozena A, Imarisio S, O'Kane CJ, Brown SD. Dyneins, autophagy, aggregation and neurodegeneration. *Autophagy* 2005;**1**(3):177–8.
96. Fader CM, Savina A, Sanchez D, Colombo MI. Exosome secretion and red cell maturation: Exploring molecular components involved in the docking and fusion of multivesicular bodies in K562 cells. *Blood Cells Mol Dis* 2005;**35**(2):153–7.
97. Saftig P, Beertsen W, Eskelinen EL. LAMP-2: a control step for phagosome and autophagosome maturation. *Autophagy* 2008;**4**(4):510–2.
98. Settembre C, Di Malta C, Polito VA, Garcia Arencibia M, Vetrini F, Erdin S, et al. TFEB links autophagy to lysosomal biogenesis. *Science* 2011;**332**(6036):1429–33.
99. Fader CM, Colombo MI. Autophagy and multivesicular bodies: two closely related partners. *Cell Death Differ* 2009 Jan;**16**(1):70–8.
100. Platta HW, Stenmark H. Endocytosis and signaling. *Curr Opin Cell Biol* 2011;**23**(4):393–403.
101. Raiborg C, Stenmark H. The ESCRT machinery in endosomal sorting of ubiquitylated membrane proteins. *Nature* 2009;**458**(7237):445–52.
102. Henne WM, Buchkovich NJ, Emr SD. The ESCRT pathway. *Dev Cell* 2011;**21**(1):77–91.
103. Rusten TE, Simonsen A. ESCRT functions in autophagy and associated disease. *Cell Cycle* 2008;**7**(9):1166–72.
104. Berg TO, Fengsrud M, Stromhaug PE, Berg T, Seglen PO. Isolation and characterization of rat liver amphisomes. Evidence for fusion of autophagosomes with both early and late endosomes. *J Biol Chem* 1998;**273**(34):21883–92.
105. Fader CM, Sanchez D, Furlan M, Colombo MI. Induction of autophagy promotes fusion of multivesicular bodies with autophagic vacuoles in k562 cells. *Traffic* 2008;**9**(2):230–50.
106. Hurley JH, Hanson PI. Membrane budding and scission by the ESCRT machinery: it's all in the neck. *Nat Rev Mol Cell Biol* 2010;**11**(8):556–66.
107. Luzio JP, Parkinson MD, Gray SR, Bright NA. The delivery of endocytosed cargo to lysosomes. *Biochem Soc Trans* 2009;**37**(Pt 5):1019–21.
108. Filimonenko M, Stuffers S, Raiborg C, Yamamoto A, Malerod L, Fisher EM, et al. Functional multivesicular bodies are required for autophagic clearance of protein aggregates associated with neurodegenerative disease. *J Cell Biol* 2007;**179**(3):485–500.
109. Hanson PI, Shim S, Merrill SA. Cell biology of the ESCRT machinery. *Curr Opin Cell Biol* 2009;**21**(4):568–74.
110. Urwin H, Authier A, Nielsen JE, Metcalf D, Powell C, Froud K, et al. Disruption of endocytic trafficking in frontotemporal dementia with CHMP2B mutations. *Hum Mol Genet* 2010;**19**(11):2228–38.
111. Bartlett BJ, Isakson P, Lewerenz J, Heriberto Sanchez H, Kotzebue RW, Cumming R, et al. p62, Ref(2)P and Ubiquitinated Proteins are Conserved Markers of Neuronal Aging, Aggregate Formation and Progressive Autophagic Defects. *Autophagy* 2011.
112. Clausen TH, Lamark T, Isakson P, Finley K, Larsen KB, Brech A, et al. p62/SQSTM1 and ALFY interact to facilitate the formation of p62 bodies/ALIS and their degradation by autophagy. *Autophagy* 2010;**6**(3).

113. Hara T, Nakamura K, Matsui M, Yamamoto A, Nakahara Y, Suzuki-Migishima R, et al. Suppression of basal autophagy in neural cells causes neurodegenerative disease in mice. *Nature* 2006;**441**(7095):885–9.
114. Komatsu M, Waguri S, Chiba T, Murata S, Iwata J, Tanida I, et al. Loss of autophagy in the central nervous system causes neurodegeneration in mice. *Nature* 2006;**441**(7095):880–4.
115. Lee JA, Beigneux A, Ahmad ST, Young SG, Gao FB. ESCRT-III dysfunction causes autophagosome accumulation and neurodegeneration. *Curr Biol* 2007;**17**(18):1561–7.
116. Belly A, Bodon G, Blot B, Bouron A, Sadoul R, Goldberg Y. CHMP2B mutants linked to frontotemporal dementia impair maturation of dendritic spines. *J Cell Sci* 2010;**123**(Pt 17):2943–54.
117. Ju JS, Weihl CC. Inclusion body myopathy, Paget's disease of the bone and fronto-temporal dementia: a disorder of autophagy. *Hum Mol Genet* 2010;**19**(R1):R38–45.
118. Boland B, Kumar A, Lee S, Platt FM, Wegiel J, Yu WH, et al. Autophagy induction and autophagosome clearance in neurons: relationship to autophagic pathology in Alzheimer's disease. *The Journal of Neuroscience: the official journal of the Society for Neuroscience* 2008;**28**(27):6926–37.
119. Metcalf D, Isaacs AM. The role of ESCRT proteins in fusion events involving lysosomes, endosomes and autophagosomes. *Biochem Soc Trans* 2010;**38**(6):1469–73.
120. Sakurai A, Maruyama F, Funao J, Nozawa T, Aikawa C, Okahashi N, et al. Specific behavior of intracellular Streptococcus pyogenes that has undergone autophagic degradation is associated with bacterial streptolysin O and host small G proteins Rab5 and Rab7. *J Biol Chem* 2010;**285**(29):22666–75.
121. Chua CE, Gan BQ, Tang BL. Involvement of members of the Rab family and related small GTPases in autophagosome formation and maturation. *Cell Mol Life Sci* 2011;**68**(20):3349–58.
122. Lin WJ, Yang CY, Li LL, Yi YH, Chen KW, Lin YC, et al. Lysosomal targeting of phafin1 mediated by Rab7 induces autophagosome formation. *Biochem Biophys Res Commun* 2011. Nov 15.
123. Nishida Y, Arakawa S, Fujitani K, Yamaguchi H, Mizuta T, Kanaseki T, et al. Discovery of Atg5/Atg7-independent alternative macroautophagy. *Nature* 2009;**461**(7264):654–8.
124. Fader CM, Sanchez DG, Mestre MB, Colombo MI. TI-VAMP/VAMP7 and VAMP3/cellubrevin: two v-SNARE proteins involved in specific steps of the autophagy/multivesicular body pathways. *Biochim Biophys Acta* 2009;**1793**(12):1901–16.
125. Farre JC, Subramani S. Rallying the exocyst as an autophagy scaffold. *Cell* 2011;**144**(2):172–4.
126. Bjorkoy G, Lamark T, Johansen T. p62/SQSTM1: a missing link between protein aggregates and the autophagy machinery. *Autophagy* 2006;**2**(2):138–9.
127. Filimonenko M, Isakson P, Finley KD, Anderson M, Jeong H, Melia TJ, et al. The selective macroautophagic degradation of aggregated proteins requires the PI3P-binding protein Alfy. *Mol Cell* 2010;**38**(2): 265–79.
128. Finley KD, Edeen PT, Cumming RC, Mardahl-Dumesnil MD, Taylor BJ, Rodriguez MH, et al. Blue cheese mutations define a novel, conserved gene involved in progressive neural degeneration. *The Journal of Neuroscience: the official journal of the Society for Neuroscience* 2003;**23**(4):1254–64.
129. Komatsu M, Waguri S, Koike M, Sou YS, Ueno T, Hara T, et al. Homeostatic levels of p62 control cytoplasmic inclusion body formation in autophagy-deficient mice. *Cell* 2007;**131**(6):1149–63.
130. Simonsen A, Birkeland HC, Gillooly DJ, Mizushima N, Kuma A, Yoshimori T, et al. Alfy, a novel FYVE-domain-containing protein associated with protein granules and autophagic membranes. *J Cell Sci* 2004;**117**(Pt 18):4239–51.
131. Manjithaya R, Nazarko TY, Farre JC, Subramani S. Molecular mechanism and physiological role of pexophagy. *FEBS Lett* 2010;**584**(7):1367–73.
132. Oku M, Sakai Y. Peroxisomes as dynamic organelles: autophagic degradation. *FEBS J* 2010;**277**(16):3289–94.
133. Platta HW, Erdmann R. Peroxisomal dynamics. *Trends Cell Biol* 2007;**17**(10):474–84.
134. Dupont N, Temime-Smaali N, Lafont F. How ubiquitination and autophagy participate in the regulation of the cell response to bacterial infection. *Biol Cell* 2010;**102**(12):621–34.
135. Kudchodkar SB, Levine B. Viruses and autophagy. *Rev Med Virol* 2009;**19**(6):359–78.
136. Noda T, Yoshimori T. Molecular basis of canonical and bactericidal autophagy. *Int Immunol* 2009;**21**(11): 1199–204.
137. Orvedahl A, Levine B. Viral evasion of autophagy. *Autophagy* 2008 Apr;**4**(3):280–5.
138. Narendra D, Tanaka A, Suen DF, Youle RJ. Parkin is recruited selectively to impaired mitochondria and promotes their autophagy. *J Cell Biol* 2008;**183**(5):795–803.

139. Nezis IP, Simonsen A, Sagona AP, Finley K, Gaumer S, Contamine D, et al. Ref(2)P, the Drosophila melanogaster homolog of mammalian p62, is required for the formation of protein aggregates in adult brain. *J Cell Biol* 2008;**180**(6):1065–71.
140. Waguri S, Komatsu M. Biochemical and morphological detection of inclusion bodies in autophagy-deficient mice. *Methods Enzymol* 2009;**453**:181–96.
141. Simonsen A, Cumming RC, Lindmo K, Galaviz V, Cheng S, Rusten TE, et al. Genetic modifiers of the Drosophila blue cheese gene link defects in lysosomal transport with decreased life span and altered ubiquitinated-protein profiles. *Genetics* 2007;**176(2)**(1):283–97.
142. Rodriguez A, Duran A, Selloum M, Champy MF, Diez-Guerra FJ, Flores JM, et al. Mature-onset obesity and insulin resistance in mice deficient in the signaling adapter p62. *Cell Metab* 2006;**3**(3):211–22.
143. Chamoux E, Couture J, Bisson M, Morissette J, Brown JP, Roux S. The p62 P392L mutation linked to Paget's disease induces activation of human osteoclasts. *Mol Endocrinol* 2009;**23**(10):1668–80.
144. Heinen C, Garner TP, Long J, Bottcher C, Ralston SH, Cavey JR, et al. Mutant p62/SQSTM1 UBA domains linked to Paget's disease of bone differ in their abilities to function as stabilization signals. *FEBS Lett* 2010;**584**(8):1585–90.
145. Rea SL, Walsh JP, Ward L, Magno AL, Ward BK, Shaw B, et al. Sequestosome 1 mutations in Paget's disease of bone in Australia: prevalence, genotype/phenotype correlation, and a novel non-UBA domain mutation (P364S) associated with increased NF-kappaB signaling without loss of ubiquitin binding. *J Bone Miner Res* 2009;**24**(7):1216–23.
146. Duran A, Serrano M, Leitges M, Flores JM, Picard S, Brown JP, et al. The atypical PKC-interacting protein p62 is an important mediator of RANK-activated osteoclastogenesis. *Dev Cell* 2004;**6**(2):303–9.
147. Hiruma Y, Kurihara N, Subler MA, Zhou H, Boykin CS, Zhang H, et al. A SQSTM1/p62 mutation linked to Paget's disease increases the osteoclastogenic potential of the bone microenvironment. *Hum Mol Genet* 2008;**17**(23):3708–19.
148. Layfield R, Shaw B. Ubiquitin-mediated signalling and Paget's disease of bone. *BMC Biochem* 2007; **8**(Suppl. 1):S5.
149. Xu J, Wu HF, Ang ES, Yip K, Woloszyn M, Zheng MH, et al. NF-kappaB modulators in osteolytic bone diseases. *Cytokine Growth Factor Rev* 2009;**20**(1):7–17.
150. Gamper C, van Eyndhoven WG, Schweiger E, Mossbacher M, Koo B, Lederman S. TRAF-3 interacts with p62 nucleoporin, a component of the nuclear pore central plug that binds classical NLS-containing import complexes. *Mol Immunol* 2000;**37**(1-2):73–84.
151. Mamidipudi V, Wooten MW. Dual role for p75(NTR) signaling in survival and cell death: can intracellular mediators provide an explanation? *J Neurosci Res* 2002;**68**(4):373–84.
152. Saitoh T, Fujita N, Jang MH, Uematsu S, Yang BG, Satoh T, et al. Loss of the autophagy protein Atg16L1 enhances endotoxin-induced IL-1beta production. *Nature* 2008;**456**(7219):264–8.
153. Stappenbeck TS, Rioux JD, Mizoguchi A, Saitoh T, Huett A, Darfeuille-Michaud A, et al. Crohn disease: a current perspective on genetics, autophagy and immunity. *Autophagy* 2011;**7**(4):355–74.
154. Xavier RJ, Huett A, Rioux JD. Autophagy as an important process in gut homeostasis and Crohn's disease pathogenesis. *Gut* 2008;**57**(6):717–20.
155. Henckaerts L, Cleynen I, Brinar M, John JM, Van Steen K, Rutgeerts P, et al. Genetic variation in the autophagy gene ULK1 and risk of Crohn's disease. *Inflamm Bowel Dis* 2011. Nov 28.
156. Kuballa P, Huett A, Rioux JD, Daly MJ, Xavier RJ. Impaired autophagy of an intracellular pathogen induced by a Crohn's disease associated ATG16L1 variant. *PLoS One* 2008;**3**(10):e3391.
157. Rioux JD, Xavier RJ, Taylor KD, Silverberg MS, Goyette P, Huett A, et al. Genome-wide association study identifies new susceptibility loci for Crohn disease and implicates autophagy in disease pathogenesis. *Nat Genet* May; **39**(5): 596–604.
158. Travassos LH, Carneiro LA, Ramjeet M, Hussey S, Kim YG, Magalhaes JG, et al. Nod1 and Nod2 direct autophagy by recruiting ATG16L1 to the plasma membrane at the site of bacterial entry. *Nat Immunol* 2007;**11**(1):55–62. 2010.
159. Levine B. Eating oneself and uninvited guests: autophagy-related pathways in cellular defense. *Cell* 2005; **120**(2):159–62.
160. Glas J, Konrad A, Schmechel S, Dambacher J, Seiderer J, Schroff F, et al. The ATG16L1 gene variants rs2241879 and rs2241880 (T300A) are strongly associated with susceptibility to Crohn's disease in the German population. *Am J Gastroenterol* 2008;**103**(3):682–91.

161. Lappalainen M, Halme L, Turunen U, Saavalainen P, Einarsdottir E, Farkkila M, et al. Association of IL23R, TNFRSF1A, and HLA-DRB1*0103 allele variants with inflammatory bowel disease phenotypes in the Finnish population. *Inflamm Bowel Dis* 2008;**14**(8):1118–24.
162. Zhang HF, Qiu LX, Chen Y, Zhu WL, Mao C, Zhu LG, et al. ATG16L1 T300A polymorphism and Crohn's disease susceptibility: evidence from 13,022 cases and 17,532 controls. *Hum Genet* 2009;**125**(5-6):627–31.
163. Jounai N, Takeshita F, Kobiyama K, Sawano A, Miyawaki A, Xin KQ, et al. The Atg5 Atg12 conjugate associates with innate antiviral immune responses. *Proc Natl Acad Sci USA* 2007;**104**(35):14050–5.
164. Lee HK, Lund JM, Ramanathan B, Mizushima N, Iwasaki A. Autophagy-dependent viral recognition by plasmacytoid dendritic cells. *Science* 2007;**315**(5817):1398–401.
165. Billmann-Born S, Lipinski S, Bock J, Till A, Rosenstiel P, Schreiber S. The complex interplay of NOD-like receptors and the autophagy machinery in the pathophysiology of Crohn disease. *Eur J Cell Biol* 2011;**90**(6-7):593–602.
166. Mole SE. The genetic spectrum of human neuronal ceroid-lipofuscinoses. *Brain Pathol* 2004;**14**(1):70–6.
167. Rakheja D, Narayan SB, Bennett MJ. Juvenile neuronal ceroid-lipofuscinosis (Batten disease): a brief review and update. *Current molecular medicine* 2007;**S7**(6):603–8.
168. Koike M, Shibata M, Waguri S, Yoshimura K, Tanida I, Kominami E, et al. Participation of autophagy in storage of lysosomes in neurons from mouse models of neuronal ceroid-lipofuscinoses (Batten disease). *Am J Pathol* 2005;**167**(6):1713–28.
169. Vesa J, Peltonen L. Mutated genes in juvenile and variant late infantile neuronal ceroid lipofuscinoses encode lysosomal proteins. *Current molecular medicine* 2002;**2**(5):439–44.
170. Cao Y, Espinola JA, Fossale E, Massey AC, Cuervo AM, MacDonald ME, et al. Autophagy is disrupted in a knock-in mouse model of juvenile neuronal ceroid lipofuscinosis. *J Biol Chem* 2006;**281**(29):20483–93.
171. Shacka JJ, Klocke BJ, Young C, Shibata M, Olney JW, Uchiyama Y, et al. Cathepsin D deficiency induces persistent neurodegeneration in the absence of Bax-dependent apoptosis. *The Journal of Neuroscience* 2006;**27**(8):2081–90.
172. Shacka JJ, Roth KA. Cathepsin D deficiency and NCL/Batten disease: there's more to death than apoptosis. *Autophagy* 2007;**3**(5):474–6.
173. Walls KC, Klocke BJ, Saftig P, Shibata M, Uchiyama Y, Roth KA, et al. Altered regulation of phosphatidylinositol 3-kinase signaling in cathepsin D-deficient brain. *Autophagy* 2007;**3**(3):222–9.
174. Di Blasi C, Jarre L, Blasevich F, Dassi P, Mora M. Danon disease: A novel LAMP2 mutation affecting the pre-mRNA splicing and causing aberrant transcripts and partial protein expression. *Neuromuscul Disord* 2008;**18**(12):962–6.
175. Saftig P, Tanaka Y, Lullmann-Rauch R, von Figura K. Disease model: LAMP-2 enlightens Danon disease. *Trends Mol Med* 2001;**7**(1):37–9.
176. Fukuda T, Roberts A, Ahearn M, Zaal K, Ralston E, Plotz PH, et al. Autophagy and lysosomes in Pompe disease. *Autophagy* 2006;**2**(4):318–20.
177. Parenti G, Andria G. Pompe disease: from new views on pathophysiology to innovative therapeutic strategies. *Current pharmaceutical biotechnology* 2011;**12**(6):902–15.
178. Raben N, Danon M, Gilbert AL, Dwivedi S, Collins B, Thurberg BL, et al. Enzyme replacement therapy in the mouse model of Pompe disease. *Molecular genetics and metabolism* 2003;**80**(1-2):159–69.
179. Levine B, Kroemer G. Autophagy in aging, disease and death: the true identity of a cell death impostor. *Cell Death Differ* 2009;**16**(1):1–2.
180. Nixon RA. Autophagy, amyloidogenesis and Alzheimer disease. *J Cell Sci* 2007;**120**(Pt 23):4081–91.
181. Ravikumar B, Rubinsztein DC. Can autophagy protect against neurodegeneration caused by aggregate-prone proteins? *Neuroreport* 2004;**15**(16):2443–5.
182. Tannous P, Zhu H, Johnstone JL, Shelton JM, Rajasekaran NS, Benjamin IJ, et al. Autophagy is an adaptive response in desmin-related cardiomyopathy. *Proc Natl Acad Sci USA* 2008;**105**(28):9745–50.
183. Zhu H, Tannous P, Johnstone JL, Kong Y, Shelton JM, Richardson JA, et al. Cardiac autophagy is a maladaptive response to hemodynamic stress. *J Clin Invest* 2007;**117**(7):1782–93.

SECTION II

ORGAN SYSTEMS

CHAPTER

3

Autophagy in the Cardiovascular System

Kazuhiko Nishida[1], *Osamu Yamaguchi*[1], *and Kinya Otsu*[1, 2]

[1]**Department of Cardiovascular Medicine, Osaka University Graduate School of Medicine, Suita, Osaka, Japan**

[2]**Correspondence to Kinya Otsu MD. PhD., Department of Cardiovascular Medicine, Osaka University Graduate School of Medicine, 2-2 Yamadaoka, Suita 565-0871, Osaka, Japan**

INTRODUCTION

Heart disease is the leading cause of death in the industrialized world[1]. Heart disease of diverse sorts culminates in heart failure, a syndrome wherein the heart is unable to meet the metabolic demands of the body[2]. Five million Americans currently suffer from chronic heart failure with a mortality of approximately 50% at five years. The heart is an organ of remarkable plasticity, and adult cardiomyocytes are terminally differentiated and replication-deficient[2]. In response to pathological stress from neurohumoral activation, hypertension or other myocardial injury, the heart initially compensates with an adaptive hypertrophic increase in cardiac mass in order to normalize ventricular wall stress. However, cardiac hypertrophy is an independent risk factor for cardiovascular mortality[3] and is a major factor for heart failure[4, 5]. Under prolonged stress, the heart undergoes apparently irreversible changes culminating in heart failure. Although several theories exist regarding the mechanism governing the transition from hypertrophy to heart failure[6], the precise mechanism remains elusive.

Autophagy has evolved as a conserving process for bulk degradation and recycling of cytoplasmic components, such as long-lived proteins and organelles[7-9]. Autophagy is controlled by autophagy-related genes (Atgs), many of which, including Atg5, Atg7 and Beclin 1, are involved in autophagosome formation. Recent studies have demonstrated a variety of physiological and pathophysiological roles in autophagy such as adaptation to nutrient deprivation, intracellular clearance of proteins and organelles, elimination of micro-organelles including mitochondria, and maintenance of endoplasmic reticulum

Autophagy in Health and Disease
http://dx.doi.org/10.1016/B978-0-12-385101-7.00003-6

(ER). Paradoxically, autophagy also appears to modulate cell death through excessive self-digestion and degradation of essential cellular constituents[10, 11]. In the heart, the level of autophagy is altered in response to stresses triggered by cardiovascular diseases such as cardiac hypertrophy[12] and heart failure[13, 14]. The multiple features of apoptosis, necrosis and autophagy have been simultaneously observed in failing human hearts[15, 16]. Since it is unclear whether autophagy is protective or detrimental, it is imperative to have a full understanding of the cell death mechanism regarding not only apoptosis and necrosis but also autophagy in cardiomyocytes.

AUTOPHAGY IN THE HEART AT A LOW BASAL LEVEL UNDER NORMAL CONDITIONS

Cardiomyocytes are mitochondria-rich, and damaged mitochondria release proapoptotic factors such as cytochrome *c*[17]. Autophagy can remove the damaged mitochondria and prevent activation of apoptosis[18, 19].

In previous reports regarding autophagy in the heart's basal state, LAMP2-deficient mice showed excessive accumulation of autophagic vacuoles and impaired autophagic degradation of long-lived proteins, resulting in cardiomyopathy[14, 20]. In our examination, autophagy also appears to play a protective role under normal or mildly stressed conditions, in the heart's so-called basal state. Temporally controlled and cardiac-specific Atg5-deficiency in tamoxifen-treated $Atg5^{flox/flox}$;$MerCreMer^{+}$ mice[21] leads to left ventricular dilatation and contractile dysfunction (Figure 3.1a)[22]. Ultra-structural analyses of Atg5-deficient hearts reveal a disorganized sarcomere structure, misalignment and aggregation of mitochondria, and aberrant concentric membranous structures (Figure 3.1b). Inactivation of Atg5 causes the accumulation of abnormal proteins and organelles and promotes ER stress and apoptosis[22].

Autophagy also plays a role in cardiac aging. Abnormal proteins and damaged mitochondria accumulate during aging[23, 24]. In cardiomyocytes, the production of reactive oxygen species from mitochondria increases with age and leads to more mitochondrial damage[25]. In our examination, the cardiac-specific Atg5-deficient mice died earlier than the wild-type mice due to cardiac dysfunction with a disorganized sarcomere structure and collapsed mitochondria[26]. These results indicate that constitutive cardiomyocyte autophagy is required for protein quality control and normal cellular structure and function under a basal state. Accumulation of abnormal proteins and organelles, especially mitochondria, may directly cause cardiac dysfunction[27].

AUTOPHAGY IN CARDIAC HYPERTROPHY

Cardiac hypertrophy is the consequence of an increase in cardiomyocyte size. Since cardiomyocytes have little or no capacity for cellular proliferation, their only means of growth is by hypertrophy. In response to stress from neurohumoral activation, hypertension or other myocardial injury, however, the heart initially compensates with an adaptive hypertrophic increase in cardiac mass in order to normalize ventricular wall stress. This increase is

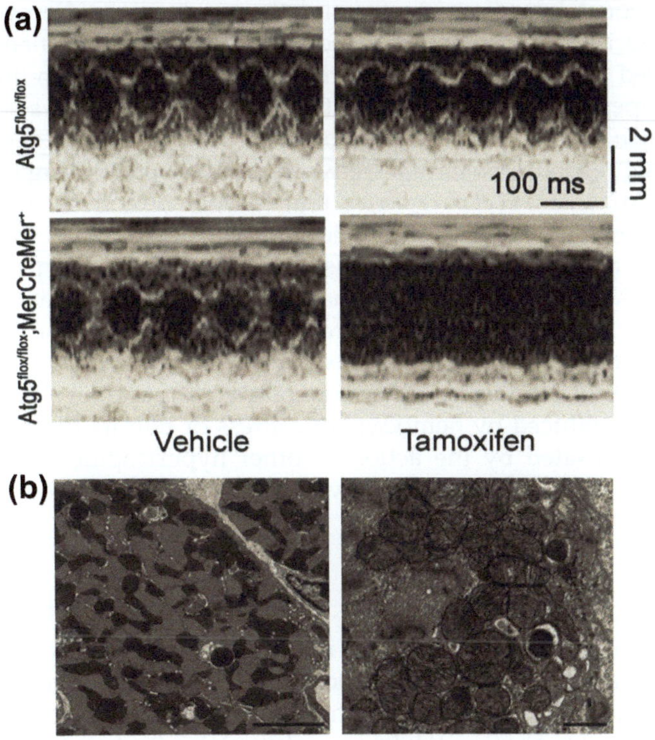

FIGURE 3.1 Cardiac dysfunction in tamoxifen-treated Atg5$^{flox/flox}$;MerCreMer^{+} mice.[22] a) Echocardiographic analysis. b) Electron micrographs of tamoxifen-treated Atg5$^{flox/flox}$;MerCreMer^{+} mouse hearts. Scale cars: 5μm in left, 1μm in right.

characterized by enhanced protein synthesis and an increase in the size and organization of cardiomyocyte sarcomeres[2, 28]. The compensated hypertrophic response is beneficial. However, prolonged stress leads to irreversible decompensation, culminating in dilatation and contractile dysfunction. This is accompanied by thinning of the ventricular walls through a combination of proteolysis and/or cardiomyocyte cell death[29].

Autophagy has been observed in hypertrophied myocardium[30]. Protein turnover is increased during hypertrophy, although in previous reports, autophagy was diminished in response to aortic stenosis[12] and isoproterenol infusion[31]. Pressure overload is accompanied by an elevated rate of myocardial protein synthesis and cardiac hypertrophy[32]. In our experiments, pressure overload due to transverse aortic constriction (TAC) induced hypertrophy without cardiac dysfunction 1 week after TAC in wild-type mice[33]. Autophagic activity was suppressed in TAC-induced hypertrophied hearts at the 1-week time point compared with that in sham-operated hearts[22]. This suggests that autophagic activity is reduced during compensative hypertrophic response[27].

We also have reported that temporally controlled and cardiac-specific Atg5-deficiency in tamoxifen-treated Atg5$^{flox/flox}$;MerCreMer^{+} mice leads to cardiac hypertrophy[22]. Knockdown of Atg7 by adenovirus expressing short hairpin RNA targeted to Atg7 inhibits autophagy in rat neonatal cardiomyocytes and then induces cardiomyocyte hypertrophy with typical characteristics[22]. Rapamycin, a potent activator of autophagy, prevents cardiac hypertrophy induced by thyroid hormone treatment[34], or aortic banding[35], and regresses existing cardiac hypertrophy[36]. This suggests that autophagy can also decrease cardiac mass and antagonize cardiac hypertrophy by increasing protein degradation. However, cardiac hypertrophic response is similar between cardiac-specific Atg5-deficient mice and control mice after TAC[22]. Zhu H *et al* reported that autophagic activity increases rapidly after severe TAC and is maintained at elevated levels for at least three weeks[37], whereas we reported that autophagic activity is suppressed in TAC-induced hypertrophied hearts as mentioned above[22]. Although mice that are haploinsufficient for Beclin 1 (Beclin $1^{+/-}$) have partial suppression of autophagic activity, cardiac hypertrophic response is similar between Beclin $1^{+/-}$ mice and control mice three weeks after TAC[37].

These results suggest that autophagy does not play an important role in regulating cardiomyocyte hypertrophy induced by hemodynamic stress or that its function in the hypertrophic process is compensated by the action of other hypertrophic signaling mechanisms. The role of autophagy in cardiac hypertrophy remains to be elucidated.

AUTOPHAGY IN HEART FAILURE

Autophagy has been observed in failing myocardium caused by dilated cardiomyopathy[13, 15, 38], by valvular disease[39], and by ischemic heart disease[40-42]. In human failing hearts with idiopathic dilated cardiomyopathy, the prevalence of autophagic, apoptotic and necrotic cells also has been observed[15]. In animal models, dead and dying cardiomyocytes showing characteristics of autophagy also have been observed. In our experiments, pressure overload due to TAC induced heart failure four weeks after TAC in wild-type mice[33]. Autophagy was up-regulated in failing wild-type hearts at the four-week time point[22, 27]. However, the question remains as to whether autophagy is a sign of failed cardiomyocyte repair or is a suicide pathway for the failing cardiomyocytes.

We examined the role of autophagy in cardiac remodeling during sustained pressure overload. Cardiac-specific Atg5 deficiency was associated with cardiac dysfunction and left ventricular dilatation one week after TAC[22]. Polyubiquitinated proteins accumulated, ER stress was increased and apoptosis was promoted in Atg5-deficient hearts. We also examined the role of autophagy against β-adrenergic stimulation in the heart. Infusion of isoproterenol for seven days in cardiac-specific Atg5 deficient mice led to greater left ventricular dilatation and contractile dysfunction compared to wild-type mice. Furthermore, adult cardiomyocytes isolated from cardiac-specific Atg5 deficient mouse hearts were more susceptible to isoproterenol than were those from control hearts. These results indicate that autophagy in failing hearts is an adaptive response to protect cells either from pressure overload or from isoproterenol stimulation[22].

Conversely, the pathological remodeling by severe pressure overload is moderately diminished in Beclin $1^{+/-}$ mice[37]. In the mice engineered for forced over-expression of Beclin 1 in cardiomyocytes, pressure overload can trigger an amplified autophagic response and pathological remodeling is more severe. This suggests that autophagy can be a maladaptive response.

These experimental outcomes lead to opposing conclusions as to whether autophagic activation by hemodynamic stress is adaptive or maladaptive as evident by several possibilities[43, 44]. First, Hill's group proposes a working model of the dual nature of autophagy in heart disease[43]. Hemodynamic stress triggers autophagy. The effect of autophagic activity depends on the level of basal autophagic flux and the amplitude and duration of autophagic activation. Autophagic activation by hemodynamic stress may be cardioprotective to a point, beyond which autophagic over-activation can exceed the threshold for autophagic cell death. Second, our Atg5 deficient mice are cardiac-specific, whereas Beclin $1^{+/-}$ mice are globally haploinsufficient. The influence of autophagic activities in non-cardiac organs must be considered. Third, Atg5 and Beclin 1 may have distinct molecule-specific functions. Since Beclin 1 is a haploinsufficient tumor suppressor gene, Beclin $1^{+/-}$ mice display an increased incidence of spontaneous malignancies[45, 46]. Considering that conventional Beclin 1-deficient mice die at 7.5 to 8.5 days of embryogenesis[45], whereas conventional Atg5- or Atg7-deficient mice survive until birth[47, 48], Beclin 1 may have additional functions beyond autophagy.

CONCLUDING REMARKS

Autophagy has evolved as a conserving process that uses bulk degradation and recycling of cytoplasmic components, such as long-lived proteins and organelles. In the heart, autophagy is beneficial for the turnover of organelles at low basal levels under normal conditions and during aging. During pathological hemodynamic stress, autophagy may not play an important role in regulating cardiomyocyte hypertrophy. However, autophagic activation by hemodynamic stress may be cardioprotective to a point, beyond which autophagic over-activation can exceed the threshold for autophagic cell death.

References

1. Lloyd-Jones D, Adams R, Carnethon M, De Simone G, Ferguson TB, Flegal K, et al, American Heart Association Statistics Committee and Stroke Statistics Subcommittee. Heart disease and stoke satistics-2009 update: a report from the American Heart Association Statistics Committee and Stroke Statistics Subcommittee. *Circulation* 2009;**119**:e21–181.
2. Hill JA, Olson EN. Cardiac plasticity. *N Engl J Med* 2008;**10**:1370–80.
3. Drazner MH, Rame JE, Marino EK, Gottdiener JS, Kitzman DW, Gardin JM, et al. Increased left ventricular mass is a risk factor for the development of a depressed left ventricular ejection fraction within five years: the Cardiovascular Health Study. *J Am Coll Cardiol* 2004;**43**:2207–15.
4. Mathew J, Sleight P, Lonn E, Johnstone D, Pogue J, Yi Q, et al, Outcomes Prevention Evaluation (HOPE) investigators. Reduction of cardiovascular risk by regression of electrocardiographic markers of left ventricular hypertrophy by the angiotensin-converting enzyme inhibitor ramipril. *Circulation* 2001;**104**: 1615–21.
5. Rame JE, Dries DL. Heart failure and cardiac hypertrophy. *Curr Treat Options Cardiovasc Med* 2007;**9**: 289–301.
6. Hilfiker-Kleiner D, Landmesser U, Drexler H. Molecular mechanisms in heart failure: focus on cardiac hypertrophy, inflammation, angiogenesis, and apoptosis. *J Am Coll Cardiol* 2006;**48**:56–66.
7. Maiuri MC, Zalckvar E, Kimchi A, Kroemer G. Self-eating and self-killing: crosstalk between autophagy and apoptosis. *Nat Rev Mol Cell Biol* 2007;**8**:741–52.
8. Levine B, Klionsky DJ. Development by self-digestion: molecular mechanisms and biological functions of autophagy. *Dev Cell* 2004;**6**:463–77.

9. Klionsky DJ, Emr SD. Autophagy as a regulated pathway of cellular degradation. *Science* 2000;**290**:1717–21.
10. Levine B, Yuan J. Autophagy in cell death: an innocent convict? *J Clin Invest* 2005;**115**:2679–88.
11. Shimizu S, Kanaseki T, Mizushima N, Mizuta T, Arakawa-Kobayashi S, Thompson CB, et al. Role of Bcl-2 family proteins in a non-apoptotic programmed cell death dependent on autophagy genes. *Nat Cell Biol* 2004;**6**:1221–8.
12. Dammrich J, Pfeifer U. Cardiac hypertrophy in rats after supravalvular aortic constriction. II. Inhibition of cellular autophagy in hypertrophying cardiomyocytes. *Virchows Arch B Cell Pathol Incl Mol Pathol* 1983;**43**:287–307.
13. Shimomura H, Terasaki F, Hayashi T, Kitaura Y, Isomura T, Suma H. Autophagic degeneration as a possible mechanism of myocardial cell death in dilated cardiomyopathy. *Jpn Circ J* 2001;**65**:965–8.
14. Nishino I, Fu J, Tanji K, Yamada T, Shimojo S, Koori T, et al. Primary LAMP-2 deficiency causes X-linked vacuolar cardiomyopathy and myopathy (Danon disease). *Nature* 2000;**406**:906–10.
15. Kostin S, Pool L, Elsasser A, Hein S, Drexler HC, Arnon E, et al. Myocytes die by multiple mechanisms in failing human hearts. *Circ Res* 2003;**92**:715–24.
16. Kostin S. Pathways of myocyte death: implications for development of clinical laboratory biomarkers. *Adv Clin Chem* 2005;**40**:37–98.
17. Gustafsson AB, Gottlieb RA. Mechanisms of apoptosis in the heart. *J Clin Immunol* 2003;**23**:447–59.
18. Decker RS, Wildenthal K. Lysosomal alterations in hypoxic and reoxygenated hearts. I. Ultrastructural and cytochemical changes. *Am J Pathol* 1980;**98**:425–44.
19. Hamacher-Brady A, Brady NR, Logue SE, Sayen MR, Jinno M, Kirshenbaum LA, et al. Response to myocardial ischemia/reperfusion injury involves Bnip3 and autophagy. *Cell Death Differ* 2007;**14**:146–57.
20. Tanaka Y, Guhde G, Suter A, Eskelinen EL, Hartmann D, Lullmann-Rauch R, et al. Accumulation of autophagic vacuoles and cardiomyopathy in LAMP-2-deficient mice. *Nature* 2000;**406**:902–6.
21. Sohal DS, Nghiem M, Crackower MA, Witt SA, Kimball TR, Tymitz KM, et al. Temporally regulated and tissue-specific gene manipulations in the adult and embryonic heart using a tamoxifen-inducible Cre protein. *Circ Res* 2001;**89**:20–5.
22. Nakai A, Yamaguchi O, Takeda T, Higuchi Y, Hikoso S, Taniike M, et al. The role of autophagy in cardiomyocytes in the basal state and in response to hemodynamic stress. *Nat Med* 2007;**13**:619–24.
23. Terman A, Dalen H, Eaton JW, Neuzil J, Brunk UT. Mitochondrial recycling and aging of cardiac myocytes: the role of autophagocytosis. *Exp Gerontol* 2003;**38**:863–76.
24. Madeo F, Tavernarakis N, Kroemer G. Can autophagy promote longevity? *Nat Cell Biol* 2010;**12**:842–6.
25. Juhaszova M, Rabuel C, Zorov DB, Lakatta EG, Sollott SJ. Protection in the aged heart: preventing the heart-break of old age? *Cardiovasc Res* 2005;**66**:233–44.
26. Taneike M, Yamaguchi O, Nakai A, Hikoso S, Takeda T, Mizote I, et al. Inhibition of autophagy in the heart induces age-related cardiomyopathy. *Autophagy* 2010;**6**:600–6.
27. Nishida K, Kyoi S, Yamaguchi O, Sadoshima J, Otsu K. The role of autophagy in the heart. *Cell Death Differ* 2009;**16**:31–8.
28. Heineke J, Molkentin JD. Regulation of cardiac hypertrophy by intracellular signalling pathways. *Nat Rev Mol Cell Biol* 2006;**7**:589–600.
29. Wang ZV, Rothermel BA, Hill JA. Autophagy in hypertensive heart disease. *J Biol Chem* 2010;**285**:8509–14.
30. Yamamoto S, Sawada K, Shimomura H, Kawamura K, James TN. On the nature of cell death during remodeling of hypertrophied human myocardium. *J Mol Cell Cardiol* 2000;**32**:161–75.
31. Pfeifer U, Fohr J, Wilhelm W, Dammrich J. Short-term inhibition of cardiac cellular autophagy by isoproterenol. *J Mol Cell Cardiol* 1987;**19**:1179–84.
32. Imamura T, McDermott PJ, Kent RL, Nagatsu M, Gt Cooper, Carabello BA. Acute changes in myosin heavy chain synthesis rate in pressure versus volume overload. *Circ Res* 1994;**75**:418–25.
33. Yamaguchi O, Higuchi Y, Hirotani S, Kashiwase K, Nakayama H, Hikoso S, et al. Targeted deletion of apoptosis signal-regulating kinase 1 attenuates left ventricular remodeling. *Proc Natl Acad Sci USA* 2003;**100**:15883–8.
34. Kuzman JA, O'Connell TD, Gerdes AM. Rapamycin prevents thyroid hormone-induced cardiac hypertrophy. *Endocrinology* 2007;**148**:3477–84.
35. Ha T, Li Y, Gao X, McMullen JR, Shioi T, Izumo S, et al. Attenuation of cardiac hypertrophy by inhibiting both mTOR and NFkappaB activation in vivo. *Free Radic Biol Med* 2005;**39**:1570–80.

36. McMullen JR, Sherwood MC, Tarnavski O, Zhang L, Dorfman AL, Shioi T, et al. Inhibition of mTOR signaling with rapamycin regresses established cardiac hypertrophy induced by pressure overload. *Circulation* 2004;**109**:3050–5.
37. Zhu H, Tannous P, Johnstone JL, Kong Y, Shelton JM, Richardson JA, et al. Cardiac autophagy is a maladaptive response to hemodynamic stress. *J Clin Invest* 2007;**117**:1782–93.
38. Knaapen MW, Davies MJ, De Bie M, Haven AJ, Martinet W, Kockx MM. Apoptotic versus autophagic cell death in heart failure. *Cardiovasc Res* 2001;**51**:304–12.
39. Hein S, Arnon E, Kostin S, Schonburg M, Elsasser A, Polyakova V, et al. Progression from compensated hypertrophy to failure in the pressure-overloaded human heart: structural deterioration and compensatory mechanisms. *Circulation* 2003;**107**:984–91.
40. Yan L, Vatner DE, Kim SJ, Ge H, Masurekar M, Massover WH, et al. Autophagy in chronically ischemic myocardium. *Proc Natl Acad Sci USA* 2005;**102**:13807–12.
41. Elsasser A, Vogt AM, Nef H, Kostin S, Mollmann H, Skwara W, et al. Human hibernating myocardium is jeopardized by apoptotic and autophagic cell death. *J Am Coll Cardiol* 2004;**43**:2191–9.
42. Depre C, Vatner SF. Cardioprotection in stunned and hibernating myocardium. *Heart Fail Rev* 2007;**12**:307–17.
43. Rothermel BA, Hill JA. Autophagy in load-induced heart disease. *Circ Res* 2008;**103**:1363–9.
44. Nishida K, Yamaguchi O, Otsu K. Crosstalk between autophagy and apoptosis in heart disease. *Circ Res* 2008;**103**:343–51.
45. Yue Z, Jin S, Yang C, Levine AJ, Heintz N. Beclin 1, an autophagy gene essential for early embryonic development, is a haploinsufficient tumor suppressor. *Proc Natl Acad Sci USA* 2003;**100**:15077–82.
46. Qu X, Yu J, Bhagat G, Furuya N, Hibshoosh H, Troxel A, et al. Promotion of tumorigenesis by heterozygous disruption of the Beclin 1 autophagy gene. *J Clin Invest* 2003;**112**:1809–20.
47. Kuma A, Hatano M, Matsui M, Yamamoto A, Nakaya H, Yoshimori T, et al. The role of autophagy during the early neonatal starvation period. *Nature* 2004;**432**:1032–6.
48. Komatsu M, Waguri S, Ueno T, Iwata J, Murata S, Tanida I, et al. Impairment of starvation-induced and constitutive autophagy in Atg7-deficient mice. *J Cell Biol* 2005;**169**:425–34.

CHAPTER

4

Autophagy in the Immune System

Michael Gurney, Girija Muralidhar, and Phyllis-Jean Linton

Donald P. Shiley BioScience Center, San Diego State University, San Diego, CA, USA

OVERVIEW OF AUTOPHAGY

Autophagy involves the sequestering of cytoplasmic constituents into a de novo-formed double-membrane vesicle called the autophagosome, and the delivery of the cargo to lysosomes for degradation[1] (Figure 4.1). The formation of the autophagosome is mediated by the coordinated action of autophagy-related proteins (Atgs). Beclin 1, the mammalian ortholog of yeast Atg6, is a critical component of a complex comprised of PI3K class III Vps34 that regulates macroautophagy through the stimulation of autophagosome nucleation[2, 3]. Beclin 1 also interacts with Bcl-2 and Bcl-x_L[4, 5], and it is believed that this interaction links autophagy and apoptosis. The formation or elongation of the autophagosome requires two ubiquitin-like conjugation systems that are regulated by Atg7: these involve the conjugation of Atg12, Atg5, and Atg16L1 and the conjugation of the Atg4-cleaved Atg8 (also known as LC3) to phosphatidylethanolamine[6]. As the autophagosome forms, Atg4 removes the cleaved Atg8 (also called LC3-II) from the external autophagosomal membrane. Eventually, the autophagosome fuses with a lysosome or an endosome via an interaction with Lamp-2, becoming an autophagolysosome or amphisome. Knocking down the proteins involved in the conjugation processes (e.g., Atg5 or Atg7) or inhibiting PI3K class III causes inhibition of macroautophagy. A non-canonical form of autophagy also exists which is independent of Atg5/7 and Beclin 1; it can be induced by Ulk1 and involves the GTPase Rab9. Its regulation and ability to function in place of canonical autophagy is incompletely understood.

Atg1/Ulk1 complex is a positive regulator of autophagosome formation. Under normal (unstressed) conditions phosphorylated Atg1 and Atg13 are bound by the mammalian target of rapamycin (mTOR) complex 1 (mTORC1), which restricts autophagosome formation to basal levels. Under conditions of stress, e.g., starvation, mTOR dissociates from the complex, freeing Atg1 and Atg13 to trigger autophagosome nucleation and elongation. As noted in Table 4.1, the deletion of autophagy genes or a reduction in their expression results in profound alterations in the phenotype and function of the immune system.

Autophagy in Health and Disease
http://dx.doi.org/10.1016/B978-0-12-385101-7.00004-8

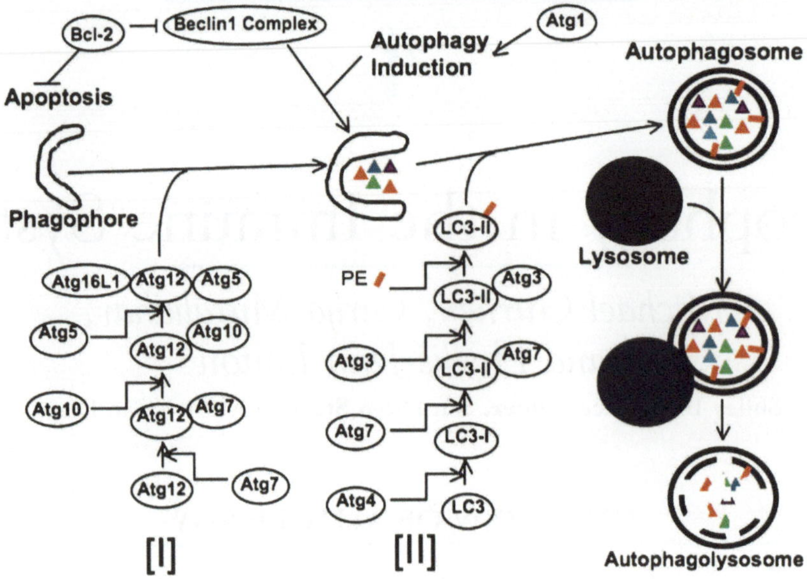

FIGURE 4.1 **Schematic Overview of Autophagy.** Stimulation of Atg1 in conjunction with the Beclin 1 complex results in the nucleation and elongation of the phagophore. Following the induction of autophagy, Beclin 1 dissociates from the complex, releasing the anti-apoptotic proteins Bcl-2 and Bcl-X_L. The formation of the autophagosome requires two ubiquitin-like conjugation systems, labeled [I] and [II]. In the first pathway, Atg7 and Atg10 are needed to create the Atg5-Atg12-Atg16L1 complex required for the elongation of the phagophore membrane. The second pathway involves the incorporation of LC3-II into the autophagosomal double membrane. Atg4 cleaves the terminal amino acid(s) from LC3, yielding LC3-I. The activity of the ubiquitin-like proteins, Atg3 and Atg7, result in the addition of phosphatidylethanolamine (in red) to LC3-I, resulting in LC3-II and subsequent incorporation of LC3-II into the autophagosomal membrane. The autophagosome encloses its cargo and undergoes lysosomal fusion mediated by LAMP2. In the autophagolysosome, lytic lysosomal-borne enzymes catalyze the degradation of the inner autophagosomal membrane in addition to the protein/organelle cargo.

AUTOPHAGY AND ANTIGEN PROCESSING/PRESENTATION

Antigen presenting cells (APCs) of the innate immune system monitor the body for the presence of foreign material as well as dying or damaged cells. These antigens are processed by APCS. Peptides that are generated by the proteasome are presented on MHC class I molecules and recognized by CD8 T cells, whereas the products of lysosomal degradation are presented on MHC class II molecules and recognized by CD4 T cells. Antigens presented on MHC class II are typically exogenous molecules that are endocytosed by APCs or endogenous proteins that reside in the secretory system. Biochemical analysis of the natural MHC class II ligands revealed that up to 20% of the eluted peptides originate from cytosolic and nuclear proteins[7, 8]. Among the peptides eluted from HLA-DR molecules of Epstein Barr virus (EBV)-transformed B cell lines were two fragments of Atg8 (LC3)[7]. This finding

TABLE 4.1 Effect of Altered Autophagy Gene Expression on the Immune System

Autophagy Gene	Phenotype/Function
Atg5 deletion in thymic epithelial cells and dendritic cells	Thymus: cellularity decreased; morphology maintained
	Positive selection: CD4 cells altered; CD8 cells unaffected
Atg5 deletion in fetal liver cells	Normal thymus development but decreased T cell numbers in periphery of repopulated mice
	Reduced T cell proliferation after TCR stimulation
Atg7 deletion in hematopoietic cells	Inability to reconstitute hematopoietic system
	Numbers of progenitor cells from multiple lineages reduced due to increased apoptosis
Beclin 1 deficiency in hematopoietic cells	Thymus cellularity reduced
	in vitro T cell development impaired
	Number of early B cells decreased in bone marrow
	Peripheral T cell numbers and proliferation intact
Atg7 deletion in T lineage cells	Marked reduction in peripheral T cell numbers due to apoptosis
	Response by CD4 effector cells: cytokine production and proliferation reduced; apoptosis unchanged
Atg5 deficiency in B lineage cells	Marked reduction in number of B-1a B cells in the peritoneum and a dramatic increase in cell death in bone marrow pro- to pre-B cells
Atg16L1 deletion	Endotoxin-stimulated macrophages produce increased amounts of IL-1β and IL-18

suggested that macroautophagy can deliver cytosolic and nuclear antigens onto MHC class II molecules[7, 9, 10, 11].

Other studies using pharmacological agents implicated autophagy as having a role in antigen presentation. These studies showed that inhibition of autophagy with 3-methyladenine abrogated or reduced MHC class II presentation of endogenous antigen, bacterial peptides, and tumor antigens[9, 12, 13]. Amino acid deprivation, which is known to induce autophagy, was also shown to effect changes in MHC class II peptide presentation[7]. These studies suggest that autophagy promotes MHC class II presentation of cytosolic antigens; however, due to the broad effects of 3-methyladenine, e.g., on membrane trafficking, lysosomal acidification, phosphorylation of signal transduction molecules and on mitochondrial permeability[14], the role of autophagy in antigen processing and presentation remained inconclusive until the studies of Paludan *et al*[10]. This study demonstrated that the viral antigen EBNA1 is processed intracellularly for MHC class II presentation to CD4 T cells via macroautophagy in EBV-transformed B cells and EBNA1-transfected EBV-negative Hodgkin's lymphoma cells[10, 15]. The endogenously synthesized viral protein EBNA1 was shown to accumulate in autophagosomes when lysosomal degradation was inhibited, and the processing of EBNA1 for MHC class II presentation was shown to be inhibited with siRNA-silencing of the autophagy gene Atg12.

MHC Class II Presentation of Endogenous Antigens

Macroautophagy is a constitutively active pathway in all MHC class II positive cells, including dendritic cells, B lymphocytes and epithelial cells[16]. Upon examination of these cells in mice transgenic for the green fluorescent protein (GFP) fused to Atg8 (GFP-LC3), at least half of all GFP-LC3 positive autophagosomes were found to frequently fuse with MHC class II loading compartments (MIICs), suggesting that macroautophagy frequently delivers autophagosome content to MIICs. In addition to continuously receiving input from the cytoplasm via macroautophagy, Schmid *et al*[16] demonstrated that the targeting of antigens to autophagosomes led to enhanced MHC class II presentation and CD4 T cell stimulation. When the matrix protein 1 (MP1) of influenza virus was fused to LC3, a 5-20 fold enhancement of MHC class II presentation by epithelial cells, EBV-transformed B lymphoblastoid cells, and dendritic cells to MP1-specific CD4 T cell clones was observed.

Chaperone mediated autophagy also has a role in delivering cytosolic antigens for MHC class II presentation[17]. Chaperone-mediated autophagy imports individual cytosolic proteins containing specific pentapeptide recognition motifs into the lysosome via a particular isoform of lysosome-associated membrane protein 2 (LAMP2a) and an accessory chaperone, the heat-shock protein HSC70[18, 19, 20]. Targeting to the chaperone-mediated autophagy pathway is intrinsic to a large fraction of self proteins, as the targeting signal (KFERQ) is present in ~30% of all cytosolic proteins[21]. Zhou *et al* have shown that overexpression of LAMP2a or HSC70 increases cytoplasmic self-antigen presentation of glutamate decarboxylase 65 (GAD65) and the mutant human immunoglobulin κ chain SMA[11]. MHC class II processing of these auto-antigens requires proteasomal and calpain-mediated degradation[22] suggesting that these proteins are processed into peptides in the cytosol and then imported to MIICs by LAMP-2a[11].

Autophagy and Tolerance

Central and peripheral tolerance in the CD4 T cell compartment is dependent upon the presentation of self-antigens in the context of MHC class II by thymic epithelial cells and dendritic cells. In the thymus, developing T cells interact with cortical and medullary epithelial cells, which promote the selection of antigen-specificities. Thymic epithelial cells are the only non-hematopoietic cell type that constitutively expresses MHC class II. Whereas MHC class II-peptide ligands are primarily generated through the processing of exogenous proteins, thymic epithelial cells are inefficient at capturing and processing extracellular antigens[23]. Therefore, it has been suggested that thymic epithelial cells rely on intracellular pathways, such as autophagy, for the processing of antigens and subsequent MHC class II loading. In agreement with this hypothesis, high levels of constitutive autophagy were observed in the thymic epithelial cells of newborn mice that transgenically express GFP-LC3[24]. These findings suggest that autophagy may enable thymic epithelial cells to present self-antigens to developing thymocytes during antigen selection. In addition,

Schmid *et al*[16] noted considerable autophagy in immature dendritic cells, which have been implicated in the induction of peripheral tolerance[25]. To determine if autophagy in thymic epithelial cells influences T cell selection, thymi from either Atg5-deficient or wild type mice were engrafted into a variety of different T cell receptor transgenic mice[26]. Although $Atg5^{-/-}$ grafts were reduced in cellularity, the morphology and compartmentalization of the thymus were maintained. However, most provocative was the finding that the depletion of Atg5 specifically in thymic epithelial cells altered the selection of certain MHC class II-restricted T cell specificities, such that severe colitis and multi-organ inflammation resulted. Altogether these findings support a role for autophagy in the generation of a self-tolerant T cell repertoire by focusing the MHC class II peptides intracellularly. No change in the selection of CD8 T cells was observed[26].

AUTOPHAGY AND DEVELOPMENT

To determine the effect of autophagy on hematopoietic stem cell (HSC) maintenance and function, bone marrow HSCs and progenitor cells were examined in mice with Atg7 conditionally depleted in the hematopoietic system[27]. The inability of $Atg7^{-/-}$ bone marrow cells or $Lin^{-}Sca\text{-}1^{+}c\text{-}Kit^{+}$ (LSK) cells, which are enriched for HSCs and early progenitors, to reconstitute the hematopoietic system of lethally irradiated mice suggests that Atg7 is essential for HSC function. In line with the loss of HSC activity, reductions in progenitors of multiple lineages (i.e., the common lymphoid progenitor, natural killer (NK) cell progenitor and myeloid progenitor compartments) were observed in hematopoietic conditional Atg7-deficient mice. The LSK cells were shown to display an increase in apoptosis as well as an accumulation of mitochondria, mitochondrial superoxide and DNA damage, suggestive of a decline in mitophagy, i.e., autophagic clearance of mitochondria. The finding that autophagy is requisite to sustain HSC function is consistent with the findings that normal quiescent HSCs are maintained in hypoxic niches[28] and autophagy plays an adaptive pro-survival role in response to hypoxia[29].

The autophagy protein Beclin 1 also plays a role in lymphocyte development. Because targeted disruption of Beclin 1 results in early embryonic lethality before the generation of the lymphoid system, Arsov *et al*[30] examined the role of Beclin 1 in lymphoid development in Beclin $1^{-/-} \rightarrow Rag1^{-/-}$ chimeric mice and control Beclin $1^{+/-} \rightarrow Rag1^{-/-}$ chimeric mice using micro-injected Beclin $1^{-/-}$ and Beclin $1^{+/-}$ embryonic stem cell (ESC) clones, respectively. [$Rag1^{-/-}$ mice cannot repopulate the lymphoid compartment and thus, lymphoid cells in these mice would be the progeny of the ESCs.] Compared to control mice, Beclin 1-deficient chimeras display a marked reduction in thymic cellularity and a decline in thymocytes that have developed into the $CD4^{+}CD8^{+}$ double positive stage. Co-cultures of Beclin 1-deficient ESCs with OP9 stromal cells that are virally transduced to express the Notch ligand Delta-like 1 (OP9-DL1) failed to sustain normal T cell development. These findings suggest that the inability to maintain normal thymic cellularity is likely the consequence of the impaired maintenance of thymocyte progenitors due to impaired Beclin 1 expression. This autophagy protein appears to play a more general role in lymphocyte development as evidenced by a reduction in the number of early B cells in the bone marrow of the Beclin 1-deficient compared to control chimeras[30]. The absence of Beclin 1 does not appear to impact peripheral

T cell numbers or their proliferative response *in vitro*. LC3 puncta were detected in Beclin 1-deficient T cells, suggesting that Beclin 1-independent autophagy may be sufficient to sustain normal peripheral T cell proliferation[30].

AUTOPHAGY AND T CELLS

Autophagy has a role in the life and death decision-making in several cell types. During nutrient deprivation and cellular stress, autophagy is elevated and functions as a pro-survival mechanism. However, when autophagy cannot keep up with the needs of the cell or when apoptosis is defective, autophagy functions in cell death. Similarly, autophagy has a role in maintaining homeostasis in T cells.

The pool of T cells in the periphery is tightly regulated. A role for autophagy in T cell survival and proliferation was first shown *in vivo* using lethally irradiated mice repopulated with hematopoietic cells from fetal livers of $Atg5^{-/-}$ mice[31]. Mice with an Atg5 deficiency had decreased T cell numbers in secondary lymphoid organs; furthermore, CD4 and CD8 T cells failed to undergo efficient proliferation after T cell receptor (TCR) stimulation[31]. In the chimeric mice, $Atg5^{-/-}$ T cells developed normally in the thymus, but failed to repopulate the periphery due to overwhelming cell death. Although these findings suggest a role for autophagy in T cell survival, one cannot formally exclude that the defective survival of $Atg5^{-/-}$ T cells is caused by: 1. the role of Atg5 in the induction of autophagy or other survival pathways, as Atg5 interacts with FADD and $Bcl\text{-}X_L$; or 2. the contribution of autophagy in other hematopoietic cells. However, a similar survival defect was obtained in mice using the Cre-LoxP technology to delete Atg7 alleles exclusively within T lymphocytes early in development[32]. These conditional knockout mice demonstrate a marked reduction (~75%) in the total number of mature CD4 and CD8 T cells, thus confirming an important cell-intrinsic role for autophagy in T cells. Consistent with the enhanced apoptosis observed in $Atg5^{-/-}$ T cells, these autophagy-deficient lymphocytes have increased Annexin V staining on the cell surface as well as enhanced activation of effector caspases. Therefore, the loss of autophagic machinery in primary T cells results in apoptotic cell death, defining autophagy as a pro-survival pathway during normal T lymphocyte homeostasis. It should be noted that in sharp contrast with the peripheral T cell defects reported for Atg5 or Atg7-deficient T cells, a Beclin 1 deficiency did not impact peripheral T cell numbers nor their proliferative response *in vitro*[30]. LC3 puncta were detected in Beclin 1-deficient T cells, suggesting that these cells retain some autophagic activity, possibly explaining the different outcomes between Beclin 1 and Atg5 and Atg7.

One interpretation is that T cells, on exit from the thymus, become exposed to nutritional stress and require autophagy to sustain them. Data in the literature demonstrate that autophagy contributes to cell survival by supporting metabolism during growth factor withdrawal. However, autophagy has other functions, including the sequestration and degradation of whole organelles. This organelle-engulfing autophagy may occur as part of a programmed cellular response to stress, such as in the ER unfolded protein response and mitochondrial depolarization. Pua *et al*[32] found that mitochondrial content is indeed developmentally regulated in T lymphocytes exiting the thymus, marking a transition from high mitochondrial content in thymocytes to lower mitochondrial content in mature

T cells. Whether mitochondria removal by autophagy, i.e., "mitophagy", is a developmentally programmed process or occurs as a response to a change in stress encountered by T cells once they enter peripheral circulation remains unknown.

Effector T Cell Responses

Activation of T cells is a highly demanding bioenergetic process and a recent study suggests that autophagy may accommodate these demands in helper T (Th) effector responses[33]. Hubbard *et al* showed that cytokine production and proliferation of Th cells after engagement of the TCR and the CD28 costimulatory molecule requires autophagosome formation and clearance[33]. Using a conditional knockout mouse model in which Atg7 is specifically deleted in T cells, effector Th cells display defective IL-2 and Interferon-γ (IFN-γ) production, reduced proliferation, with no significant increase in apoptosis and a decline in ATP generation. Importantly, when an exogenous energy source was provided to the Atg7 deleted Th cells, cytokine production was restored. Examination of the autophagosome cargo revealed differences in the resting vs. activated cells in that mitochondria and other organelles were absent in the autophagosomes of activated Th cells. As suggested by Hubbard *et al*[33], the need for increased ATP production provided by the mitochondria could explain why these organelles are spared from autophagic degradation during T cell activation. Instead, autophagy turns to degradation of cytosolic soluble components to ensure a sufficient energetic and metabolic output.

Excessive autophagy has been linked to the death of effector T cells. Th2 polarized CD4 T cells display more GFP-LC3 positive autophagosomes than Th1 cells[34]. In this study[34], Th2 cells become more resistant to cell death induced by growth factor withdrawal when autophagy is blocked using pharmacological inhibitors or RNA silencing of Atg7 or Atg6 (Beclin 1). This finding suggests that Th2 cells either die by macroautophagy or need the autophagy machinery to undergo apoptotic death upon growth factor withdrawal.

T Cell Death

The dual role of autophagy in life and death decisions is not unexpected given the extensive interplay between autophagy and apoptosis[35, 36, 37]. Studies with HIV-1 suggest that the virus not only uses the autophagy machinery for replication but it also uses autophagy in the depletion of uninfected bystander CD4 T cells[38, 39]. Espert *et al*[39] demonstrated that the binding of HIV-1 envelope protein to CXCR4 on uninfected CD4 T cells leads to cell death that is reversed by pharmacological autophagy inhibitors or siRNA against Atg7 and Beclin 1. Future studies are required to determine if this mode of cell death contributes to the progressive decline in CD4 T cell numbers that occurs in patients with AIDS.

Apoptosis is a process by which cells are induced to commit suicide, resulting in nuclear contraction, DNA fragmentation, changes in mitochondrial membrane permeability, and the compartmentalization of cytoplasmic components, forming apoptotic bodies and blebs[40]. Apoptosis is controlled by one of two upstream regulatory pathways: the intrinsic and extrinsic[41]. Interestingly, the intrinsic and extrinsic pathways associate with autophagy proteins, thus providing a link between autophagy and apoptosis[42; 43].

Intrinsic Pathway

The intrinsic pathway is initiated by signals of intracellular stress or damage, such as DNA damage or ER stress, and is regulated by the Bcl-2 family, resulting in apoptosome formation and apoptosis[42, 44]. The Bcl-2 protein family is known to control apoptosis by either promoting or inhibiting apoptosis. The anti-apoptotic members, Bcl-2 and Bcl-X_L (Bcl-2/X_L) were shown to associate with the autophagy protein, Beclin 1[3], and this association functions to regulate autophagy and apoptosis[3, 44]. It is thought that under stress Bcl-2/X_L decreases Beclin 1-mediated autophagosomal nucleation[3], and that disruption of the Beclin 1/Bcl-2/X_L complex through either posttranslational modification, gene expression or phosphorylation allows Beclin 1-induced autophagy concomitant with Bcl-2/X_L induced anti-apoptosis. Thus, it has been suggested that the interaction between Beclin 1 and Bcl-2/X_L may fine-tune the cellular control of autophagy[3].

Extrinsic Pathway

The extrinsic pathway is initiated by the ligation of death receptors residing on the cell surface, followed by the activation of the cytoplasmic caspase cascade and the formation of the apoptosome[37]. Death receptors possess cytoplasmic death domains that promote the recruitment of adapter molecules such as FADD (Fas Associated Death Domain) and assist in the assembly of complexes that promote survival or apoptosis through caspase-8[37]. A non-catalytic caspase-8-like protein along with caspase-8 (and 10 in humans) are recruited to a death-induced signaling complex (DISC) that promotes either survival or apoptosis depending on the intracellular context[45]. In support of autophagy playing a role in this pathway, Yu *et al*[46] showed that knocking down caspase-8 resulted in increased autophagy and Pyo *et al*[47] showed that Atg5 plays a role in IFN-γ-induced autophagic cell death by interacting with FADD. Further studies are needed to understand the specific role of autophagy.

AUTOPHAGY AND B CELLS

Autophagy is required for the efficient development of all B cells and for the maintenance of B-1a B cell numbers[48]. Using Atg5-deficient mice, Miller *et al*[48] observed a dramatic reduction in the number of B-1a B cells in the peritoneum, fewer B-2 B cells in the secondary lymphoid tissue and a marked increase in cell death in bone marrow cells transitioning from the pro- to pre-B cell stage. Although the requirement for Atg5 in bone marrow B cell development could explain the decrease in peripheral B-2 B cells, the requirement for Atg5 in the survival of B-2 B cells in the periphery could not be ruled out. However, the finding of similar B-2 B cell numbers and reduced B-1a B cell numbers in mice with the targeted Atg5 deletion specific to mature B cells suggest that autophagy is not required for the maintenance of B-2 B cells but is required for the maintenance of B-1a B cells.

As mentioned above, Beclin 1 was found to be essential at early stages of both T and B cell development[30]. Although it is very clear that Atg5 and Atg7 have a role throughout lymphocyte development, Beclin 1 dependent processes are selectively required in progenitor cells and early developmental stages. Whether these differences reflect the role of Beclin 1 in autophagy or in other autophagy-independent function in cell survival/differentiation has yet to be determined.

In addition to B cell development, autophagy plays a role in responses by mature B cells. It is known that crosslinking the B cell receptor (BCR) with antigen can induce B cells to undergo apoptosis or B cell activation. Recently, it was found that crosslinking the BCR of primary B cells and B cell lines with either antigen or anti-BCR antibody leads to the accumulation of cytoplasmic LC3 puncta characteristic of autophagosome formation[49]. Furthermore, it was shown that the ligation of BCR along with costimulation through CD40 not only rescued B cells from apoptosis but blocked autophagy. Because B cells are able to interact with CD40 ligand (CD40L) after CD40L-expressing T cells recognize antigen presented by B cells, these findings led to the model that BCR crosslinking induces autophagosome formation, MHC class II loading and presentation to T cells, and that CD40L-CD40 interaction blocks further autophagy and rescues cells from apoptosis[50]. In support of this model, it was shown that BCR ligation also translocates Toll-like receptor 9 (TLR9) to autophagosomes where TLR9 appears to interact with its ligands[51]. Altogether, these findings suggest that autophagy plays a key role in B cell activation by connecting BCR signaling to costimulatory signaling.

AUTOPHAGY REGULATION BY IMMUNE SIGNALS

Not only does autophagy enhance innate and adaptive immune responses, but macroautophagy is also induced as an effector mechanism of innate and adaptive immune cells through IFNs and members of the tumor necrosis factor (TNF) family. Immune signaling molecules shown to positively regulate autophagy in some contexts include double-stranded-RNA-dependent protein kinase R[52], IFN-γ (and its downstream effector immunity-related GTPases)[47, 53-56], members of the TNF family[57-59], and CD40-CD40L interaction[60]. Autophagy has been shown to be negatively regulated by the Th2-type cytokines IL-4 and IL-13 in non-immune cell lines[61, 62].

Both type I and II IFNs modulate autophagy. Type I IFN can induce macroautophagy and prevent HSV-1 neurovirulence through the activation of double-stranded-RNA-dependent protein kinase R[63]. IFN-γ has been reported to enhance *Mycobacterium tuberculosis* degradation by macroautophagy in infected cells[53]. The IFN-γ-induced autophagy and antimycobacterial activity in macrophages requires immunity-related GTPases (IRGs), which are also known as p47 GTPases, and include the mouse LRG-47 and the human IRGM proteins[53, 56]. Whereas IFN-γ readily induces the upregulation of macroautophagy in mouse macrophages, this effect is difficult to detect in human macrophages. This is most likely because murine IRGs are IFN-γ-inducible whereas IFN-γ does not regulate the expression of human IRGM although autophagy must be induced for IRGM to exert its action. The mechanism by which IRGM promotes autophagy not known.

TNFα, TNF-related apoptosis-inducing ligand (TRAIL) and CD40L are members of the TNF family and upregulate autophagy. TNFα was shown to enhance macroautophagy in cells lacking NF-κB activation[57]. TRAIL induces autophagy in human epithelial cells[64]. The induction of CD40 signaling through interaction with CD40L was shown to induce macroautophagy-mediated fusion of phagosomes containing *Toxoplasma gondii* with lysosomes in macrophages[60].

AUTOPHAGY, AUTOIMMUNITY AND INFLAMMATION

Autophagy and inflammation are highly inter-related processes. Pathogen associated molecular patterns (PAMPs) as well as damage associated molecular proteins (DAMPs) trigger pathogen recognition receptors (PRRs) to coordinately activate transcription factors that lead to the expression of inflammatory cytokines, chemokines, type I IFNs and antimicrobial genes. Autophagy is integrated in the PRR pathways. Signaling through TLR, one family of PRRs, was shown to induce autophagy in macrophages/monocytes[65, 66]. Similarly, Cooney *et al*[67] showed that stimulation of Nod 2 (nucleotide-binding oligomerization domain 2), an intracellular sensor that recognizes unique bacterial polypeptides, induces the formation of autophagosomes, thus promoting antigen presentation on MHC class II molecules by human dendritic cells. Moreover, a Nod 2 deficiency had previously been shown to result in enhanced inflammation. Thus, aberrant activation of innate immune responses utilizing these pathways may result in the development of inflammatory disorders, such as that observed in Crohn's disease.

Very recent data suggest that the specific form of macroautophagy that catabolizes mitochondria, i.e., mitophagy, is a critical negative regulator of the activation of the NLRP3 inflammasome[68-70]. Zhou *et al*[70] and Nakahira *et al*[68] demonstrate that after an inflammatory stimulus, the accumulation of damaged mitochondria leads to the significant increase in mitochondrial reactive oxygen species (ROS) production, which in turn enhances inflammasome activation and the subsequent proteolytic maturation of the inflammatory cytokines, IL-1β and IL-18. The secretion of IL-1β and IL-18 by macrophages stimulated with specific DAMPs (such as lipopolysaccharides (LPS) and ATP) is mediated by caspase-1 and controlled by inflammasomes[69]. Pharmacologic inhibition of autophagy by treatment of macrophages with 3-methyladenine resulted in the accumulation of damaged mitochondria and increased mitochondrial ROS[70]. The activation of the NLRP3 inflammasome in a ROS dependent manner was shown by the ablation of IL-1β secretion as a consequence of elevated mitochondrial ROS in *Nlrp3*$^{-/-}$ and *Asc*$^{-/-}$ cells. In a separate study[68], the disruption of autophagy (by genetic deletion of LC3B or Beclin 1) compromised mitochondrial homeostasis leading to the accumulation of mitochondrial ROS, translocation of mitochondrial DNA to the cytosol, and the activation of caspase-1 followed by secretion of IL-1β and IL-18. The cytosolic translocation of mitochondrial DNA was dependent upon the NLRP3 inflammasome and mitochondrial ROS and contributed to the secretion of IL-1β and IL-18 in response to LPS and ATP. The emerging picture is that autophagy regulates NLRP3-dependent inflammation by maintaining overall mitochondrial "health".

A strong genetic link has recently been established between autophagy and Crohn's disease. Genome wide scans have identified a strong association between a non-synonymous single-nucleotide polymorphism (SNP) in the autophagy gene ATG16L1 and susceptibility to Crohn's disease[71-73]. Atg16L1 forms a complex with Atg12-Atg5 conjugates. This Atg12-Atg5-Atg16L1 complex recruits an Atg3-LC3 intermediate to the isolation membrane. Cells lacking Atg16L1 are deficient in the conjugation of LC3 to phosphatidylethanolamine, which is essential for the elongation of the autophagosomes[74]. Atg16L1 is an essential component of autophagy under both nutrient rich and nutrient starved conditions. Using Atg16L1-deficient mice, Saitoh *et al*[74] demonstrated that the loss of Atg16L1 enhances endotoxin-induced inflammatory immune responses. Commensal bacteria are thought to be a major causative agent of inflammatory bowel disease. When intestinal epithelial cells are damaged, commensal bacteria are able to pass through the epithelial layer and stimulate PRRs, leading to the induction of intestinal inflammation. TLRs, one family of PRRs, detect microbial components and induce the production of inflammatory cytokines such as TNF-α, IL-1β, IL-6, IL-12, and IL-18 by macrophages/dendritic cells, resulting in the infiltration of activated lymphocytes into the intestine[75, 76]. Macrophages lacking Atg16L1 produce high amounts of IL-1β and IL-18 in response to lipopolysaccharide, a ligand for TLR4. Macrophages from Atg7-deficient mice also show enhanced production of IL-1β, indicating the importance of autophagy in the regulation of the inflammatory response. Stimulation of TLR family members other than TLR3 fails to enhance production of IL-1β by Atg16L1-deficient macrophages. TRIF (Toll/IL-1 receptor domain-containing adaptor inducing IFN-β), the adaptor protein involved in TLR3/4 signaling pathways, also mediates IL-1β production in Atg16L1-deficient macrophages. In Atg16L1-deficient macrophages, the enhanced IL-1β production is induced by TRIF-dependent generation of ROS and is blocked by ROS scavengers. Excess ROS production has been shown to lead to the activation of caspase-1, an essential protease for IL-1β production[70]. Mice with hematopoietic cells lacking Atg16L1 are highly susceptible to dextran sulfate sodium-induced colitis, which is relieved by the injection of anti-IL-1β and anti-IL-18 antibodies. These findings indicate that the loss of Atg16L1, an essential component of the autophagic machinery, results in the production of inflammatory cytokines in response to endotoxin. The source of the ROS has yet to be determined. Candidate sources of ROS include NADPH oxidase[77] and old/damaged mitochondria; the latter is especially interesting in light of the finding that loss of autophagy results in increased ROS production in immune cells, possibly from damaged mitochondria[32, 78].

Recent findings suggested that autophagy may play a role in the development of multiple sclerosis (MS). MS is an inflammatory, autoimmune central nervous system disorder characterized by T cell-mediated demyelination. Disease progression has been linked to increased proliferation of autoreactive T cells as well as resistance of autoreactive T cells to apoptosis. Alirezaei *et al*[79] examined the expression of Atg5 using both a mouse model of autoimmune demyelination as well as blood and brain tissues from MS cases. Studies with the experimental autoimmune encephalomyelitis (EAE) mice revealed a strong correlation between Atg5 expression and clinical disability wherein the upregulation and post-translational modification of Atg5 positively correlated with EAE severity. Similarly, Atg5 expression was significantly elevated in lymphocytes from individuals with active relapsing-remitting MS compared to non-diseased controls and immunofluorescence microscopy of brain tissue sections from relapsing-remitting MS cases revealed encephalitogenic T cells as a source of Atg5 expression in MS brain samples. Together these data suggest that increased T cell expression of Atg5 may contribute to inflammatory demyelination in MS.

SUMMARY

Recent advances have shown that autophagy plays a fundamental and indispensable role in regulating immune cell survival/homeostasis at various stages in development and the immune response. As such, one might predict that compromised autophagic function might result in decreased resistance to infection and increased incidence of inflammatory disorders. Indeed, as one ages, autophagy declines and concomitant changes in immune function are noted, e.g., increased incidence of inflammatory diseases such as Alzheimer's and decreased immune responsiveness to pathogens. As studies that dissect the role of autophagy in various immune response parameters in health and disease continue to unfold, future studies that determine the efficacy of intervention with treatments that modulate autophagy will be of great interest.

References

1. Klionsky DJ. Autophagy: from phenomenology to molecular understanding in less than a decade. *Nat Rev Mol Cell Biol* 2007;**8**:931–7.
2. He C, Klionsky DJ. Regulation mechanisms and signaling pathways of autophagy. *Annu Rev Genet* 2009;**43**:67–93.
3. He C, Levine B. The Beclin 1 interactome. *Curr Opin Cell Biol* 2010;**22**:140–9.
4. Liang XH, Kleeman LK, Jiang HH, Gordon G, Goldman JE, Berry G, et al. Protection against fatal Sindbis virus encephalitis by beclin, a novel Bcl-2-interacting protein. *J Virol* 1998;**72**:8586–96.
5. Pattingre S, Tassa A, Qu X, Garuti R, Liang XH, Mizushima N, et al. Bcl-2 antiapoptotic proteins inhibit Beclin 1-dependent autophagy. *Cell* 2005;**122**:927–39.
6. Ohsumi Y, Mizushima N. Two ubiquitin-like conjugation systems essential for autophagy. *Semin Cell Dev Biol* 2004;**15**:231–6.
7. Dengjel J, Schoor O, Fischer R, Reich M, Kraus M, Muller M, et al. Autophagy promotes MHC class II presentation of peptides from intracellular source proteins. *Proc Natl Acad Sci USA* 2005;**102**:7922–7.
8. Dongre AR, Kovats S, deRoos P, McCormack AL, Nakagawa T, Paharkova-Vatchkova V, et al. In vivo MHC class II presentation of cytosolic proteins revealed by rapid automated tandem mass spectrometry and functional analyses. *Eur J Immunol* 2001;**31**:1485–94.
9. Brazil MI, Weiss S, Stockinger B. Excessive degradation of intracellular protein in macrophages prevents presentation in the context of major histocompatibility complex class II molecules. *Eur J Immunol* 1997;**27**: 1506–14.
10. Paludan C, Schmid D, Landthaler M, Vockerodt M, Kube D, Tuschl T, et al. Endogenous MHC class II processing of a viral nuclear antigen after autophagy. *Science* 2005;**307**:593–6.
11. Zhou D, Li P, Lin Y, Lott JM, Hislop AD, Canaday DH, et al. Lamp-2a facilitates MHC class II presentation of cytoplasmic antigens. *Immunity* 2005;**22**:571–81.
12. Dorfel D, Appel S, Grunebach F, Weck MM, Muller MR, Heine A, et al. Processing and presentation of HLA class I and II epitopes by dendritic cells after transfection with in vitro-transcribed MUC1 RNA. *Blood* 2005;**105**:3199–205.
13. Nimmerjahn F, Milosevic S, Behrends U, Jaffee EM, Pardoll DM, Bornkamm GW, et al. Major histocompatibility complex class II-restricted presentation of a cytosolic antigen by autophagy. *Eur J Immunol* 2003;**33**:1250–9.
14. Mizushima N. Methods for monitoring autophagy. *Int J Biochem Cell Biol* 2004;**36**:2491–502.
15. Munz C, Bickham KL, Subklewe M, Tsang ML, Chahroudi A, Kurilla MG, et al. Human CD4(+) T lymphocytes consistently respond to the latent Epstein-Barr virus nuclear antigen EBNA1. *J Exp Med* 2000;**191**:1649–60.
16. Schmid D, Pypaert M, Munz C. Antigen-loading compartments for major histocompatibility complex class II molecules continuously receive input from autophagosomes. *Immunity* 2007;**26**:79–92.
17. Massey AC, Zhang C, Cuervo AM. Chaperone-mediated autophagy in aging and disease. *Curr Top Dev Biol* 2006;**73**:205–35.

18. Agarraberes FA, Terlecky SR, Dice JF. An intralysosomal hsp70 is required for a selective pathway of lysosomal protein degradation. *J Cell Biol* 1997;**137**:825–34.
19. Chiang HL, Terlecky SR, Plant CP, Dice JF. A role for a 70-kilodalton heat shock protein in lysosomal degradation of intracellular proteins. *Science* 1989;**246**:382–5.
20. Cuervo AM, Dice JF. A receptor for the selective uptake and degradation of proteins by lysosomes. *Science* 1996;**273**:501–3.
21. Cuervo AM. Autophagy: many paths to the same end. *Mol Cell Biochem* 2004;**263**:55–72.
22. Lich JD, Elliott JF, Blum JS. Cytoplasmic processing is a prerequisite for presentation of an endogenous antigen by major histocompatibility complex class II proteins. *J Exp Med* 2000;**191**:1513–24.
23. Klein L, Roettinger B, Kyewski B. Sampling of complementing self-antigen pools by thymic stromal cells maximizes the scope of central T cell tolerance. *Eur J Immunol* 2001;**31**:2476–86.
24. Mizushima N, Yamamoto A, Matsui M, Yoshimori T, Ohsumi Y. In vivo analysis of autophagy in response to nutrient starvation using transgenic mice expressing a fluorescent autophagosome marker. *Mol Biol Cell* 2004;**15**:1101–11.
25. Steinman RM, Hawiger D, Nussenzweig MC. Tolerogenic dendritic cells. *Annu Rev Immunol* 2003;**21**:685–711.
26. Nedjic J, Aichinger M, Emmerich J, Mizushima N, Klein L. Autophagy in thymic epithelium shapes the T-cell repertoire and is essential for tolerance. *Nature* 2008;**455**:396–400.
27. Mortensen M, Soilleux EJ, Djordjevic G, Tripp R, Lutteropp M, Sadighi-Akha E, et al. The autophagy protein Atg7 is essential for hematopoietic stem cell maintenance. *J Exp Med* 2011;**208**:455–67.
28. Takubo K, Goda N, Yamada W, Iriuchishima H, Ikeda E, Kubota Y, et al. Regulation of the HIF-1alpha level is essential for hematopoietic stem cells. *Cell Stem Cell* 2010;**7**:391–402.
29. Semenza GL. HIF-1: upstream and downstream of cancer metabolism. *Curr Opin Genet Dev* 2010;**20**: 51–6.
30. Arsov I, Adebayo A, Kucerova-Levisohn M, Haye J, MacNeil M, Papavasiliou FN, et al. A role for autophagic protein Beclin 1 early in lymphocyte development. *J Immunol* 2011;**186**:2201–9.
31. Pua HH, Dzhagalov I, Chuck M, Mizushima N, He YW. A critical role for the autophagy gene Atg5 in T cell survival and proliferation. *J Exp Med* 2007;**204**:25–31.
32. Pua HH, Guo J, Komatsu M, He YW. Autophagy is essential for mitochondrial clearance in mature T lymphocytes. *J Immunol* 2009;**182**:4046–55.
33. Hubbard VM, Valdor R, Patel B, Singh R, Cuervo AM, Macian F. Macroautophagy regulates energy metabolism during effector T cell activation. *J Immunol* 2010;**185**:7349–57.
34. Li C, Capan E, Zhao Y, Zhao J, Stolz D, Watkins SC, et al. Autophagy is induced in CD4+ T cells and important for the growth factor-withdrawal cell death. *J Immunol* 2006;**177**:5163–8.
35. Kroemer G, Jaattela M. Lysosomes and autophagy in cell death control. *Nat Rev Cancer* 2005;**5**:886–97.
36. Levine B, Yuan J. Autophagy in cell death: an innocent convict? *J Clin Invest* 2005;**115**:2679–88.
37. Walsh CM, Edinger AL. The complex interplay between autophagy, apoptosis, and necrotic signals promotes T-cell homeostasis. *Immunol Rev* 2010;**236**:95–109.
38. Denizot M, Varbanov M, Espert L, Robert-Hebmann V, Sagnier S, Garcia E, et al. HIV-1 gp41 fusogenic function triggers autophagy in uninfected cells. *Autophagy* 2008;**4**:998–1008.
39. Espert L, Denizot M, Grimaldi M, Robert-Hebmann V, Gay B, Varbanov M, et al. Autophagy is involved in T cell death after binding of HIV-1 envelope proteins to CXCR4. *J Clin Invest* 2006;**116**:2161–72.
40. Krysko DV, Vanden Berghe T, D'Herde K, Vandenabeele P. Apoptosis and necrosis: detection, discrimination and phagocytosis. *Methods* 2008;**44**:205–21.
41. Danial NN, Korsmeyer SJ. Cell death: critical control points. *Cell* 2004;**116**:205–19.
42. Salazar M, Carracedo A, Salanueva IJ, Hernandez-Tiedra S, Lorente M, Egia A, et al. Cannabinoid action induces autophagy-mediated cell death through stimulation of ER stress in human glioma cells. *J Clin Invest* 2009;**119**:1359–72.
43. Yu L, Lenardo MJ, Baehrecke EH. Autophagy and caspases: a new cell death program. *Cell Cycle* 2004;**3**:1124–6.
44. Dunkle A, He YW. Apoptosis and autophagy in the regulation of T lymphocyte function. *Immunol Res* 2010;**49**:70–86.
45. Siegel RM. Caspases at the crossroads of immune-cell life and death. *Nat Rev Immunol* 2006;**6**:308–17.
46. Yu L, Alva A, Su H, Dutt P, Freundt E, Welsh S, et al. Regulation of an ATG7-beclin 1 program of autophagic cell death by caspase-8. *Science* 2004;**304**:1500–2.

47. Pyo JO, Jang MH, Kwon YK, Lee HJ, Jun JI, Woo HN, et al. Essential roles of Atg5 and FADD in autophagic cell death: dissection of autophagic cell death into vacuole formation and cell death. *J Biol Chem* 2005;**280**:20722–9.
48. Miller BC, Zhao Z, Stephenson LM, Cadwell K, Pua HH, Lee HK, et al. The autophagy gene ATG5 plays an essential role in B lymphocyte development. *Autophagy* 2008;**4**:309–14.
49. Watanabe K, Ichinose S, Hayashizaki K, Tsubata T. Induction of autophagy by B cell antigen receptor stimulation and its inhibition by costimulation. *Biochem Biophys Res Commun* 2008;**374**:274–81.
50. Watanabe K, Tsubata T. Autophagy connects antigen receptor signaling to costimulatory signaling in B lymphocytes. *Autophagy* 2009;**5**:108–10.
51. Chaturvedi A, Dorward D, Pierce SK. The B cell receptor governs the subcellular location of Toll-like receptor 9 leading to hyperresponses to DNA-containing antigens. *Immunity* 2008;**28**:799–809.
52. Talloczy Z, Jiang W, Virgin HWt, Leib DA, Scheuner D, Kaufman RJ, et al. Regulation of starvation- and virus-induced autophagy by the eIF2alpha kinase signaling pathway. *Proc Natl Acad Sci USA* 2002;**99**:190–5.
53. Gutierrez MG, Master SS, Singh SB, Taylor GA, Colombo MI, Deretic V. Autophagy is a defense mechanism inhibiting BCG and Mycobacterium tuberculosis survival in infected macrophages. *Cell* 2004;**119**:753–66.
54. Inbal B, Bialik S, Sabanay I, Shani G, Kimchi A. DAP kinase and DRP-1 mediate membrane blebbing and the formation of autophagic vesicles during programmed cell death. *J Cell Biol* 2002;**157**:455–68.
55. Ling YM, Shaw MH, Ayala C, Coppens I, Taylor GA, Ferguson DJ, et al. Vacuolar and plasma membrane stripping and autophagic elimination of Toxoplasma gondii in primed effector macrophages. *J Exp Med* 2006;**203**:2063–71.
56. Singh SB, Davis AS, Taylor GA, Deretic V. Human IRGM induces autophagy to eliminate intracellular mycobacteria. *Science* 2006;**313**:1438–41.
57. Djavaheri-Mergny M, Amelotti M, Mathieu J, Besancon F, Bauvy C, Souquere S, et al. NF-kappaB activation represses tumor necrosis factor-alpha-induced autophagy. *J Biol Chem* 2006;**281**:30373–82.
58. Jia G, Cheng G, Gangahar DM, Agrawal DK. Insulin-like growth factor-1 and TNF-alpha regulate autophagy through c-jun N-terminal kinase and Akt pathways in human atherosclerotic vascular smooth cells. *Immunol Cell Biol* 2006;**84**:448–54.
59. Subauste CS, Andrade RM, Wessendarp M. CD40-TRAF6 and autophagy-dependent anti-microbial activity in macrophages. *Autophagy* 2007;**3**:245–8.
60. Andrade RM, Wessendarp M, Gubbels MJ, Striepen B, Subauste CS. CD40 induces macrophage anti-Toxoplasma gondii activity by triggering autophagy-dependent fusion of pathogen-containing vacuoles and lysosomes. *J Clin Invest* 2006;**116**:2366–77.
61. Petiot A, Ogier-Denis E, Blommaart EF, Meijer AJ, Codogno P. Distinct classes of phosphatidylinositol 3′-kinases are involved in signaling pathways that control macroautophagy in HT-29 cells. *J Biol Chem* 2000;**275**:992–8.
62. Wright K, Ward SG, Kolios G, Westwick J. Activation of phosphatidylinositol 3-kinase by interleukin-13. An inhibitory signal for inducible nitric-oxide synthase expression in epithelial cell line HT-29. *J Biol Chem* 1997;**272**:12626–33.
63. Orvedahl A, Alexander D, Talloczy Z, Sun Q, Wei Y, Zhang W, et al. HSV-1 ICP34.5 confers neurovirulence by targeting the Beclin 1 autophagy protein. *Cell Host Microbe* 2007;**1**:23–35.
64. Mills KR, Reginato M, Debnath J, Queenan B, Brugge JS. Tumor necrosis factor-related apoptosis-inducing ligand (TRAIL) is required for induction of autophagy during lumen formation in vitro. *Proc Natl Acad Sci USA* 2004;**101**:3438–43.
65. Blander JM, Medzhitov R. Regulation of phagosome maturation by signals from toll-like receptors. *Science* 2004;**304**:1014–8.
66. Delgado MA, Elmaoued RA, Davis AS, Kyei G, Deretic V. Toll-like receptors control autophagy. *Embo J* 2008;**27**:1110–21.
67. Cooney R, Baker J, Brain O, Danis B, Pichulik T, Allan P, et al. NOD2 stimulation induces autophagy in dendritic cells influencing bacterial handling and antigen presentation. *Nat Med* 2010;**16**:90–7.
68. Nakahira K, Haspel JA, Rathinam VA, Lee SJ, Dolinay T, Lam HC, et al. Autophagy proteins regulate innate immune responses by inhibiting the release of mitochondrial DNA mediated by the NALP3 inflammasome. *Nat Immunol* 2011;**12**:222–30.
69. Schroder K, Zhou R, Tschopp J. The NLRP3 inflammasome: a sensor for metabolic danger? *Science* 2010;**327**:296–300.

70. Zhou R, Yazdi AS, Menu P, Tschopp J. A role for mitochondria in NLRP3 inflammasome activation. *Nature* 2011;**469**:221–5.
71. Hampe J, Franke A, Rosenstiel P, Till A, Teuber M, Huse K, et al. A genome-wide association scan of non-synonymous SNPs identifies a susceptibility variant for Crohn disease in ATG16L1. *Nat Genet* 2007;**39**:207–11.
72. Prescott NJ, Fisher SA, Franke A, Hampe J, Onnie CM, Soars D, et al. A nonsynonymous SNP in ATG16L1 predisposes to ileal Crohn's disease and is independent of CARD15 and IBD5. *Gastroenterology* 2007;**132**:1665–71.
73. Rioux JD, Xavier RJ, Taylor KD, Silverberg MS, Goyette P, Huett A, et al. Genome-wide association study identifies new susceptibility loci for Crohn disease and implicates autophagy in disease pathogenesis. *Nat Genet* 2007;**39**:596–604.
74. Saitoh T, Fujita N, Jang MH, Uematsu S, Yang BG, Satoh T, et al. Loss of the autophagy protein Atg16L1 enhances endotoxin-induced IL-1beta production. *Nature* 2008;**456**:264–8.
75. Beutler B. Microbe sensing, positive feedback loops, and the pathogenesis of inflammatory diseases. *Immunol Rev* 2009;**227**:248–63.
76. Kawai T, Akira S. The roles of TLRs, RLRs and NLRs in pathogen recognition. *Int Immunol* 2009;**21**:317–37.
77. Dostert C, Petrilli V, Van Bruggen R, Steele C, Mossman BT, Tschopp J. Innate immune activation through Nalp3 inflammasome sensing of asbestos and silica. *Science* 2008;**320**:674–7.
78. Stephenson LM, Miller BC, Ng A, Eisenberg J, Zhao Z, Cadwell K, et al. Identification of Atg5-dependent transcriptional changes and increases in mitochondrial mass in Atg5-deficient T lymphocytes. *Autophagy* 2009;**5**:625–35.
79. Alirezaei M, Fox HS, Flynn CT, Moore CS, Hebb AL, Frausto RF, et al. Elevated ATG5 expression in autoimmune demyelination and multiple sclerosis. *Autophagy* 2009;**5**:152–8.

CHAPTER

5

Autophagy in the Gastrointestinal Tract

Leticia A.M. Carneiro[1] *and Leonardo H. Travassos*[2]

[1]**Instituto de Microbiologia Professor Paulo de Góes and** [2]**Instituto de Biofísica Carlos Chagas Filho, Universidade Federal do Rio de Janeiro, 21941-902, Rio de Janeiro, Brazil**

INTRODUCTION

The gastrointestinal (GI) tract from the esophagus through the rectum, including stomach, duodenum, small and large intestines, represents the largest environmentally exposed surface in the human body. It contains the largest repertoire of immune functions and is only second to the liver in terms of metabolic activity. Much of its surface is designed to be highly exposed and absorptive since it is the site where food is converted into nutrients and metabolites that provide energy, cofactors amd biosynthetic substrates for our cells. On the other hand, the intestinal mucosa is exposed to the largest accumulation and diversity of commensal and pathogenic microbial life in mammalian hosts and, thus, must be constantly scrutinized for the presence of toxic or pathogenic components. The intestinal epithelium prevents commensal bacteria from invading the mucosa through the formation of mechanical barriers and the production of antimicrobial peptides. Indeed, the epithelium of the small intestine and colon create a complex and very dynamic chemical and physical barrier between host and microbes. Furthermore, by virtue of its specific location at the frontier of immune system, the intestinal epithelium plays a key role in the maintenance of immune homeostasis in the gut. Since it is exposed to an enormous range of nutrients, factors derived from both host and microbial metabolism as well as signals derived from the complex and highly dynamic and immune compartment, the GI tract is constantly subject to potential inflammation and infection.

From birth onwards, the GI tract surfaces are colonized by complex polymicrobial communities (collectively referred to as "microbiota") that have a profound impact on the architecture and function of these tissues. This is illustrated by the observation that the

Autophagy in Health and Disease
http://dx.doi.org/10.1016/B978-0-12-385101-7.00005-X

immune composition and function of the intestine and other organs are significantly altered in germ-free mice[1]. The GI tract of a human adult has an estimated concentration of 10^{12} microorganisms per milliliter of luminal content and harbors at least 1800 genera and between 15,000 and 36,000 species of bacteria[2]. In particular, the small and large intestine present the highest concentration and most diverse microflora that have important functions including nutrient absorption, vitamin production and metabolism of xenobiotic compounds. This normal homeostatic and mutually beneficial co-evolved relationship between the gut microbiota and the host immune system relies on a fine balance and its disruption may result in initiation and/or propagation of inflammatory processes. The intestinal epithelium, as an interface between the microbiota and immune system, plays a critical regulatory function on both the microbiota and immune components of the intestine, both innate and adaptive, and has a major impact on idiopathic inflammatory bowel diseases (IBD) such as Crohn's disease and ulcerative colitis.

Autophagy is a highly conserved process throughout the eukaryotic domain through which cells are able to turnover stable or aggregated proteins, remove surplus or damaged organelles and eliminate intracellular microbes. Autophagy has recently gained considerable attention as an essential contributor to human health and disease such as neurodegeneration, cancer and metabolic syndrome[3]. It is becoming increasingly clear from genetic inactivation studies that the type of disease resulting from autophagy defects depends on the type of tissue. For example, in the brain, autophagy prevents the accumulation of ubiquitinated proteins and disposes of proteine aggregates and damaged organelles implicated and also prevents neurodegeneration in sufferers of Huntington's and Parkinson's diseases[4, 5]. In the liver, in addition to suppressing the accumulation of protein aggregates and lipid autophagic activity attenuates oxidative stress, cell death, inflammation and cancer [6, 7]. As we will see in the next section, autophagy also plays an important role in the homeostasis of the GI tract, and defects in the autophagic machinery/pathway have been placed at the core of IBD pathogenesis, in particular Crohn's disease (CD).

AUTOPHAGY IN THE HEALTHY GI TRACT

Homeostasis is the process by which organisms strive to maintain a constant internal environment despite perturbations in their external conditions. In this context, autophagy can fulfill several roles as a cytoplasmic homeostatic process. In the absence of stress, basal autophagy serves a housekeeping function providing a routine "garbage disposal" service to cells by eliminating and recycling damaged and/or long-lived components that could otherwise become toxic. In starvation, it provides an internal source of energy generation for survival—this is a key feature since even multicellular organisms that possess sophisticated multiorgan systems are unable to secure a constant supply of nutrients to all their cells. In order to guarantee its own survival, cells capture their own cytoplasm and organelles and consume them in lysosomes. The resulting breakdown products input to cellular metabolism and are used to generate energy and to build new proteins and membranes[8].

As the major site where nutrients are absorbed, the intestinal epithelium alternates periods when it is loaded with a vast array of nutrient molecules that must be taken up and redistributed (following meals) with periods of relatively lower nutrient abundance. The effects of these alternating cycles on the regulation of autophagy in the epithelium itself are just beginning to be elucidated, especially regarding the availability of amino acids. Amino acids have a central role in intestinal homeostasis as illustrated by the action of arginine, glutamine and leucine in mucosal cell migration and restitution[3]. Amino acid availability also greatly influences the rate of autophagy within intestinal epithelial cells[3]. For example, glutamine increases the levels of both basal and stress-induced autophagy promoting cell survival. Conversely, glutamine starvation results in increased caspase-3 and poly (ADP ribose) activity after stress conditions. Together with the observations that inhibit mTOR or p38 MAP kinase reduction, these data collectively suggest that autophagy protects cells undergoing glutamine restriction from apoptosis[9]. More recently, it has been shown that glutamine uptake via the SLC1A5 transporter is a prerequisite for essential amino acid signaling to mTORC1. This is because glutamine serves as a substrate for the leucine transporter SLC7A5/SLC3A2, an obligate exchanger that requires an intracellular subtrate to allow leucine import that, in turn, leads to the activation of mTORC1[10]. Thus, under amino acid rich conditions, mTORC1 inhibits the cellular catabolism of autophagy. When the amino acid supply lowers, autophagy is activated to allow the cells to recycle their own components to generate amino acids to maintain homeostasis.

Autophagy may also have a relevant role on other characteristic features of the intestinal epithelium like the process of constant epithelial renewal and shedding of apical epitheliam cells. In order to maintain the integrity of the epithelial barrier, enterocytes are constantly shed and renewed, forming an "epithelial escalator" from near the base of the crypt to the villus tip where they are shed from the gut surface. This process must be tightly controlled to avoid nutrient and fluid loss as well as to prevent bacterial entry. Thus, it is predictable that cells try to recover and/or recycle as many nutrients as possible before being shed. Although studies of autophagy have not been performed on these cells, it is plausible that cells consume their stores of nutrients prior to being shed. The shedding of the cells itself must follow a series of organized events in order to prevent loss of epithelial integrity—junctions between viable epithelial cells must be maintained until the last moment to prevent osmotic leakage and bacterial invasion. Once again, it is still not known whether autophagy is required for this process or not, but there are plausible reasons to believe it might be since focal adhesion interacting proteins have been implicated in autophagy, and loss of focal adhesion has been observed during starvation of cultured cells. Finally, in models of detachment-induced cell death, autophagy is induced following detachment from the extracellular matrix and its inhibition results in increased levels of apoptosis[3]. Therefore, defects in autophagy may alter the state of shed epithelial cells, as well as affect the permeability of the epithelium they leave behind. Indeed, there are significant abnormalities associated with autophagic defects in mouse models of cell shedding. In addition, autophagy is also considered a type of cell death, in addition to apoptosis and necrosis. Cell death plays an important role during embryogenesis and metamorphosis in multicellular organisms, and enables the regulation and maintenance of cell populations in all organs depending upon physiological and pathological conditions[11,12]. Thus, autophagy might also contribute to proper function of the intestinal epithelium as a mechanism of cell death.

The intestinal epithelium is composed of four principal terminally differentiated cell types:

1. absorptive cells or enterocytes
2. mucus-producing goblet cells
3. a complex population of enteroendocrine cells
4. Paneth cells that synthesize lysozyme plus defensins[13].

How autophagy, or defects in the autophagic machinery, affects each one of these cell types is related to their different functions. Paneth cells, the most highly secretory cell type in the intestinal epithelium, are found at the bottom of ileal crypts and are thought to control intestinal microbiota by secreting granule contents including antimicrobial peptides, lysozyme, α-defensins and secretory phospholipase A2. These antimicrobials are believed to protect neighboring stem cells from microbial insults and keep gut microbiota under check[14]. It has been recently shown that autophagy is critical for normal secretory function of these cells as mice with impaired autophagy present an overall normal morphology of the ileum and colon but display profound abnormalities specifically in Paneth cells[15]. Defects in autophagy result in defective degranulation of Paneth cells accompanied by a variety of gene-expression changes, including many related to mucosal homeostasis and healing, as well as strong pro-inflammatory signatures demonstrating that autophagy is crucial to preserve cellular function and prevents expression of damage and inflammatory markers (discussed in more detail in the next sections). In enterocytes, autophagy seems to be a key player in the innate immune defense against invading microbes as it serves as a mechanism to confine and degrade intracellular bacteria (also discussed in more detail in the next section).

AUTOPHAGY IN GI TRACT DISEASES

As previously mentioned, autophagy is a mechanism that allows cells to degrade large structures such as organelles and protein aggregates, preventing the buildup of these components that can cause chronic tissue damage and disease. In addition to eliminating toxic components from the cytoplasm, autophagy's metabolic role also contributes to ensure consistent availability of internal nutrients, enabling cells to survive periods of poor external nutrition in good health[8]. Thus, it is not surprising that autophagy is crucial for the maintenance of homeostasis in living organisms and that impaired autophagy contributes to the pathogenesis of several diseases.

Inflammatory Bowel Diseases

The idiopathic inflammatory bowel diseases (IBD) are a group of chronic debilitating inflammatory disorders affecting the intestine and that have been empirically defined by clinical, pathological, endoscopic and radiological features[16]. Crohn's disease (CD) and ulcerative colitis (UC) are the two major forms of IBD with a combined prevalence of about 150-200 cases per 100,000 in Western countries. While UC is a continuous mucosal inflammatory disorder of the colon, CD may affect any segment of the intestine and is generally notable for transmural, discontinuous ulceration with granuloma formation. These two conditions, DC and UC, might be considered as subtypes that lie at different ends of the IBD spectrum and, even though certain characteristics are used to differentiate them, in current clinical practice there remains considerable overlap of treatment regimens between these conditions[17].

Several aspects of intestinal mucosal dysfunction have been implicated in the etiology of IBD, including loss of mucosal integrity, intolerance towards the intestinal microbiota and abnormal immune sensing, signaling, differentiation and processing of either self or non-self antigens[3]. Although the precise causes of IBD remain elusive, a huge body of evidence accumulated in the last decades converge to a model of multifactorial disease in which individuals with a genetic predisposition in whom an environmental or microbial trigger cause an abnormal immune response that persists as a chronic intestinal inflammation. Still, many questions await clarification, including what factor(s) trigger disease onset and what factor(s) affect the distribution of inflammation within the gut. A number of recent studies in well defined model systems are just beginning to elucidate the fundamental importance of the interplay between the intestinal epithelium and microbiota, environmental factors and immune system and how the host's genetics influence the outcome of these interactions. As we will see in the following sections, dysregulation of autophagy could be implicated in the pathogenesis of IBD in several ways:

1. by disrupting Atg16L1-mediated control of endotoxin-induced pro-inflammatory signaling in macrophages
2. by altering Paneth cell homeostasis and function
3. by affecting the control of intracellular bacteria replication
4. potentially via autophagy related lymphocyte homeostasis.

Crohn's Disease

CD is a chronic and debilitating inflammatory condition of the gastrointestinal tract mostly characterized by uncontrolled inflammation in the distal ileum, though the colon can be involved as well. The distinguishing features of inflammation in CD are that it is transmural and includes frequent lymphoid aggregates that are present even in the outer muscle layer and mesothelial lining as well as non-caseating granulomas composed of fused monocyte-derived cells. The transmural inflammation in CD leads to atrophy of ileal villi, fibrosis, hypertrophy of smooth muscle and autonomic nerve cells in the outer layer of the bowel wall and is associated with fistula formation with adjacent tissues. Peak incidence is in early adult life, although any age can be affected and a majority of affected individuals progress to relapsing and chronic disease. Prevalence in Western populations is 100-150/100,000. There is no cure for CD and treatment is focused on limiting the recurrence and severity of the symptoms, usually with empirical corticosteroid therapy and immunosuppression; intestinal resectional surgery is frequently required[16].

CD is a complex polygenic trait whereby multiple genetic and non-genetic factors contribute to disease susceptibility. CD has been one of the most tractable multigenic disorders for discovery of susceptibility genes and, to date, more than 30 genetic loci have been associated with increased susceptibility to CD[18]. This suggests that what we call CD may actually represent a spectrum of disorders that present with similar gross phenotypes. The most widely accepted model is one in which, in a genetically susceptible host, a commensal or environmental bacteria or viral agent triggers a breach of the intestinal epithelial barrier that separates the lumenal contents from the mucosal immune system. This primary event then leads to inflammatory flares in the gut mucosa. Subsequently, failures to properly downregulate the immunological response and establish epithelial restitution and repair leave the gut primed for future flare-ups, perhaps triggered by progressively less significant environmental insults[19].

In recent years, a number of related lines of investigation have generated substantial advances in our understanding of the molecular pathogenesis of CD. Genetic associations demonstrate that multiple cell types and pathways are involved in generating the disease phenotype. In the next section, we will discuss the role of autophagy in these different aspects in pathophysiology of CD.

The Genetic Component

Epidemiological evidence for a genetic contribution to CD has been accumulating for decades – about 15% of the patients with CD have an affected family member which gives a relative risk to siblings of 17 times. Additionally, CD studies in twins have demonstrated approximately 50% concordance in monozygotic twins compared to less than 10% in dizygotics[20]. However, it was the advent of rapid, relatively cheap and flexible genotyping technologies developed in the last few years that opened up the possibility of recruiting large cohorts of patients and healthy controls and analysing their genetics in an unbiased and systematic manner. The genome-wide association studies (GWAS) enabled the examination of the majority of common genetic variation for roles in disease susceptibility. Such studies test hundreds of thousands of markers, known as single nucleotide polymorphism (SNPs) in thousands of individuals, seeking markers at which there are statistically significant differences in allele frequency between cases and controls[3]. Since genetic variants in GWAS are spread throughout the genome rather than selected to test a specific set of genes, these studies have proven invaluable in identifying biological pathways not previously suspected of playing a role in a disease's pathogenesis. This powerful new approach has allowed a deeper analysis of complex multigenic diseases. For instance, in the case of CD, the number of associated genetic variants has grown from two (NOD2 and the IBD5 locus) to three; two in just five years. As a consequence, not only our knowledge about this disease has greatly improved but also several cohesive themes have emerged and are now being tested in model systems. Among these themes, these recent findings strongly imply that autophagy is an important biological pathway in CD pathogenesis. To date, two genes within the autophagic pathway have been identified and replicated as being associated with CD. The presence of two such genes, Atg16L1 and IRGM, in just 32 associations points toward a strong role for autophagy in the pathogenesis of CD. In addition, the product of the NOD2 gene, which had been previously implicated as a susceptibility gene for CD, has been recently shown to have a key role in anti-bacterial autophagy.

NOD2

Before the GWAS era, NOD2 (also designated IBD1 or CARD15) was the first susceptibility gene identified in CD through fine mapping within linkage intervals[21]. The three most common mutations (Arg702Trp, Gly908Arg and Leu1007fs) account for about 80% of NOD2 variants associated with CD[22]. Two of the mutant alleles, Arg702Trp and Gly908Arg, specify single amino acid substitutions within the LRR, whereas the third, Leu1007fs, results in a frame-shift mutation caused by a cytosine insertion (3020insC) that removes the last 33 amino acids of the NOD2 polypeptide. Individuals who are either homozygotes or compound heterozygotes for any one of the three common identified germline variations of NOD2 have as much as a 40-fold increased likelihood of developing CD[21]. Patients harboring NOD2 variants often have an ileal, inflammatory, fibrosing phenotype of CD, with stricturing complications and slightly earlier age of onset[23]. However, the mechanism by which NOD2-mediated functions contribute to intestinal immune homeostasis and how dysregulation of these functions in individuals with disease-associated NOD2 polymorphisms contribute to the increased propensity to develop CD are not yet fully understood.

NOD2 is an innate immunity pattern recognition molecule (PRM) that senses muramylpeptides released from bacterial peptidoglycan, more specifically muramyldipeptide (MDP). It contributes to immune defense by various mechanisms such as cytokine production, upregulation of defensins, killing of intracellular microbes (e.g., *Salmonella* spp, *Listeria monocytogenes*, *Streptococcus pneumoniae* and mycobacterial species) and maintenance of the epithelial barrier integrity[24]. More recently, NOD2 was also implicated in anti-viral responses by triggering activation of interferon regulatory factor 3 (IRF3) and production of interferon-β (IFNβ) by direct sensing of ssRNA[25]. Its implication in CD pathogenesis positioned innate immunity as a crucial element for disease. The *L1007fsinsC* mutation, the most prevalent in CD patients, has been shown to abrogate PG and MDP sensing[26] which supports the concept that patients with Crohn's disease carrying NOD2 mutations have a defect that is associated with failure of microbial eradication at mucosal surfaces.

Upon activation, NOD2 recruits the serine/threonine kinase RIP2 to mount an inflammatory response that is largely dependent on NF-κB and induces adaptive immune responses with the objective of removing the attack. Activation of NF-κB drives the transcription of target genes that include the pro-inflammatory cytokines, chemokines and antimicrobial products[27]. Early work on the role of NOD2 in CD demonstrated that monocytes with the *Leu1007fs* mutation are impaired in IL-1β secretion upon stimulation with PG or MDP. Similarly, these cells display a reduced cytokine production (TNFα, IL-1β, CXCL8/IL-8 and IL-10) following co-stimulation with MDP and TLR ligands, but not with TLR ligands alone. Finally, impaired anti-inflammatory IL-10 secretion in Leu1007fs human monocytes stimulated by TLR2 ligands or whole bacteria has been reported[28]. In a murine model, ectopic expression of the NOD2fs protein was shown to interfere with transcription of the IL-10 promoter, a proposed mechanism to account for the reduced IL-10 secretion[29]. Together, these results suggest that NOD2 mutations would result in impaired inflammatory/immune response.

However, the results above suggest that CD-linked NOD2 mutations are loss-of-function mutations, which is not consistent with the common clinical observations of elevated levels of circulating pro-inflammatory cytokines and presence of activated NF-κB in the lamina propria of CD patients. Indeed, the analysis of dendritic cells derived from such patients reveal increased expression of certain inflammatory genes, including IL-1β, IL-6 and tumor necrosis factor α (TNFα), upon incubation with MDP[30]. Two hypotheses may contribute to reconciling this apparent paradox:

1. defective NOD2 signaling may lead to uncontrolled proliferation of microorganisms in the intestine, with an increase in bacterial burden and prolonged inflammation. In this model,

an impaired capacity to produce antimicrobial peptides by NOD2-deficient Paneth cells could lead to an increased microbial burden in the intestine, promoting the onset of CD. Indeed, in mice deficient in NOD2, Paneth cells express significantly fewer antimicrobial peptides (cryptidins), and these mice are impaired in the eradication of *L. monocytogenes* upon oral administration[31]. Similarly, patients with Crohn's disease with a NOD2 mutation exhibit strongly diminished expression of the human variant of these defensins, or;

2. defective NOD2 signalling could dampen antiinflammatory signals, such as the secretion of IL-10 in response to MDP or whole bacteria. Based on these findings, it was hypothesized that a defect in clearance of certain bacteria may lead to a perpetual, albeit insufficient, mucosal immune response that results in chronic inflammation. This also highlights the importance of microbial mucosal interaction in the pathogenesis of IBD.

It has been proposed that NOD2 signaling could limit TLR2-driven activation of nuclear factor-κB (NF-κB) and IL-12 expression, while conversely, NOD2-deficiency or expression of the human Leu1007fs mutation increased TLR2-mediated activation of NF-κB and secretion of IL-12 and IFNγ. As a result, NOD2-deficiency would lead to exacerbated Th1 responses, which is a hallmark of CD. Moreover, another study from the same group reported that NOD2-deficiency leads to deregulated TLR2 signaling and induction of antigen-specific colitis[32, 33]. Using a murine model in which the human Leu1007fs mutation has been introduced by knock-in, Maeda *et al* reported the unexpected observation that macrophages harbouring such a mutation exhibit a gain-of-function phenotype with increased NF-κB and more efficient IL-1β secretion upon MDP stimulation[34]. Although these data would fit with the global increased inflammation observed in Crohn's patients, they are at odds with those obtained with human monocytes harbouring the Leu1007fs mutation or with macrophages from NOD2-deficient mice that have been shown to be impaired in MDP sensing and synergy with TLRs.

Many patients who are homozygotes for NOD2 mutations remain healthy, and even those who develop the disease are commonly asymptomatic for the first 10-15 years of their lives suggesting that additional factors, both genetic and environmental, act in conjunction with NOD2 mutations for the development of the disease. Recently, the role of NOD2 in innate immune defense has been expanded to autophagy[21]. With the recently reported observations that NOD2 and Atg16L1 form a complex at the cytoplasma membrane, the hypothesis of autophagy deficiency in IBD patients is highlighted as any mutation in NOD2 or autophagy related proteins would result in diminished and ineffective autophagic response to intracellular pathogens. A first study showed a synergy between NOD2 and Atg16L1 in the recognition and wrapping of bacteria at an early time of the autophagic process[35], and another complementary study further extended the interplay between NOD2 and Atg16L1 to the initiation of adaptive immune response (discussed in detail below)[36].

IRGM

Further strenghthening the relevance of autophagy in CD pathogenesis, the landmark Wellcome Trust Case Control Consortium (WTCCC) GWAS identified association between variants in a key autophagy gene, the immunity-related GTPase family M (IRGM) and CD susceptibility, using a panel of 2000 cases of CD and 3000 controls[37]. This association was replicated in subsequent independent studies[38, 39].

These recent GWAS for CD identified SNPs, flanking and within the IRGM gene, as being strongly correlated with disease. Sequencing of this gene in samples from CD patients and healthy controls, however, did not identify any causal amino acid changes. More recent and detailed studies of the IRGM gene and its surrounding genomic region identified the functional CD-associated polymorphisms to be in perfect linkage with a large (20kb) upstream deletion of a potential regulatory sequence, suggesting that genetic variation conferring susceptibility to CD could operate via modulation of IRGM gene expression[40]. Indeed, analysis of allele-specific expression showed a complex regulatory picture with different tissues and cell lines heterozygous for the two haplotypes (CD-risk and CD-protective) exhibit varying levels of expression of each allele. Thus, one possibility is that regulation of IRGM is specific to cell type and altered regulation resulting from the CD-associated deletion might be reflected by functional deficits in specific cell types. Polymorphisms in IRGM seems to predispose to the fistulizing behaviour observed in CD and appears to be specifically associated with CD and not other forms of IBD[3].

Even though less is known about the IRGM-related autophagic pathways when compared to the Atg pathways described below, recent studies are starting to shed light on its role in antibacterial autophagy. The IRGM protein is highly expressed in response to infection and stimulates the early stages of autophagy. This seems to be an effective mechanism against intracellular bacteria since mice rendered deficient in the homologue of IRGM gene have impaired ability to eliminate the intracellular pathogens *Toxoplasma gondii*, *Listeria monocytogenes* and *Salmonella typhimurium*[41, 42]. In a murine tuberculosis model, macrophages infected with the Bacillus Calmette-Guérin are able to clear intracellular bacteria through IRGM-dependent autophagy. In both human and mouse macrophage cell lines, IRGM-mediated autophagy generated large mycobacteria-containing vacuoles acquiring lysosomal markers and vacuole acidification, indicating that these bacterial vacuoles were fused with lysosomes to effect antibacterial control. Of note, experiments in human macrophages reported that IRGM silencing leads to markedly prolonged survival of *M. tuberculosis*[43, 44].

More recently, IRGM was shown to restrict the intracellular growth of a pathotype of *E. coli*, known as adherent-invasive *E. coli* (AIEC), that is frequently found colonizing ileal lesions in CD patients (described in more detail below). It was observed that cells lacking IRGM presented more and significantly larger clusters of bacteria growing in the cytoplasm when compared to wild type controls or even if compared to other autophagy-impaired cells such as Atg5-deficient mouse embryonic fibroblasts (MEFs) or Atg16L1 depleted human epithelial cell lines CD[45]. This last observation suggests that the role of IRGM may not be restricted to autophagy. Indeed, this protein has been described to be associated also to the membrane of phagosomes.

Atg16L1

In one of the first GWAS, the group of Dr. Ramnik Xavier and the NIDDK IBD Genetics Consortium examined approximately 300,000 SNPs in 1000 CD patients and in 1000 healthy control individuals and identified an association between Atg16L1 variants and CD[46]. The same causal variant in Atg16L1 had also been identified in an independent screen of about 7000 common non-synonymous coding variants in a German CD study[47]. In a subsequent study of CD that was part of the WTCCC GWAS, the Atg16L1 association was confirmed[37]. Interestingly, the Atg16L1 disease variant is frequent (50% in European-derived populations) and it is still unclear why disease occurs in only a small proportion of persons carrying such a common risk allele of disease susceptibility gene[18].

The Atg16L1 risk-associated polymorphism corresponds to a SNP that encodes a non-synonymous amino acid change, a threonine to alanine substitution (T300A, Ala197Thr, rs2241880), in the protein sequence at position 300. Interestingly, as observed with *NOD2* and *IRGM* risk variants, Atg16L1 risk-associated polymorphism has been implicated in CD but not UC and is usually associated with ileal disease[39].

The Atg16L1 protein functions in autophagy as part of a multimeric protein complex with Atg5 and Atg12. This complex is not only required for autophagosome formation but is also responsible for proper subcellular localization of the autophagy machinery[48]. In mouse models in which the Atg16L1 gene is disrupted, autophagy is abolished, indicating that Atg16L1, like its homologous yeast Atg16, is indeed a bonafide autophagy protein[49]. Mammalian Atg16L1 displays three distinct domains:

1. the N-terminal domain that is homologous to yeast Atg16 and that interacts with Atg5;
2. the coil-coiled domain (CCD) necessary for Atg16L1 oligomerization and in association withAtg5-Atg12;
3. a C-terminal region with seven WD repeat domain—WD repeats are normally associated with protein-protein or protein-membrane interactions. The fact that this domain is absent in yeast Atg16 suggests that it might be involved in a more complex regulation of mammalian autophagy compared to the yeast or it could have an important role in the autophagic clearance of pathogens, fulfilling a function absent in yeast[18].

Currently, functional studies are looking into understanding how this polymorphism impacts development of CD. As we will see in the following sections, the risk-associated polymorphism or lack of Atg16L1 can affect different cell types in different ways, from limiting inflammatory responses in myeloid cells to fighting intracellular bacteria in epithelial cells, to maintaining normal secretory function in the specialized Paneth cells, all of which can be implicated in CD pathogenesis.

Thus, these GWAS breakthroughs have expanded the role of innate immunity components in CD beyond the already implicated NOD2 to include autophagy based on association with Atg16L1, an autophagy execution gene, and an autophagy-linked factor, IRGM, involved in clearing bacteria. The strong association of NOD2 and Atg16L1 polymorphisms with CD and their combined increased risk for the disease suggest an intricate interplay of these two factors in the biology of CD. Yet these genetic components individually, or in combination, confer limited risk. It is not clear why some individuals with a given polymorphism acquire disease whereas others remain unaffected. The concept that environmental factors including infections trigger disease in individuals with certain genetic backgrounds is broadly recognized.

The Intestinal Epithelium Component

Paneth Cells

Paneth cells are the most highly secretory of the four lineages of epithelial cells that form the intestinal mucosa. Their main function is to produce, package and export a variety of anti-microbial proteins and peptides including α-defensins, angiogenin-4, lysozyme and secretory phospholipase A2. Moreover, Paneth cells also produce inflammatory cytokines in both constitutive (TNFα) and induced-upon-injury (IL-17) manner. Hence, these cells are predicted to regulate the complex intestinal microbiota by affecting both its composition and adherence to the epithelium, as well as to participate in the clearance of pathogens. Indeed, NOD2-, CD1s- and MyD88-deficient mice all present structural and/or functional alterations in Paneth cells accompained by altered composition of the microbiota and enhanced adherence of the microbiota to the intestinal epithelium[50]. Paneth cells have been previously suspected to play a role in CD[16] and corroborating this is the fact that several primary human genetic risk factors for CD have been recently shown to affect Paneth cell function including NOD2, Atg16L1 and XBP1[15, 31, 51].

One of the ways by which autophagy is implicated in intestinal epithelium homeostasis is through its selective effects on the cell biology and specialized regulatory properties of Paneth cells. The picture that has emerged from recent studies using mice with either hypomorphic expression of Atg16L1 or reduced intestinal epithelial Atg5 or Atg7 expression, is that compromised autophagy impacts on Paneth cell function in more than one way[15, 52]. On the one hand, impaired autophagy results in loss of function phenotype, with diminished production and delivery of antimicrobial proteins resulting in alteration in the microbiota. On the other hand, it might also be seen as a gain of function phenotype with increased inflammatory cytokine production affecting the mucosal immune system. Accordingly, in humans, CD patients who carry the Atg16L1 risk allele show similar concomitant loss and gain of function defects in their Paneth cells[15], indicating that within the intestinal epithelium, Paneth cells have a unique sensitivity to autophagy gene disruption.

The loss of function in mice expressing a hypomorphic variant of Atg16L1 is charaterized by distinctive Paneth cell defects, including notable lack of lysozyme in the mucus, decreased numbers of aberrant and disorganized granules resulting in defective degranulation, degenerating mitochondria, increased numbers of cytoplasmic vesicles and, frequently, absence of the apical microvilli. Importantly, similar phenotypes were observed in mice lacking Atg5 or Atg7 expression in the intestinal epithelium as well as in CD patients, confirming both that the defect is related to autophagy and that it is relevant to human disease[15]. Indeed, ileal specimens from CD patients homozygous for the Atg16L1 risk allele display Paneth cell morphogical abnormalities remarkably similar to those observed in Atg16L1^{HM} mice. In addition, Atg16L1 hypomorphic cells also present a gain of function whereby they express increased levels of inflammatory cytokines. This was demonstrated by microarray analysis in Paneth cells from Atg16L1^{HM} mice that revealed a notable signature of genes involved in peroxisome proliferator-activated receptor (PPAR) pathways, adipocytokine signaling and aspects of lipid metabolism. Of note, the adipocytokines leptin and adiponectin that had been previously reported to be increased in CD patients were amongst the most highly enriched transcripts[53, 54]. Finally, Paneth cells from CD patients with the Atg16L1 risk allele also presented increased staining for leptin protein compared to similar cells in controls[15].

These findings demonstrate similar features between the pathology and trancriptional profile of Paneth cells from Atg16L1^{HM} mice and Paneth cell abnormalities observed in CD patients with the risk allele of Atg16L1. Thus, in addition to its role in preserving the function of the granule exocytosis pathway, Atg16L1 might also contribute to homeostasis by serving as a brake for the expression of pro-inflammatory genes in Paneth cells.

An intriguing question is why impaired autophagy results in such dramatic and specific effects on Paneth cells while other IEC lineages and stem cells do not exhibit obvious morphological abnormalities. Interesting clues came from a series of recent studies showing that Paneth cells are particularly sensitive to endoplasmic reticulum (ER) stress. Paneth cells, in contrast to other IEC lineages, contain abundant ER and these studies uncovered an important link between cell-intrinsic ER stress and intestinal inflammation. In a murine model, a conditional deletion of a transcriptional factor relevant for ER homeostasis, XBP1, in the intestinal epithelium results in spontaneous development of small intestine inflammation displaying several features of human IBD such as crypt abscesses, leukocyte infiltration and ulcerations[51].

Cells respond to ER stress by engaging the unfolded protein response (UPR), a primitive cellular pathway that is activated upon accumulation of unfolded or misfolded protein within the ER[55]. Highly secretory cells such as Paneth cells and goblet cells are particularly susceptible to ER stress and are exceedingly dependent upon properly functioning UPR to maintain cellular viability and homeostasis. As such, these cells are highly susceptible to a variety of secondary or cell extrinsic factors that further stress the secretory requirements of the cell or adversely influence the environment and function of the secretory pathway of the cell including defects in components of the UPR itself (XBP1, ARG2, ORMDL3) or in genes that encode proteins reliant upon a robust secretory pathway (MUC2, HLAB27) or even factors that can disturb the UPR (availability of nutrients and their metabolic products, oxygen supply, microbial products or cytokines and other inflammatory mediators among others)[56, 57].

It seems that autophagy and ER stress pathways are intersecting and interdependent pathways. The loss of key autophagy proteins may disrupt autophagy-mediated organelle turnover and subsequently disruption of homeostasis in Paneth cells abundant ER. Indeed, Atg16L1-deficient Paneth cells display degeneration of mitochondria and replacement of ER with excess vesicular structures[15]. Furthermore, with the potential role for autophagic machinery in controlling ER stress and exocytosis, it is possible that defects in autophagy disproportionately affect the function of secretory cells such as Paneth cells and goblet cells. On the other hand, it has been shown in *in vitro* model systems that UPR IRE1-mediated JNK activation or PERK-mediated eIF2α phosphorylation can both regulate autophagy[58].

In an alternative hypothesis, ER stress and autophagy pathways might dialogue indirectly through modulation of the microbiota. As we will discuss below, there is a bidirectional interaction between autophagy and the intestinal microbiota. The same holds true for ER stress pathways – XBP1 in particular. On the one hand, it has been shown that XBP1, through its effects in antimicrobial activity of the epithelium may have substantial impact on the composition of intestinal microbiota[51]. On the other hand, the microbiota can interfere with ER stress pathway. For example, bacteria such as *Streptomyces* spp. secrete small molecules named macrocytic lactams which directly inhibit XBP1 activation[59]. Thus, an abnormal ER stress pathway may result in alterations of the microbial milieu influencing other risk factors such as autophagy and NOD2, and, as a corollary, abnormalities in the microbial milieu that may emanate from abnormalities in risk factors such as autophagy and NOD2 may adversely affect ER stress pathways.

Paneth cells dysfunction and aberrant cytokine responses were also previously described in IBD patients harboring NOD2 and Atg16L1 polymorphisms. In addition,

NOD2-, Atg16L1- and XBP1-deficient mice all share a phenotypic impairment of Paneth cell function[15, 31, 51]. The fact that innate immunity-, autophagy- and ER-associated risk factors converge on Paneth cells suggest that these pathways may be synergistically related to inflammation and this is probably a relevant factor in the development and/or perpetuation of intestinal inflammation. Future research should define their relative contribution to the pathogenesis of Crohn's disease.

The Inflammatory Component

Innate immunity and its pattern recognition molecules (PRMs) are major players in the inflammatory response. PRMs are germ line-encoded proteins that have been shown to detect a vast array of pathogen-associated molecular patterns (PAMPs) from diverse microbes including lipopolysaccharides (LPS), peptidoglycan (PG), lipoproteins, flagellin and nucleic acids, as well as danger signals or damage-associated molecular patterns (DAMPs) such as ATP, uric acid and HMGB1. Prominent among these receptors are the Toll-like-receptors (TLRs), the NOD-like receptors (NLRs, that include NOD2) and the RIG-I-like receptors (RLRs) families. Upon the detection of PAMPs, various cell types can trigger a potent protective response that includes abundant cytokine secretion, recruitment of phagocytic cells, secretion of antimicrobial peptides and instruction of dendritic cells to promote the generation of an adaptive immune response[27]. However, possible signaling pathways leading to autophagy activation downstream of PRMs remain poorly understood. A potential involvement of TLR signaling in autophagy induction has been proposed based on studies that reported upregulation of LC3 puncta in macrophages following stimulation with several TLR agonists including mycobacterial lipoproteins (TLR2), poly(I:C) (TLR3), LPS (TLR4) and ssRNA (TLR7). In some instances, this PAMP-induced autophagy improved the capacity of the cells to eliminate intracellular pathogens[60-63]. In addition, it has been reported that TLR4 adaptor proteins, TRIF and MyD88, physically interact with the autophagic protein Beclin-1 to promote LPS-induced autophagosome formation[62]. However, the role of TLRs in induction of autophagy remains a matter of some debate as in recent studies LPS stimulation did not increase the number of endogenous LC3 puncta in mouse primary macrophages. In addition, Atg16L1-dependent formation of bacteria-containing autophagosomes in macrophages following infection with *Salmonella typhimuruim* does not require engagement of TLRs as it occurs also in the absence of MyD88 and TRIF adaptor molecules[49, 64].

In mammals, protein degradation systems are crucial for a tight control of inflammatory responses. Indeed, it appears that autophagy has a role in limiting inflammation and, therefore, it is not surprising that defects in the autophagic machinery/pathways are associated with diseases characterized by excessive inflammation. There is increasing evidence showing that basal and induced autophagy play important roles under pathophysiological conditions such as dysregulation of the immune response, neurodegeneration and hepatic dysfunction. Yet the mechanisms by which autophagy limits inflammation are less clear.

Aberrant activation of PRM-induced signaling pathways, often as a result of disruption of regulatory mechanisms, leads to massive inflammation and may cause septic shock or autoimmune diseases. In order to investigate the role of Atg16L1 and autophagy in response to PAMPs, the group of Dr. Shizuo Akira generated mice with the Atg16L1 gene deleted for the CCD domain (ΔCCD). In contrast to the Atg16L1^{HM} mouse model in which autophagy is reduced but not completely eliminated, MEFs from Atg16L1-ΔCCD mice are null for autophagy. However, these mice are not viable and die within one day of birth, a phenomenon that had been previously observed in mice deficient in other Atg proteins, such as Atg5 and Atg7[65, 66]. Saitoh *et al* tested fetal liver-derived macrophages isolated from Atg16L1-ΔCCD mice for the production of pro-inflammatory cytokines in response to LPS or Gram-negative commensal bacteria (*E. coli, Enterobacter aerogenes* and *Klebsiella pneumoniae*). While TNFα and IL-6 levels were normal, they observed abnormally high pro-IL-1β processing, resulting in elevated IL-1β release. Like Atg16L1-deficient macrophages, Atg7-deficient macrophages also presented increased production of IL-1β showing that autophagy in general and not Atg16L1 deficiency per se is responsible for the phenotype. While investigating the molecular mechanisms underlying enhanced LPS-induced IL-1β release in Atg16L1-deficient macrophages, the authors observed comparable levels of IL-1β mRNA as well as of immature pro-IL-1β protein following LPS stimulation in these and wild type macrophages, indicating that the differences are post-translational[49]. Processing of pro-IL-1β to mature IL-1β (the active form) is dependent on the activation of caspase-1[67]. In Atg16L1-deficient macrophages, LPS stimulation induces cleavage of pro-caspase-1 to its activated form, caspase-1, resulting in enhanced IL-1β and IL-18 (also processed by caspase-1) production. Next, the authors analysed the involvement of TLR-dependent signaling in IL-1β production by Atg16L1-deficient macrophages[49]. Upon activation, TLRs recruit the adaptor proteins MyD88 or TRIF (or both) to initiate signaling cascades leading to the activation of NF-κB and MAP kinases and transcription of type I IFNs[69]. The activation of NF-κB and the MAP kinase p38 in response to LPS stimulation was similar in wild type and Atg16L1-deficient macrophages, indicating that the difference in IL-1β production is not due to disruption of these signaling pathways. Interestingly, enhanced IL-1β release by Atg16L1-deficient macrophages was observed in response to the TLR4 ligand, which signals both through MyD88 and TRIF, but not to the ligands for TLR2 or TLR5, that signal only via MyD88. These findings prompted them to assess the involvement of TRIF signaling in the autophagic control of IL-1β production. To this end, they generated Atg16L1-TRIF double-deficient mice and found that macrophages isolated from these mice failed to produce IL-1β in response to LPS stimulation due to lack of caspase-1 activation. Finally, as ROS have been implicated in activation of caspase-1 induced by ATP, silicas, uric acid crystal and asbestos[70], its potential involvement in enhanced IL-1β production by Atg16L1-deficient macrophages was analysed. The authors show that not only ROS levels are higher in Atg16L1-deficient macrophages in comparison to wild type cells but also that ROS scavengers block IL-1β production by Atg16L1-deficient macrophages[49]. Together, these findings suggest that ROS accumulate in Atg16L1-deficient macrophages following LPS stimulation and induce TRIF-dependent production of IL-1β, indicating that autophagy has an important role in buffering oxidative stress and regulating inflammation.

Abnormally high levels of IL-1β and IL-18 are associated with the development of intestinal inflammation[70]. As Atg16L1 is a susceptibility gene for CD, the observations described above led to the question as to whether the inability of Atg16L1-deficient macrophages to limit IL-1β and IL-18 production could be implicated in the development of disease. To analyse this issue, the authors used lethally irradiated mouse chimeras reconstituted with Atg16L1 ΔCCD embryonic liver cells in an experimental model of colitis. Strikingly, mice expressing the truncated Atg16L1 on the hematopoietic compartment had increased sensitivity to experimental colitis induced by the chemical damaging agent dextran sodium sulfate (DSS) and died following severe weight loss after DSS treatment. These animals presented significantly elevated IL-1β and IL-18 in the sera relative to controls. Histological analysis showed a more severe inflammation in the distal colons of chimeric mice lacking Atg16L1 in hematopoietic cells than seen in control mice, with larger areas of ulceration and increased infiltration of lymphocytes. Importantly, death was avoided and pathology was reversed by anti-IL-1β and anti-IL-18 neutralizing antibodies, demonstrating that excessive production of these cytokines due to a defect in the autophagic machinery results in the development of severe colitis[49].

A subsequent study further analysed the mechanisms of autophagic control of caspase-1 activation and subsequent release of IL-1β and demonstrated that depletion of the autophagic proteins LC3B or beclin-1 promoted the accumulation of dysfunctional mitochondria and release of mitochondrial DNA (mtDNA) into the cytosol in response to LPS and ATP in macrophages. This resulted in enhanced activation of caspase-1 and secretion of IL-1β and IL-18. Release of mtDNA into the cytosol depended on the NALP3 inflammasome and mitochondrial ROS. Additionally, LC3B-deficient mice produced more caspase-1-dependent cytokines in two sepsis models and were susceptible to LPS-induced mortality. These results suggest that autophagic proteins regulate NLRP3-dependent inflammation by preserving mitochondrial integrity[71].

A more recent study confirmed a role for autophagy in regulating the production of IL-1β in antigen-presenting cells. The authors show that, after treatment of macrophages with TLR ligands, pro-IL-1β is specifically sequestered into autophagosomes, while further activation of autophagy with rapamycin induces the degradation of pro-IL-1β and blocks secretion of the mature cytokine. Similar to the studies described above, in this case also, inhibition of autophagy promoted the processing and secretion of IL-1β in a NLRP3- and TRIF-dependent manner and inhibition of ROS also reduced the effects of impaired autophagy. Furthermore, induction of autophagy in mice *in vivo* with rapamycin reduced serum levels of IL-1β in response to challenge with LPS. Thus, this study proposes that autophagy controls the production of IL-1β through at least two separate mechanisms; by targeting pro-IL-1β for lysosomal degradation and by regulating activation of the NLRP3 inflammasome[73].

The data described above implicate Atg16L1 and autophagy in the regulation of endotoxin-induced inflammatory response as macrophages deficient in different components of the autophagic machinery are hyperreactive to LPS stimulation and produce increased amounts of IL-1β and IL-18. These findings demonstrate an essential role for autophagy not only in regulation of inflammation, especially in the gut, where exposure to LPS is constant, but also for proper healing and homeostasis. These insights strengthen the notion that subtle defects in autophagy are probably more evident in organs where homeostasis imposes large and/or extremely dynamic loads in the autophagic machinery. Can elevated IL-1β help us understand CD pathogenesis? The effects of the truncated Atg16L1 on IL-1β are reminiscent of CD-associated NOD2 variants, that also elicit elevated IL-1β I in a murine model[74]. The components of IL-1β signaling are often abnormal in intestinal tissue specimens from IBD patients, and increased IL-1β levels may, among other effects increase epithelial barrier susceptibility, possibly enhancing microbial product translocation[70, 74].

Fujishima *et al* investigated the involvement of autophagy in the regulation of inflammatory responses specifically in the intestinal epithelium. In mice with a conditional deletion in

Atg7 in the small intestine epithelium, the production of TNFα and IL-1β mRNA following stimulation with LPS given orally was enhanced when compared with control small intestine tissues. In addition, the level of NF-κB activation in response to LPS was also elevated in Atg7-deficient IECs (Fujishima *et al*, Ach Biochem Bioph, 2010). The (NF-κB), a transcription factor that regulates the expression of several genes involved in inflammation such as the pro-inflammatory cytokines IL-8 and (TNFα), is intimately associated with the pathogenesis of several inflammatory disorders, including IBD[75]. As a master regulator of inflammation, the activation of NF-κB is a rate limiting step for various inflammatory responses[76]. The nuclear translocated NF-κB is often detected in the mucosae of IBD patients suggesting that it might contribute to disease progression[77]. The relationship between autophagic pathways and NF-κB signaling has not yet been completely elucidated but the emerging picture shows these two events on opposite sides of the balance, i.e., there seems to be an inverse correlation between them. Recent studies have demonstrated that activation of NF-κB blocks autophagy and, conversely, induction of autophagy limits NF-κB activation[78]. These antagonistic pathways can be concomitantly induced upon stimulation with microbes (or their products) or foreign substances and they may counter balance each other. It is plausible that simultaneous induction of autophagy along with NF-κB activation may serve to limit excessive inflammatory responses and help maintain tissue homeostasis. Thus, understanding the interplay between autophagy and NF-κB-dependent inflammatory responses might prove valuable for elucidating the mechanisms underlying the pathogenesis of diseases that have a strong inflammatory component. Autophagy may become a target for therapeutic intervention for inflammatory disorders including IBD.

Altogether, these data showing that disruption of different components of the autophagic machinery results in excessive inflammation suggest that autophagy can attenuate LPS-induced inflammation in macrophages and IECs to help maintain intestinal homeostasis.

The Immunological Component

Early work in the study of IBD in general, and CD in particular, focused on alterations in the activation and polarization of T cells (Th1, Th2 and, more recently, Th17) and their resultant cytokine profiles. With the identification of the innate immune sensor NOD2 as the first susceptibility gene for CD, much attention was drawn to the role of the innate immunity in initiating, modulating and perhaps driving the abnormal inflammatory response observed in CD. As a consequence, significant advances in our understanding of the relationship between autophagy and innate immunity and their roles in CD have been made. However, there remains a substantial gap in our knowledge of the role of autophagy in the adaptive immune system.

Antigen-presenting cells (APCs) are a crucial link between innate and adaptive immune responses. These cells, in particular dendritic cells (DCs), are capable of presenting exogenous (e.g., bacteria-derived) and endogenous antigens as well as the co-stimulatory signals necessary to activate T cells to initiate the adaptive immune response. Autophagy is known to be required for optimal MHC class-II antigen cross-presentation in thymic development and impaired autophagy in a specific APC in the thymus results in the development of colitis[79]. Even though MHC class-II presentation is typically associated with exogenous antigens, autophagosomes can engulf cytosolic components and thereby deliver cellular and microbial degradation products to MHC class-II compartments. Indeed, processing and MHC class-II presentation of cytoplasmic influenza or measles virus by DCs require lysosome acidification but not transporter associated with antigen processing (TAP) transporters or protein trafficking through the Golgi apparatus, indicating direct delivery of proteins to lysosomes from the cytosol[80, 81]. In addition, autophagy is required for viral recognition by plasmacytoid DCs and has an important role in vaccination by enhancing antigen presentation[82]. However, the question remains of how intracellular microbes, as opposed to self antigens, are specifically recognized and targeted by the autophagic machinery for MHC class-II presentation. A recent study sheds new light upon this mechanism by showing that both NOD2 and Atg16L1 are required in DCs for the induction of autophagy and proper induction of antigen-presentation and bacterial handling pathways. As evidence to support this, it was shown that stimulation of DCs with the NOD2 ligand, MDP, increases the levels of MHC class-II surface expression as a result of redistribution from within the cell. The increase in MHC class-II surface expression correlated with LC3 re-localization to MHC class-II compartments. DCs in which Atg5, Atg7 or Atg16L1 have been ablated via siRNA failed to relocate MHC class-II to the cell surface following MDP stimulation. This was associated with loss of the localization of MHC class-II to autophagic compartments. In addition, in DCs expressing the CD-associated NOD2 variant infected with either *Salmonella* or adherent-invasive *E. coli* (AIEC) isolated from the mucosa of CD patients, failure to translocate bacteria to lysosomes resulted in increased bacterial proliferation. Furthermore, CD4+ T cell proliferation assays following infection with *Salmonella typhimurium* showed that both NOD2- and Atg16L1-deficient DCs had reduced antigen-presentation capacity strengthening the role of NOD2 in autophagy-mediated MHC class-II antigen presentation. Finally, in DCs isolated from CD patients and harboring the CD-associated variants of NOD2 (L1007fs) or Atg16L1 (Atg16L1 *300A) a similar loss of MHC class-II expression was observed. Functionally, this results in curtailed antigen presentation and inability to generate CD4+T cell responses to bacteria-derived antigens following infection. Of note, in contrast with what was observed in epithelial cells and fibroblasts, in DCs, autophagy mediated by NOD2 seems to require an intact RIP-2 dependent signaling pathway[36]. Thus, altogether, these data highlight the contribution of NOD2-mediated autophagy to the adaptive immune response in response to pathogens via MHC class-II presentation in DCs.

In the gut, DCs act as sentinels and capture bacteria that breach the epithelial barrier and invade the intestinal mucosa. They are also constantly sampling the intestinal luminal environment by extending dendrites in between epithelial cells and into the lumen. Finally, in the particular immune network of the gut, antigens are also captured by the specialized M cells, that continuously sample and deliver luminal antigens to DCs, and macrophages are also considered professional APCs[83, 84]. In addition, several factors present in the intestine, that can either be released from other cell types (e.g. thymic stromal lymphopoietin, TSLP), from bacteria (e.g. adenosine 5′-triphosphate, ATP) or nutrients (e.g. retinoic acid) can condition the function of intestinal DCs[83, 85, 86]. Furthermore, there are several subsets of DCs in the intestine and each one of them plays a different role in intestinal homeostasis and IBD pathogenesis. Some DC subsets, in particular those that express CD103, maintain tolerance towards the commensal intestinal microbiota, while other subsets have been implicated in the generation of effector immune responses against potential pathogen[83, 84]. One DC subset in particular is found in increased numbers in intestinal granulomas of a mouse CD model as well as in the inflamed colon of CD patients. This unique DC subset shares some phenotypic and functional features with macrophages and produces abundant amounts of IL-23 in response to bacterial products[85, 87]. Even though the involvement of autophagy in DC function has become increasingly apparent, it is still not known if it plays a specific role in each of these subsets, and in particular in those involved in IBD pathogenesis. Autophagy may also have a critical role in the differentiation of T cells, as induction of autophagy results in down-regulation of lysosomal cathepsins and altered patterns of peptide processing in DCs and, thus, has a potential to change the nature of peptides presented by these cells to $CD4^+$ T cells. Thus, given that degradation systems such as autophagy are important not only for the elimination of intracellular pathogens but also influence adaptive immune responses, it is likely that autophagy has a role in modulating antigen presentation and, subsequently, immune activation in the intestinal compartment influencing both healthy and disease states.

Even though they are not considered professional APCs, intestinal epithelial cells (IECs) possess some attributes of APCs such as the ability to sample soluble proteins via fluid-phase pinocytosis or selective receptor mediated routes, and to express MHC class-I and class-II. Autophagy might be particularly important for these cells. In an inflammatory setting, in the presence of pro-inflammatory cytokines such as IFNγ, epithelial cells have increased capacity to present antigens due to increased MHC class-II expression coupled with increased delivery of LC3-positive autophagosomes to MHC class-II compartments[89]. Another aspect to be considered is the potential role of these cells in the phenomenon of oral tolerance that refers to the suppression of all aspects of systemic immune responses to an antigen, even thymus-dependent antigens, when it is administrated orally. In general, T-cell mediated responses and tissue immunopathology are very sensitive to low doses of orally administered antigens. Despite its profund effects in immunity and its obvious practical and clinical importance, the mechanisms of oral tolerance are still largely unknown. As IECs can present antigens in the context of MHC class-I and class-II molecules[90, 91] but, in general, lack co-stimulatory signaling required for full T cell activation[92], they are good candidates for tolerogenic antigen presenting cells *in vivo*. Apparently, this is not the only means by which IECs can suppress mucosal T cell responses, as some primary colonic epithelial cells do express co-stimulatory molecules but they still interfere with CD4+ T cell activation by professional APCs[92]. It is plausible that perturbations in the processing and presentation of self antigens could impair the ability of IECs to maintain oral tolerance. In this sense, autophagy could be involved as it is involved in intracellular processing for MHC class-II loading. Additionally, substantial levels of autophagy have been observed in thymic epithelial cells as well as in immature DCs that are involved in central and peripheral T cell tolerance induction, respectively[88]. Finally, induction of autophagy by microbial infection can potentially increase processing not only of microbial but also host and dietary antigens by MHC class-II pathways which could alter the spectrum of peptides presented to the immune system. As a result, dysregulation of oral tolerance mechanisms could certainly contribute to the uncontrolled inflammation characteristic of IBD.

Adaptive immune responses are affected by a combination of resident and recruited cell populations that comprise mucosal B cells secreting IgA and IgG, a complex mixture of T cells that are dominated by a Th1, Th17 or Th2 phenotype and T_{reg} cells. Abnormal cytokine responses are a feature of both CD and UC. In simplistic terms, UC seems to exhibit a Th2-type-like cytokine profile, while a Th1 profile has been implicated in the pathogenesis of CD[93]. Recent studies, however, suggest a more complex and significant overlap between the two major forms of IBD. Another $CD4^+$ T cell lineage, Th17, has been defined and is characterized by the production of the eponymic cytokine IL-17 (IL-17A) as well as IL-17F, IL-21, IL-22 and IL-26 but not IFNγ or IL-4[94]. The development of the Th17 phenotype is promoted by IL-23 and suppressed by transcription factors required for Th1 and Th2 cells. For many years, a role in the development of CD was attributed to IL-12 p70 (a heterodimer of p40 and p35 subunits) based on the beneficial effects of anti-p40 treatment on CD patients and experimental models. However, it is becoming increasingly clear that IL-23 (heterodimer of p40 and p19 subunits) rather than IL-12 contributes to the development of CD[94]. Furthermore, polymorphisms in the gene encoding the IL-23 receptor are negatively associated with CD development[96]. A pathogenic role of IL-17 has been demonstrated in CD models and intestine-specific subsets of Th17 cells, based on differential production of cytokines and expression of surface markers, and were identified in the inflamed colon of CD patients[97]. Even though it is not known if autophagy has different effects on these functionally different T cell subsets, recent evidence linking autophagy and production of/response to key cytokines implies that the immune dysregulation characteristic of CD may result from aberrant autophagic processes.

Once T cells are activated, autophagy plays an important role in immune regulation through its role in proliferation and survival of these cells[97]. Characteristic double-membrane vesicles are observed in both human and murine activated T cells and autophagy is thought to enhance survival and proliferation of memory T cell through clearance of superfluous mitochondria[79, 97, 98]. Colitis might develop from dysregulated homeostasis of memory T cells. Indeed, in a murine model, implantation of Atg5-deficient thymus in an athymic recipient mouse results in systemic inflammation, including colitis, indicating that autophagy is required for appropriate selection of self antigen-specific $CD4^+$ T cells[79]. As it is important to initiate adaptive immune responses, autophagy is also a key to preventing the responding effector cells from causing harmful effects by directly or indirectly participating in the differentiation of several T and B regulatory subtypes.

B cells represent the major lymphocyte population in the mucosal immune compartment. The importance of these cells in mucosal immunity can be illustrated by the fact that more than 80% of the total of activated B cells are located within mucosal tissues[99]. Even though these cells have not been considered a major source of cytokines, especially when compared to T cells, emerging evidence show that not only do they produce a vast array of cytokines but also that specific subsets of cytokine-producing B cells may have a role in controling the development of IBD[100, 101]. One of these subsets, named Breg for regulatory B cell, produces IL-10 under intestinal inflammatory conditions and contributes to improvement of ongoing colitis seen in several IBD models. A role for B_{regs} in ameliorating colitis is further supported by reports showing an exacerbation of IBD concomitant to reduction in IL-10 production after B cell depletion therapy using a humanized mouse anti-human CD20 monoclonal antibody (known commercially as rituximab). Another intestinal B cell subset relevant to IBD is the unique IgM+B2 subset that is characterized by very high expression of MHC class-II[102, 103]. This B subset develops in an antigen-independent manner and significantly expands in the context of inflammation and contributes to host defense against enteric microorganisms by producing IL-12 p70. By drawing a parallel with peritoneal B1 B cells that have the ability to suppress colitis by also producing natural IgM against enteric microorganisms and where homeostasis is known to be Atg5-dependent[104], one can speculate that autophagy may also be relevant for homeostasis of IgM+B2 B cells, but this remains to be investigated. Despite the role of autophagy in MHC class-II antigen presentation and a known role for Atg5 in B cell development and homeostasis, the effect of perturbations in autophagy upon B cell repertoires is still largely unknown.

In addition to exogenous antigen presentation, a major function of autophagy is the regulation of self responses[105]. In this sense, autophagy contributes to the clearance of dying cells that represent the major source of self antigens. It is well documented that during development the clearance of cell corpses is dependent on autophagy. Cells with defective autophagy are not able to release lysophosphatidylcholine to recruit phagocytes or present phosphatidylserine in their surfaces to trigger their engulfment by the phagocytes. During infection and/or inflammation of the gut, prematurely shed and/or dying cells are released and must also be cleared by phagocytes. The failure to promptly engulf and clear dead or dying cells is strongly associated with diverse autoimmune disorders, such as arthritis and lupus, and may take on special immunological significance during gut infection and inflammation. Indeed, past findings in IBD models showed that impaired clearance of apoptotic cells resulted in exacerbation of chronic colitis. It is likely that clearance of dead cells in these situations is also dependent on defined membrane signals and, thus, might be affected by reduced levels of autophagy. Combined with increased leakage of soluble factors from the gut lumen during inflammation and tissue disruption, this may lead to enhanced autoantigen exposure and subsequently increased risk of autoimmunity. On the other hand, efficient autophagy targeting pathogens and cell corpses combined may allow a more rapid immune detection of pathogens since cells with intact autophagy not only are more efficient in presenting pathogen-derived antigens to immune cells but also enhance their own uptake by phagocytes upon their death. Thus, the participation of autophagy in these different processes is likely to result in the reduction of the spread of pathogens as well as in a faster adaptive immune response, both leading to a more prompt resolution of inflammation.

Even though there is still much to be done to elucidate how polymorphisms in autophagy genes influence CD pathogenesis, it is clear that autophagy is paramount to the mucosal immune system as it is required for the entire immune processes involved in the pathogenesis of CD from controlling the function to the development of many different cell subsets in the intestine[106].

The Microbial Component

It is essential for the maintenance of gut homeostasis that microbes that gain entry into intestinal tissues are promptly confined and removed, not only to limit microbial replication and spread but also to avoid the induction of unnecessary and potentially damaging inflammation. In this context, autophagy is crucial for confining and degrading microbial intruders and limiting the extent of immune recognition and activation. Thus, defects in autophagy and its antimicrobial activity may predispose individuals to IBD. For example, the alterations observed in autophagy-impaired Paneth cells increases epithelial exposure to luminal microbiota and allow the microbiota to gain access to the base of the crypts. This, in turn, induces inflammation and promotes epithelial turnover but, in the absence of fully functional autophagy, junctional regulation may be compromised resulting in improper coordination of cell shedding. This destabilizes the epithelium resulting in exposure of the basolateral surfaces of the epithelial cells accompanied by depletion of mucus and the epithelial layer itself—all factors that predispose to infection with pathogens as well as commensal microbiota. In this way, epithelial cells and phagocytes have to respond to higher burdens of bacterial invasion/attachment and epithelial breakthrough. If the autophagic machinery is unable to keep up with the bacterial challenge, these epithelial cells and phagocytes may become infected and perpetuate the inflammatory cycle[3]. In summary, another way autophagy seems to affect gastrointestinal homeostasis is by removing bacteria that make it into tissues and degrading any non-self immuno-stimulatory molecules to maintain intestinal tolerance.

A recent study showed a crucial role of the interplay between NOD2 and Atg16L1, two relevant genes for CD pathogenesis, in autophagy-mediated clearance of intracellular bacteria by lysosomal degradation. NOD2 (as well as NOD1) was shown to direct Atg16L1 to bacterial entry sites on the plasma membrane, initiating autophagy to remove invading pathogens. Initially, it was demonstrated that NOD agonists induce autophagy in human and mouse cells. In mice, exposure to NOD1 or NOD2 ligands via intraperitoneal injection following thioglycholate priming shows NOD-dependent induction of autophagy within macrophages isolated from wild type mice but not NOD1- or NOD2-deficient mice, respectively. However, treatment with rapamycin or stimulation with LPS resulted in normal activation in either NOD1- or NOD2-deficient mice. Using *Shigella flexneri* and *Listeria monocytogenes* infection of mouse embryonic fibroblasts (MEFs, wild type and NOD1-deficient) it was demonstrated that autophagic control of both *S. flexneri* and *L monocytogenes* infection was dependent upon NOD1, but independent of RIP2 and NF-κB[35]. As previously mentioned, Atg16L1 is essential to specify the site for LC3 conjugation during autophagosome formation[35]. The NOD proteins are immunoprecipitated with Atg16L1, and a fraction of these proteins also co-localized at the plasma membrane when overexpressed[35]. It had been previously shown that, upon infection with invasive bacteria, NOD2 localizes to the bacterial entry foci at the plasma membrane from where it recruits RIP2 to initiate pro-inflammatory signaling pathways[35]. In this study, both endogenous and overexpressed NOD2 and Atg16L1 (and overexpressed NOD1) were shown to be co-localized preferentially at sites of bacterial entry following short periods of infection with *S. flexneri*, indicating that NOD2 (and NOD1) directly interact with Atg16L1 to recruit the autophagic machinery to the site of bacterial entry and initiate anti-bacterial autophagy. Notably, in the epithelial cells and MEFs used in this study, the RIP2 adaptor protein was dispensable for NOD2-induced autophagy[35]. This would provide the cell with a mechanism to promptly confine and remove invading bacteria without the need to activate signaling pathways or synthesis of new components. Thus, activation of NOD2 at the plasma membrane can initiate two independent innate immune mechanisms to fight infection: RIP2-dependent NF-κB activation and anti-bacterial autophagy. Importantly, the CD-associated NOD2 frame-shift mutation results in a protein that no longer localizes to the plasma membrane and, as a consequence, it retains Atg16L1 in the cytosol resulting in impaired wrapping of invading bacteria by autophagosomes. Finally, it was shown that human cells homozygous for the CD-associated Atg16L1 polymorphism show an impaired autophagic induction upon treatment with MDP, the NOD2 ligand, suggesting that NOD2 and Atg16L1 localization to the plasma membrane is required for optimal innate immune signaling and efficient autophagic response[35]. These results show that mutations in either NOD2 or Atg16L1 can compromise anti-bacterial autophagy. Given that these are two of the most important genetic susceptibility factors associated with CD, it is likely that this defect is relevant to CD pathogenesis.

Microbiota

It is clear from the ongoing discussion that the host's genetic make-up is one of the essential elements associated with risk of developing CD. The challenge now is to determine what the other risk factors are that serve as triggers in the susceptible host. The host microbiota is a strong contender. Indeed, studies directed towards the identification of environmental factors implicate gut resident commensal microbiota rather than pathogens as drivers of dysregulated immunity and susceptibility to CD[35]. The examination of how genetics influences the host responses to commensal microbiota at the intestinal epithelial interface has revealed a number of previously unappreciated pathways that are of considerable importance to mucosal homeostasis and inflammatory diseases. Two hypotheses have arisen regarding the role of microbiota in CD:

1. it could result from an excessive/dysregulated immune response to the normal intestinal commensal microbiota or;
2. it could be a consequence of an imbalance in the composition of the microbiota leading to generalized or localized dysbiosis.

The enteric microbiota is composed of more than 15,000 species of bacteria which can reach a density of more than 10^{12} microorganisms per ml of luminal content in the colon. The luminal contents of the colon account for 99.9% of the known human indigenous microbiota[2]. Colonization of mucosal surfaces including the GI tract, starts at birth and evolves until early childhood. The adult microbiota for each individual is unique and stable over time. The commensal intestinal bacteria are essential for intestinal epithelial cell repair and homeostasis as well as for the development of the host immune system, but they might also be critical factors in the onset and/or course of CD.

Several lines of evidence point to the role of the intestinal microbiota in the onset or perpetuation of CD:

1. Intestinal bacteria are essential for the development of intestinal inflammation since they are required for the onset of inflammation in numerous knock-out models of IBD[111, 112, 113] (including the Atg16L1^{HM} mice discussed below).
2. In animal models, severity of colitis is decreased with the use of antibiotics and no sign of colitis is observed in animals kept in germ-free conditions[115].
3. Generalized or localized dysbiosis (microbial imbalance) is often observed in CD patients and might represent a breakdown in the balance between putative protective versus harmful bacteria.
   Indeed, molecular biology analysis of the luminal bacterial composition in CD patients provided evidence that in these patients intestinal microbiota is less diverse with predominance of *Proteobacteria* and *Bacteroidetes* over the Bifidobacteria, Lactobacilli and *Firmicutes* that are frequently observed in healthy individuals[112, 113]. In particular, a recent study has implicated diminished levels of *Faecalibacterium prausnizii* as being associated with higher risk of post-operative recurrence of ileal CD[111] (Sokol *et al*, PNAS, 2008 - http://www.ncbi.nlm.nih.gov/pubmed/18936492). In addition, these studies

demonstrated that in CD patients intestinal microbiota is more abundant, forming a thick biofilm on the surface of the intestinal mucosa[111].

4. CD develops predominantly at anatomical sites (terminal ileum and colon) where there is a dramatic increase in commensal bacteria mass.
5. In CD patients, exposure of the terminal ileum post-surgically to luminal contents is associated with increased inflammation and, conversely, diversion of the fecal stream is associated with improvement of disease[115].
6. Some CD patients improve upon antibiotic treatment[116].
7. Administration of probiotic bacteria can also contribute to the supression of colitis[111, 117].

Before recognition of the role of autophagy in CD, it was known that the microbiota was an essential component in CD pathogenesis but no singular cellular pathway that could confer the immune defects seen in patients had been identified. Now, autophagy integrates numerous aspects of CD pathogenesis including mutations in NOD2, innate immune activation with aberrant inflammation, defects in bacterial handling and antigen presentation and homeostasis of the epithelial barrier. Thus, autophagy provides a defined pathway with which to study the interaction of host and gut microbiota and how this is perturbed in disease states. The use of new molecular microbiology tools to analyze variations in the gut microbiota of patients with known predisposing Atg16L1, IRGM and NOD2 polymorphisms will allow us to determine the extent of correlation between colonization with particular microbial species and CD. A better definition of these processes and the identification of the nature of such perturbations and how to effect their restoration is the key to understanding the pathophysiology of this inflammatory disorder and routes to therapy.

Pathogenic Microorganisms

CD exibits features that resemble those of microbial infections in the gut. Indeed, some characteristic pathological elements of CD, including aphthous ulcers of the mucosa, mural abscesses and macrophage and epithelioid cell granulomas also occur in well recognized infectious diseases such as shigellosis, salmonellosis and yersinial enterocolitis that have as a common theme the invasion ability and intracellular life style of the bacteria involved[118]. This has prompted many investigations into finding a potential causative infectious agent for CD. In addition, the implication of the NOD2-autophagy axis in CD has strengthened the idea of microbial agents as triggers for CD onset due to impaired function of the host defense against intracellular bacteria. Indeed, one potential mechanism by which autophagy contributes to intestinal homeostasis is through elimination of microbes through the fusion of microbe-containing autophagosomes with degradative lysosomal compartments. To be implicated as a primary causal infectious agent in CD pathogenesis, a microorganism would probably require an initial invasive ability and also an inherent mechanism for chronic persistance. The enteric microbiota is normally not sufficiently invasive or lacks virulence factors that are able to initiate disease progression.

Several established bacterial pathogens were tentatively implicated as CD-associated pathogens, including *Listeria monocytogenes* and *Yersinia* spp. but evidence has been lacking despite some circumstancial observations. These bacteria were suggested to play a role in CD based on the so-called "cold chain hypothesis"—epidemiologic assessment of familial environmental risk factors related to the western lifestyle, diet, bacteria and domestic hygiene that pointed to refrigeration as a potential risk factor for CD. As psychotropic bacteria (capable of growing between -1°C and 10°C), *L. monocytogenes* and *Yersinia* spp. appeared as good candidates[119]. In addition, several studies demonstrated the presence of *Y. enterocolitica* and *Y. pseudotuberculosis* in intestinal samples of CD[111, 120] patients and there are a number of aspects of yersiniosis that are reminiscent of the inflammatory reaction seen in CD, such as the presence of granulomas, which make the differential diagnosis difficult[121]. Even though *L. monocytogenes* has been identified in CD lesions by immunohistochemical analysis, PCR analysis generated conflicting results[109, 112, 122]. Interestingly, risk factors for CD also favor *L. monocytogenes* infections:

1. NOD2-deficient mice presents increased susceptibility to *L. monocytogenes* infection due to decreased levels of cryptins (defensin homologues) secreted by Paneth cells[31].
2. *In vitro* systems demonstrate that *L. monocytogenes* is efficiently targeted by autophagy by both the Ub-p62-LC3 and NOD-Atg16L1 pathways and disruption of these pathways results in increased intracellular bacterial growth[123].
3. Mice deficient in LRG-47, the homologue of human *IRGM* gene, are unable to eliminate *L. monocytogenes*[41]. However, Atg16L1^{HM} mice do not have increased susceptibility to *L. monocytogenes* given orally[15].

Mycobacterium avium subspecies *paratuberculosis* (MAP) remained for many years as a putative infectious agent with a pathogenic role in CD. This bacteria causes spontaneous granulomatous enterocolitis in cattle accompanied by diarrhea and wasting, making this obligate intracellular pathogen a credible etiologic agent for CD due to the clinical similarities between the disease in ruminants and human CD[124]. In the past few decades, several dozen published studies have provided evidence that both supports and disputes the association of MAP with CD[122, 125, 126]. Thus, to date, there remains no definitive evidence that MAP plays a causal role in human CD. However, the identification of mutations in the NOD2 gene in CD patients and the GWAS that in the past few years implicated defects in autophagy as a risk factor for CD, warrant additional research to confirm or not the putative role of MAP as an etiological agent in CD. This is because MAP can selectively infect hosts with defective innate immune killing, such as those with NOD2 mutations or defects in autophagy. The autophagic machinery can effectively eliminate intracellular bacteria[43, 44] and the NOD2 pathway has been implicated in the innate immune response to *Mycobacterium* infection as NOD2-deficient macrophages produce an impaired cytokine response to *M. tuberculosis*[121]. So far, however, no relation was found between NOD2 mutations and positive MAP serology.

E. coli coexits as commensal with its mammalian hosts and rarely causes disease except in immunocompromised hosts or whenever mucosal barriers are breached. Some *E. coli* strains, however, have acquired specific virulence factors that allow them to cause a broad spectrum of diseases and, therefore, are considered pathogens. There are at least six well described pathotypes of *E. coli* that cause disease in humans[127]. Recently, independent studies in North America and Europe reported the presence of mucosa-associated *E. coli* with invasive properties in CD patients that does not belong to any of the established pathotypes[128-131]. This new CD-associated *E. coli* pathotype was termed AIEC, for adherent-invasive *E. coli.* AIEC is partiularly associated with ileal CD due to abnormal expression of the specific host receptor CEACAM6 (Carcinoembryonic antigen-related cell adhesion molecule 6), that allows adherence of the bacteria to the intestinal mucosa. The significantly increased expression of CEACAM6 in the ileum observed in uninvolved ileal mucosa of CD patients compared to healthy controls suggest that these patients are genetically predisposed to over-express that molecule. Interestingly, even though stimulation with TNFα and IFNγ increases the levels of CEACAM6 expression in cells that already express it, these cytokines do not induce CEACAM6 expression in cells that do not already express it. Additionally, AIEC infection also upregulates CEACAM6 indicating that these bacteria can promote their own colonization in CD patients[118]. Thus, the high prevalence of AIEC strains associated with the ileal lesions on mucosa of CD patients is related to an abnormal expression of a specific host receptor recognized by bacterial lectin-like surface adhesins in a genetically predisposed host gut segment. Colonization by AIEC and the inflammatory milieu would then lead to an amplification loop of colonization and inflammation.

During infection, a subpopulation of intracellular AIEC is found in LC3-positive autophagosomes. Functional autophagy limits AIEC intracellular replication as evidenced by the enhanced AIEC replication in IRGM or Atg16L1 deficient cells. Furthermore, rescue of Atg16L1 with the *300T protective variant resulted in a more efficient control of AIEC replication with phenotype similar to the control cells while the CD-associated *300A variant was unable to mediate full phenotypic rescue, despite equal levels of protein expression. Interestingly, autophagy deficiency did not interfere with the ability of any other *E. coli* strain (including non-pathogenic, environmental, commensal or pathogenic strain involved in gastroenteritis) to survive or replicate inside the cells suggesting a crucial and specific role of autophagy in restraining AIEC associated with ileal CD[45]. However, it is not clear whether colonization with AIEC represent a proximal cause of inflammation or is actually a consequence of the disease. In murine models, inflammation-induced shifts in the intestinal microbiota have been observed following infection with *Citrobacter rodentium* or *Salmonella typhimurium*, both of which promote growth of *Enterobacteriaceae*[132, 133]. *E. coli* overgrowth could also result from the abnormal expression of host molecules acting as receptors for bacterial adhesion. As mentioned in a previous section, the large CD risk-conferring deletion at the IRGM locus also results in a functional alteration in the ability of epithelial cells to initiate and sustain autophagy to control AIEC replication. Thus, the combination of polymorphisms in autophagy genes with the presence of pathogenic AIEC colonizing the intestinal mucosa may have a significant impact on the outcome of intestinal inflammation.

Until recently, viral infections were not considered to be among the major risk factors contributing to CD development as the incidence of clinical flare-ups of the disease do not correlate with outbreaks of viral infection. However, viruses could be operating subclinically. For example, it is possible that impaired autophagy or defects in NLR-mediated sensing of nucleic acids can result in abnormal persistence of viruses normally cleared by the host's immune system. Alternatively, viral infection of the host microbiome could play a part. Finally, another possibility could be abnormal handling of endogenus retroviruses as has been shown for other inflammatory diseases[133, 134]. Importantly, a recent study demonstrated that a specific viral infection in a susceptible host (Atg16L1^{HM} mice) results in profound strctural and functional alterations in Paneth cells rendering these mice prone to develop an aberrant inflammation that presents several features that are characteristic of human CD (discussed below).

Despite all this evidence showing that some pathogens are more commonly found in CD patients than in healthy individuals, to date, no single bacterial species could be implicated as a causal infectious agent in CD. Nevertheless, it is possible that in a host with impaired defense mechanisms against intracellular bacteria, such as autophagy, the multiplications of persistent bacteria (e.g., MAP or AIEC) is no longer regulated and could trigger uncontrolled intestinal inflammation.

The Interplay Between the Different Components

As we have seen, much progress has been made in our understanding of CD pathogenesis and how genetic, immunological, microbial and environmental factors all contribute to the development of CD. The model of complex disease in which an aberrant immunologic/inflammatory response to an environmental and/or microbial stimulus in a genetically susceptible host is widely recognized but experimental evidence has been lacking.

The group of Dr. Herbert Virgin presented a multi-hit model of IBD triggered by a viral infection only in the presence of a mutation in a disease susceptibility gene. They used the $Atg16L1^{HM}$ mice model that present similar Paneth cell abnormalities to the ones observed in Paneth cells from CD patients homozygous for the risk allele of Atg16L1, including aberrant packaging and exocytosis of antimicrobial granules and a gain-of-function transcriptional profile in which transcripts associated with lipid metabolism, proinflammatory cytokines, and other pathways were enriched. Strikingly, when $Atg16L1^{HM}$ mice were rederived into an enhanced barrier facility, alterations in Paneth cells were no longer observed, suggesting a role for an infectious agent present in the conventional mouse facility. Indeed, mice kept in the conventional facility were infected with an enteric virus that is prevalent in mouse facilities around the world, murine norovirus (MNV). Accordingly, oral infection of the animals kept in the "clean facility" with a specific strain of MNV capable of persistent infection resulted in the same alteration in Paneth cells in the $Atg16L1^{HM}$ mice but not in wild-type mice. The treatment with low doses of dextran sulfate triggered an inflammatory response in the colon of non-infected $Atg16L1^{HM}$ mice as well as in the colon of infected and non-infected wild-type mice. In the infected $Atg16L1^{HM}$ mice, however, the response to injury was fundamentally altered to include pathologies resembling aspects of human CD in both colon and ileum. Importantly, the $Atg16L1^{HM}$ mice did not present a higher viral burden than the wild type mice following treatment with DSS[135]. The aberrant intestinal injury response described in virally-infected $Atg16L1^{HM}$ mice was dependent on three factors associated with Crohn's disease pathogenesis in humans—TNFα, IFNγ, and commensal bacteria. This common role of two cytokines and commensal bacteria suggest that the human disease and the observed pathologies in mice share an underlying mechanism, and supports the relevance of the virus-plus-susceptibility gene interaction.

These results provide a specific example of how a genetic factor and an environmental agent, each innocuous by itself, can have profound effects on the host when combined and also provide one explanation for why persons in clinical trials may display widely variable responses to treatment for a single pathologically defined disease. The data presented in this study demonstrate the combinatory effect of several factors—host's genetic make-up + exposure to a specific virus + mucosal barrier damage caused by a toxic agent + commensal microbiota—for the establishment of an abnormal inflammatory response displaying several characteristics of CD in humans. This type of model should help us understand the pathogenesis of complex diseases and how the immune system communicates with the environment. A notable difference however, between the Atg16L1^{HM} mouse model and human CD is that, in the mouse, the genetic risk is mimicked by reducing Atg16L1 expression rather than by expressing the actual human gene variant. As previously mentioned, in human CD, the Atg16L1 risk-associated variant (T300A) displays a single amino acid change in the WD repeat. This repeat is a recent evolutionary acquisition, absent from the equivalent autophagy gene of organisms such as yeast, and its function is still unknown. Thus, understanding the interaction between viruses and the autophagy pathway in cells expressing the T300A variant should be informative. Considering that the Atg16L1 *300A protein is unstable in infected epithelial cell lines, a feature not seen with the protective Atg16L1*300T, one might speculate that the Atg16L1 *300A variant presents a hypomorphic phenotype. Thus the Atg16L1^{HM} mice may recapitulate some aspects of this Atg16L1 CD risk allele. In cultured epithelial cells, the Atg16L1*300A risk-associated variant has no effect on housekeeping autophagy but is implicated in impaired autophagic response to intracellular bacteria[136].

Finally, the observations in this study provides a basis for understanding how a common allele can be linked to an infrequent severe disease, and why mice carrying mutations in human disease susceptibility genes do not always spontaneously reproduce human pathology.

CONCLUDING REMARKS

Recent years have seen a revolution in the ability of investigators to identify the genetic determinants of IBD and indeed all complex diseases. Progress has been made possible by endeavors such as the human genome project and advances in genotyping technologies that make hypothesis-free methods of testing the majority of the human genome for association with disease a tractable proposition. Collectively, the studies described here re-affirm several long-term or recent notions about CD and provide insights into the role of autophagy in this disease, an aspect that will become more complete once the effects of the autophagy factors are revealed. However, the complexity of the autophagic response and its connection to multiple other cellular pathways are likely to make it difficult to readily understand its role in IBD pathogenesis. The significant challenge for patients and their clinicians is that their IBD remains incurable and current medical treatments are generally limited to modulating inflammation and promoting mucosal healing. Novel therapeutic strategies are likely to flow from a better understanding of the immune, microbial, environmental and genetic elements that contribute to IBD. In this context, it is hoped that an improved understanding of pathogenic mechanisms, for example, by studying the genetic basis of CD and other forms of IBD, will lead to improved therapies and possibly preventive strategies in individuals identified as being at risk.

References

1. Lee YK, Mazmanian SK. Has the microbiota played a critical role in the evolution of the adaptive immune system? *Science* 2010;**330**(6012):1768–73.
2. Frank DN, et al. Molecular-phylogenetic characterization of microbial community imbalances in human inflammatory bowel diseases. *Proceedings of the National Academy of Sciences of the United States of America.* 2007;**104**(34):13780–5.

3. Huett A, Xavier RJ. Autophagy at the gut interface: mucosal responses to stress and the consequences for inflammatory bowel diseases. *Inflamm Bowel Dis* 2010;**16**(1):152–74.
4. Hara T, et al. Suppression of basal autophagy in neural cells causes neurodegenerative disease in mice. *Nature* 2006;**441**(7095):885–9.
5. Komatsu M, et al. Loss of autophagy in the central nervous system causes neurodegeneration in mice. *Nature* 2006;**441**(7095):880–4.
6. Komatsu M, et al. Homeostatic levels of p62 control cytoplasmic inclusion body formation in autophagy-deficient mice. *Cell* 2007;**131**(6):1149–63.
7. Mathew R, et al. Autophagy suppresses tumorigenesis through elimination of p62. *Cell* 2009;**137**(6):1062–75.
8. Rabinowitz JD, White E. Autophagy and metabolism. *Science* **330**(6009):1344-8.
9. Sakiyama T, Musch MW, Ropeleski MJ, Tsubouchi H, Chang EB. Glutamine increases autophagy under Basal and stressed conditions in intestinal epithelial cells. *Gastroenterology* 2009;**136**(3):924–32.
10. Nicklin P, et al. Bidirectional transport of amino acids regulates mTOR and autophagy. *Cell* 2009;**136**(3):521–34.
11. Schock F, Perrimon N. Molecular mechanisms of epithelial morphogenesis. *Annu Rev Cell Dev Biol* 2002;**18**:463–93.
12. Tettamanti G, et al. Programmed cell death and stem cell differentiation are responsible for midgut replacement in Heliothis virescens during prepupal instar. *Cell Tissue Res* 2007;**330**(2):345–59.
13. Wice BM, Gordon JI. A strategy for isolation of cDNAs encoding proteins affecting human intestinal epithelial cell growth and differentiation: characterization of a novel gut-specific N-myristoylated annexin. *The Journal of cell biology* 1992;**116**(2):405–22.
14. Ouellette AJ. Paneth cells and innate mucosal immunity. *Curr Opin Gastroenterol* 2007;**26**(6):547–53.
15. Cadwell K, et al. A key role for autophagy and the autophagy gene Atg16l1 in mouse and human intestinal Paneth cells. *Nature* 2008;**456**(7219):259–63.
16. Xavier RJ, Podolsky DK. Unravelling the pathogenesis of inflammatory bowel disease. *Nature* 2007;**448**(7152):427–34.
17. Ramjeet M, et al. 'Nodophagy': New crossroads in Crohn disease pathogenesis. *Gut Microbes* 2010;**1**(5):307–15.
18. Xavier RJ, Huett A, Rioux JD. Autophagy as an important process in gut homeostasis and Crohn's disease pathogenesis. *Gut* 2008;**57**(6):717–20.
19. Sewell GW, Marks DJ, Segal AW. The immunopathogenesis of Crohn's disease: a three-stage model. *Current opinion in immunology* 2009;**21**(5):506–13.
20. Stappenbeck TS, et al. Crohn disease: A current perspective on genetics, autophagy and immunity. *Autophagy* 2011;**7**(4):355–74.
21. Hugot JP, et al. Association of NOD2 leucine-rich repeat variants with susceptibility to Crohn's disease. *Nature* 2001;**411**(6837):599–603.
22. Rescigno M, Nieuwenhuis EE. The role of altered microbial signaling via mutant NODs in intestinal inflammation. *Curr Opin Gastroenterol* 2007;**23**(1):21–6.
23. Yamamoto S, Ma X. Role of NOD2 in the development of Crohn's disease. *Microbes Infect* 2009;**11**(12):912–8.
24. Shaw MH, Kamada N, Warner N, Kim YG, Nunez G. The ever-expanding function of NOD2: autophagy, viral recognition, and T cell activation. *Trends in immunology* 2011;**32**(2):73–9.
25. Sabbah A, et al. Activation of innate immune antiviral responses by NOD2. *Nat Immunol* 2009;**10**(10):1073–80.
26. Inohara, et al. NOD-LRR proteins: role in host-microbial interactions and inflammatory disease. *Annu Rev Biochem* 2005;**74**:355–83.
27. Carneiro L, et al. NOD-like proteins in inflammation and disease. *J Pathol* 2008;**214**(2):136–48.
28. Netea MG, et al. NOD2 mediates anti-inflammatory signals induced by TLR2 ligands: implications for Crohn's disease. *Eur J Immunol* 2004;**34**(7):2052–9.
29. Noguchi E, et al. A Crohn's disease-associated NOD2 mutation suppresses transcription of human IL10 by inhibiting activity of the nuclear ribonucleoprotein hnRNP-A1. *Nat Immunol* 2009;**10**(5):471–9.
30. Brant SR, McGovern DP. NOD2, not yet: con. *Inflamm Bowel Dis* 2005;**11**(5):507–9.
31. Kobayashi KS, et al. NOD2-dependent regulation of innate and adaptive immunity in the intestinal tract. *Science* 2005;**307**(5710):731–4.
32. Watanabe T, Kitani A, Murray PJ, Strober W. NOD2 is a negative regulator of Toll-like receptor 2-mediated T helper type 1 responses. *Nat Immunol* 2004;**5**(8):800–8.

33. Watanabe T, et al. Nucleotide binding oligomerization domain 2 deficiency leads to dysregulated TLR2 signaling and induction of antigen-specific colitis. *Immunity* 2006;**25**(3):473–85.
34. Maeda S, et al. NOD2 mutation in Crohn's disease potentiates NF-kappaB activity and IL-1beta processing. *Science* 2005;**307**(5701):734–8.
35. Travassos LH, et al. NOD1 and NOD2 direct autophagy by recruiting Atg16L1 to the plasma membrane at the site of bacterial entry. *Nat Immunol* 2010;**11**(11):55–62.
36. Cooney R, et al. NOD2 stimulation induces autophagy in dendritic cells influencing bacterial handling and antigen presentation. *Nat Med* 2010;**16**(1):90–7.
37. Genome-wide association study of 14,000 cases of seven common diseases and 3,000 shared controls. *Nature* 2007;**447**(7145):661–78.
38. Parkes M, et al. Sequence variants in the autophagy gene IRGM and multiple other replicating loci contribute to Crohn's disease susceptibility. *Nat Genet* 2007;**39**(7):830–2.
39. Rioux JD, et al. Genome-wide association study identifies new susceptibility loci for Crohn disease and implicates autophagy in disease pathogenesis. *Nat Genet* 2007;**39**(5):596–604.
40. McCarroll SA, et al. Deletion polymorphism upstream of IRGM associated with altered IRGM expression and Crohn's disease. *Nat Genet* 2008.
41. Collazo CM, et al. Inactivation of LRG-47 and IRG-47 reveals a family of interferon gamma-inducible genes with essential, pathogen-specific roles in resistance to infection. *The Journal of experimental medicine* 2001;**194**(2):181–8.
42. Henry SC, et al. Impaired macrophage function underscores susceptibility to Salmonella in mice lacking Irgm1 (LRG-47). *J Immunol* 2007;**179**(10):6963–72.
43. Singh SB, et al. Human IRGM induces autophagy to eliminate intracellular mycobacteria. *Science* 2006;**313**(5792):1438–41.
44. Gutierrez MG, et al. Autophagy is a defense mechanism inhibiting BCG and Mycobacterium tuberculosis survival in infected macrophages. *Cell* 2004;**119**(6):753–66.
45. Lapaquette P, et al. Crohn's disease-associated adherent-invasive E. coli are selectively favoured by impaired autophagy to replicate intracellularly. *Cellular microbiology* 2010;**12**(1):99–113.
46. The International HapMap Project. *Nature* 2003;**426**(6968):789–96.
47. Hampe J, et al. A genome-wide association scan of nonsynonymous SNPs identifies a susceptibility variant for Crohn disease in Atg16L1. *Nat Genet* 2007;**39**(7730):207–11.
48. Levine B, Mizushima N, Virgin HW. Autophagy in immunity and inflammation. *Nature* 2011;**469**(7330):323–35.
49. Saitoh T, et al. Loss of the autophagy protein Atg16L1 enhances endotoxin-induced IL-1beta production. *Nature* 2008;**456**(7219):264–8.
50. Porter EM, Bevins CL, Ghosh D, Ganz T. The multifaceted Paneth cell. *Cell Mol Life Sci* 2002;**59**(1):156–70.
51. Kaser A, et al. XBP1 links ER stress to intestinal inflammation and confers genetic risk for human inflammatory bowel disease. *Cell* 2008;**134**(5):743–56.
52. Cadwell K, Patel KK, Komatsu M, Virgin HWt, Stappenbeck TS. A common role for Atg16L1, Atg5 and Atg7 in small intestinal Paneth cells and Crohn disease. *Autophagy* 2009;**5**(2):250–2.
53. Barbier M, et al. Overexpression of leptin mRNA in mesenteric adipose tissue in inflammatory bowel diseases. *Gastroenterol Clin Biol* 2003;**27**(11):987–91.
54. Yamamoto K, et al. Production of adiponectin, an anti-inflammatory protein, in mesenteric adipose tissue in Crohn's disease. *Gut* 2005;**54**(6):789–96.
55. Kaser A, Martinez-Naves E, Blumberg RS. Endoplasmic reticulum stress: implications for inflammatory bowel disease pathogenesis. *Curr Opin Gastroenterol* 2010;**26**(4):318–26.
56. Schroder M, Kaufman RJ. The mammalian unfolded protein response. *Annu Rev Biochem* 2005;**74**:739–89.
57. Todd DJ, Lee AH, Glimcher LH. The endoplasmic reticulum stress response in immunity and autoimmunity. *Nature reviews* 2008;**8**(9):663–74.
58. Kaser A, Blumberg RS. Endoplasmic reticulum stress and intestinal inflammation. *Mucosal Immunol* 2010;**3**(1):11–6.
59. Tashiro E, et al. Trierixin, a novel Inhibitor of ER stress-induced XBP1 activation from Streptomyces sp. 1. Taxonomy, fermentation, isolation and biological activities. *J Antibiot (Tokyo)* 2007;**60**(9):547–53.
60. Delgado MA, Elmaoued RA, Davis AS, Kyei G, Deretic V. Toll-like receptors control autophagy. *The EMBO journal* 2008;**27**(7):1110–21.

61. Sanjuan MA, et al. Toll-like receptor signalling in macrophages links the autophagy pathway to phagocytosis. *Nature* 2007;**450**(7173):1253–7.
62. Shin DM, et al. Mycobacterial lipoprotein activates autophagy via TLR2/1/CD14 and a functional vitamin D receptor signalling. *Cellular microbiology* 2010;**12**(11):1648–65.
63. Shi CS, Kehrl JH. MyD88 and Trif target Beclin 1 to trigger autophagy in macrophages. *The Journal of biological chemistry* 2008;**283**:33175–82.
64. Akira S, Uematsu S, Takeuchi O. Pathogen recognition and innate immunity. *Cell* 2006;**124**(4):783–801.
65. Komatsu M, et al. Impairment of starvation-induced and constitutive autophagy in Atg7-deficient mice. *The Journal of cell biology* 2005;**169**(3):425–34.
66. Kuma A, et al. The role of autophagy during the early neonatal starvation period. *Nature* 2004;**432**(7020):1032–6.
67. Dinarello CA. IL-1: discoveries, controversies and future directions. *Eur J Immunol* **40**(3):599-606.
68. Kawai T, Akira S. The role of pattern-recognition receptors in innate immunity: update on Toll-like receptors. *Nat Immunol* **11**(5):373-84.
69. Schroder K, Tschopp J. The inflammasomes. *Cell* 2010;**140**(6):821–32.
70. Al-Sadi R, Ye D, Dokladny K, Ma TY. Mechanism of IL-1beta-induced increase in intestinal epithelial tight junction permeability. *J Immunol* 2008;**180**(8):5653–61.
71. Nakahira K, et al. Autophagy proteins regulate innate immune responses by inhibiting the release of mitochondrial DNA mediated by the NALP3 inflammasome. *Nat Immunol* 2007;**12**(3):222–30.
72. Harris J, et al. Autophagy controls IL-1β secretion by targeting pro-IL-1β for degradation. *The Journal of biological chemistry* 2011;**286**(11):5987–97.
73. Kanneganti TD, et al. Pannexin-1-mediated recognition of bacterial molecules activates the cryopyrin inflammasome independent of Toll-like receptor signaling. *Immunity* 2007;**26**(4):433–43.
74. Deretic V, Master S, Singh S. Autophagy gives a nod and a wink to the inflammasome and Paneth cells in Crohn's disease. *Dev Cell* 2008;**15**(5):641–2.
75. Baker RG, Hayden MS, Ghosh S. NF-kappaB, inflammation, and metabolic disease. *Cell Metab* **13**(1): 11-22.
76. Sha WC. Regulation of immune responses by NF-kappa B/Rel transcription factor. *The Journal of experimental medicine* 1998;**187**(2):143–6.
77. Schreiber S, Nikolaus S, Hampe J. Activation of nuclear factor kappa B inflammatory bowel disease. *Gut* 1998;**42**(4):477–84. 78.
78. Fujishima Y et al. Autophagy in the intestinal epithelium reduces endotoxin-induced inflammatory responses by inhibiting NF-kappaB activation. *Arch Biochem Biophys* **506**(2):223-35.
79. Nedjic J, Aichinger M, Emmerich J, Mizushima N, Klein L. Autophagy in thymic epithelium shapes the T-cell repertoire and is essential for tolerance. *Nature* 2008;**455**(7211):396–400.
80. Jaraquemada D, Marti M, Long EO. An endogenous processing pathway in vaccinia virus-infected cells for presentation of cytoplasmic antigens to class II-restricted T cells. *The Journal of experimental medicine* 1990;**172**(3):947–54.
81. Malnati MS, et al. Processing pathways for presentation of cytosolic antigen to MHC class II-restricted T cells. *Nature* 1992;**357**(6380):702–4.
82. Jagannath C, et al. Autophagy enhances the efficacy of BCG vaccine by increasing peptide presentation in mouse dendritic cells. *Nat Med* 2009;**15**(3):267–76.
83. Coombes JL, Powrie F. Dendritic cells in intestinal immune regulation. *Nature reviews* 2008;**8**(6):435–46.
84. Yoshida M, et al. Neonatal Fc receptor for IgG regulates mucosal immune responses to luminal bacteria. *The Journal of clinical investigation* 2006;**116**(8):2142–51.
85. Kamada N, et al. Unique CD14 intestinal macrophages contribute to the pathogenesis of Crohn disease via IL-23/IFN-gamma axis. *The Journal of clinical investigation* 2008;**118**(6):2269–80.
86. Mora JR. Homing imprinting and immunomodulation in the gut: role of dendritic cells and retinoids. *Inflamm Bowel Dis* 2008;**14**(2):275–89.
87. Mizoguchi A, et al. Dependence of intestinal granuloma formation on unique myeloid DC-like cells. *The Journal of clinical investigation* 2007;**117**(3):605–15.
88. Schmid D, Pypaert M, Munz C. Antigen-loading compartments for major histocompatibility complex class II molecules continuously receive input from autophagosomes. *Immunity* 2007;**26**(1):79–92.

89. Hershberg RM, et al. Highly polarized HLA class II antigen processing and presentation by human intestinal epithelial cells. *The Journal of clinical investigation* 1998;**102**(4):792–803.
90. Mayer L, et al. Expression of class II molecules on intestinal epithelial cells in humans. Differences between normal and inflammatory bowel disease. *Gastroenterology* 1991;**100**(1):3–12.
91. Sanderson IR, Ouellette AJ, Carter EA, Walker WA, Harmatz PR. Differential regulation of B7 mRNA in enterocytes and lymphoid cells. *Immunology* 1993;**79**(3):434–8.
92. Cruickshank SM, et al. Colonic epithelial cell mediated suppression of CD4 T cell activation. *Gut* 2004;**53**(5):678–84.
93. Strober W, Fuss I, Mannon P. The fundamental basis of inflammatory bowel disease. *The Journal of clinical investigation* 2007;**117**(3):514–21.
94. Korn T, Bettelli E, Oukka M, Kuchroo. VK IL-17 and Th17 Cells. *Annual review of immunology* 2009;**27**:485–517.
95. Duerr RH, et al. A genome-wide association study identifies IL23R as an inflammatory bowel disease gene. *Science* 2006;**314**(5804):1461–3.
96. Kleinschek MA, et al. Circulating and gut-resident human Th17 cells express CD161 and promote intestinal inflammation. *The Journal of experimental medicine* 2009;**206**(3):525–34.
97. Pua HH, Guo J, Komatsu M, He YW. Autophagy is essential for mitochondrial clearance in mature T lymphocytes. *J Immunol* 2009;**182**(7):4046–55.
98. Stephenson LM, et al. Identification of Atg5-dependent transcriptional changes and increases in mitochondrial mass in Atg5-deficient T lymphocytes. *Autophagy* 2009;**5**(5):625–35.
99. Brandtzaeg P, Johansen FE. Mucosal B cells: phenotypic characteristics, transcriptional regulation, and homing properties. *Immunol Rev* 2005;**206**:32–63.
100. Fillatreau S, Gray D, Anderton SM. Not always the bad guys: B cells as regulators of autoimmune pathology. *Nature reviews* 2008;**8**(8):391–7.
101. Mizoguchi A, Bhan AK. A case for regulatory B cells. *J Immunol* 2006;**176**(2):705–10.
102. El Fassi D, et al. Ulcerative colitis following B lymphocyte depletion with rituximab in a patient with Graves' disease. *Gut* 2008;**57**(5):714–5.
103. Goetz M, et al. Exacerbation of ulcerative colitis after rituximab salvage therapy. *Inflamm Bowel Dis* 2007;**13**:1365–8.
104. Shimomura Y, et al. Regulatory role of B-1 B cells in chronic colitis. *Int Immunol* 2008;**20**(6):729–37.
105. Yoshimori T. Autophagy: paying Charon's toll. *Cell* 2007;**128**(5):833–6.
106. Munz C. Enhancing immunity through autophagy. *Annual review of immunology* 2009;**27**:423–49.
107. Dianda L, et al. T cell receptor-alpha beta-deficient mice fail to develop colitis in the absence of a microbial environment. *Am J Pathol* 1997;**150**(1):91–7.
108. Rath HC, et al. Normal luminal bacteria, especially Bacteroides species, mediate chronic colitis, gastritis, and arthritis in HLA-B27/human beta2 microglobulin transgenic rats. *The Journal of clinical investigation* 1996;**98**(4):945–53.
109. Sellon RK, et al. Resident enteric bacteria are necessary for development of spontaneous colitis and immune system activation in interleukin-10-deficient mice. *Infection and immunity* 1998;**66**(11):5224–31.
110. Wirtz S, Neurath MF. Mouse models of inflammatory bowel disease. *Adv Drug Deliv Rev* 2007;**59**(11):1073–83.
111. Sartor RB. Microbial influences in inflammatory bowel diseases. *Gastroenterology* 2008;**134**(2):577–94.
112. Seksik P, et al. Review article: the role of bacteria in onset and perpetuation of inflammatory bowel disease. *Aliment Pharmacol Ther* 2006;**3**(Suppl. 24):11–8.
113. Tamboli CP, Neut C, Desreumaux P, Colombel JF. Dysbiosis in inflammatory bowel disease. *Gut* 2004;**53**(1):1–4.
114. Sokol H, et al. Faecalibacterium prausnitzii is an anti-inflammatory commensal bacterium identified by gut microbiota analysis of Crohn disease patients. *Proceedings of the National Academy of Sciences of the United States of America* 2008;**105**(43):16731–6.
115. Rutgeerts P, et al. Effect of faecal stream diversion on recurrence of Crohn's disease in the neoterminal ileum. *Lancet* 1991;**338**(8779):771–4.
116. Sands BE. Inflammatory bowel disease: past, present, and future. *J Gastroenterol* 2007;**42**(1):16–25.
117. Reiff C, Kelly D. Inflammatory bowel disease, gut bacteria and probiotic therapy. *Int J Med Microbiol* 2010;**300**(1):25–33.
118. Glasser AL, Darfeuille-Michaud A. Abnormalities in the handling of intracellular bacteria in Crohn's disease: a link between infectious etiology and host genetic susceptibility. *Arch Immunol Ther Exp (Warsz)* 2008;**56**:237–44.

119. Hugot JP, Alberti C, Berrebi D, Bingen E, Cezard JP. Crohn's disease: the cold chain hypothesis. *Lancet* 2003;**362**(9400):2012–5.
120. Sasaki M, et al. Invasive Escherichia coli are a feature of Crohn's disease. *Lab Invest* 2007;**87**(10):1042–54.
121. Rumsey J, Valentine JF, Naser SA. Inhibition of phagosome maturation and survival of Mycobacterium avium subspecies paratuberculosis in polymorphonuclear leukocytes from Crohn's disease patients. *Med Sci Monit* 2006;**12**(4): BR130–9.
122. Feller M, et al. Mycobacterium avium subspecies paratuberculosis and Crohn's disease: a systematic review and meta-analysis. *Lancet Infect Dis* 2007;**7**(9):607–13.
123. Ogawa M, et al. Autophagy targeting of Listeria monocytogenes and the bacterial countermeasure. *Autophagy* 2011;**7**(3):90–117.
124. Chiodini RJ. Crohn's disease and the mycobacterioses: a review and comparison of two disease entities. *Clin Microbiol Rev* 1989;**2**(1):90–117.
125. Naser SA, Ghobrial G, Romero C, Valentine JF. Culture of Mycobacterium avium subspecies paratuberculosis from the blood of patients with Crohn's disease. *Lancet* 2004;**364**(9439):1039–44.
126. Selby W, et al. Two-year combination antibiotic therapy with clarithromycin, rifabutin, and clofazimine for Crohn's disease. *Gastroenterology* 2007;**132**(7):2313–9.
127. Kaper JB, Nataro JP, Mobley HL. Pathogenic Escherichia coli. *Nat Rev Microbiol* 2004;**2**(2):123–40.
128. Baumgart M, et al. Culture independent analysis of ileal mucosa reveals a selective increase in invasive Escherichia coli of novel phylogeny relative to depletion of Clostridiales in Crohn's disease involving the ileum. *ISME J* 2007;**1**(5):403–18.
129. Eaves-Pyles T, et al. Escherichia coli isolated from a Crohn's disease patient adheres, invades, and induces inflammatory responses in polarized intestinal epithelial cells. *Int J Med Microbiol* 2008;**298**(5-6):397–409.
130. Martin HM, et al. Enhanced Escherichia coli adherence and invasion in Crohn's disease and colon cancer. *Gastroenterology* 2004;**127**(1):80–93.
131. Darfeuille-Michaud A, et al. High prevalence of adherent-invasive Escherichia coli associated with ileal mucosa in Crohn's disease. *Gastroenterology* 2004;**127**(2):412–21.
132. Lupp C, et al. Host-mediated inflammation disrupts the intestinal microbiota and promotes the overgrowth of Enterobacteriaceae. *Cell Host Microbe* 2007;**2**(2):119–29.
133. Stecher B, et al. Salmonella enterica serovar typhimurium exploits inflammation to compete with the intestinal microbiota. *PLoS Biol* 2007;**5**(7):2177–89.
134. Simmons A. Crohn's disease: Genes, viruses and microbes. *Nature* 2010;**466**(7307):699–700.
135. Cadwell K, et al. Virus-plus-susceptibility gene interaction determines Crohn's disease gene Atg16L1 phenotypes in intestine. *Cell* 2010;**141**(7):1135–45.
136. Kuballa P, et al. Impaired autophagy of an intracellular pathogen induced by a Crohn's disease associated Atg16L1 variant. *PLoS ONE* 2008;**3**(10). e3391.

CHAPTER

6

Autophagy in the Homeostasis of Pancreatic β-Cells

Gilad Twig[1, 2], *Guy Las*[1], *and Orian S. Shirihai*[1]

[1]Department of Medicine, Section of Molecular Medicine, Boston University School of Medicine, Boston, MA 02118

[2]Department of Medicine and Talpiot Medical Leadership Program, The Chaim Sheba Medical Center, Tel-Hashomer, Israel

INTRODUCTION

Basal autophagy occurs at a low rate in all eukaryotes as a quality control mechanism to clear and recycle large damaged or redundant components of the cell including organelles, protein aggregates, ribosomes and lipids. Upon induction of stress such as amino acid starvation, ROS exposure or ER stress, autophagy is stimulated to protect the cell by increasing the availability of amino acids and by clearing accumulated damaged components. Autophagy is not exclusively linked to cell survival. Under chronic stress, it is sometimes associated with cell death ("autophagic cell death"), though the meaning of autophagy in this case is still controversial[1]. Indeed in many cases, autophagy-associated cell death is the result of a blockage of autophagic flux rather than an induction of autophagosome formation.

Steady state ("housekeeping") autophagy has been shown to be important for the physiology as well as for the viability of pancreatic β-cells. In samples from type 2 diabetic patients, signs of impaired autophagy were shown to include reduced levels of various lysosomal genes, which are involved in protein degradation and in fusion with the autophagosomes, including cathepsins B and D, and LAMP2[2]. Two separate studies reported impaired glucose tolerance in mice harboring specific β-cell ATG7 deletion, due to both an apoptosis-induced decrease in β-cell mass and to impaired β-cell function[3, 4]. The latter manifested as impaired basal- and high-glucose-stimulated insulin secretion. This defect in insulin secretion is attributed not only to a decrease in β-cell mass, but also to a reduced number of insulin granules. Morphological analysis showed swollen mitochondria, distended Golgi and ER, and accumulation of ubiquitinated proteins[3, 4]. This effect was reproduced in different *in vitro* models utilizing 3-MA, an inhibitor

Autophagy in Health and Disease
http://dx.doi.org/10.1016/B978-0-12-385101-7.00006-1

of early stage of autophagy[5, 6]. In addition, treatment of INS1 β-cells with 3-MA (an inhibitor of autophagy) but not epoxomicin (an inhibitor of the proteasome) resulted in the accumulation of ubiquitinated aggregates[6]. Taken collectively, these results found that at steady state, autophagy is critical for pancreatic β-cell function independently of the proteasome system.

The Importance of Selective Autophagy of Mitochondria in β-Cells (Mitophagy)

Mitophagy refers to the selective removal of mitochondria by the autophagic machinery. Mitophagy appears to be a universal route for the degradation of dysfunctional mitochondria. Under certain physiological settings, mitophagy can also eliminate functional mitochondria as seen during erythroid differentiation[7], oocyte fertilization[8] or during starvation[9]. Mitophagy seems to be of special importance compared to the removal of other organelles for two reasons:

1. mitochondria are one of the main sources of reactive oxygen species (ROS) generation (therefore they are also the immediate targets of ROS damage)
2. dysfunctional mitochondria that are not degraded can produce higher amounts of ROS, be more susceptible to the release of cytochrome c and apoptosis-inducing factor, and thereby, amplify ROS damage[10, 11].

These two arguments are of special importance when dealing with a cell that is highly-sensitive to ROS with low anti-oxidant capacity, such as the β-cell.

In β-cells, mitochondria are arranged in a dense, web-like morphology, where they function as fuel sensors, tightly regulating insulin secretion in response to varying glucose concentrations[12, 13]. Maintenance of the quality of mitochondria in β-cells is therefore of utmost importance requiring the continuous activity of an intricate mechanism of selection that we have recently reported[14-16]. Quality control of mitochondria involves a cycle of constant fusion and fission between mitochondria (a process termed *mitochondrial dynamics*), followed by selective mitophagy of those mitochondria that are depolarized and unable to re-fuse (Figure 6.2).

The opposing processes of fusion and fission determine the architecture of the entire mitochondrial population of the cell and thereby affect nearly every aspect of mitochondrial functions. In β-cells, the balance between these processes determine the susceptibility of the cell to nutrient-induced apoptosis, respiratory chain capacity, Atp production and insulin secretion[12, 14, 17]. In addition, fusion and fission events *per se* (and not only their resultant effect on architecture) are suggested to impact mitochondrial homeostasis. Fusion events mediate the exchange of matrix and membranous components between mitochondria, whereas fission events split mitochondria to a size that is suitable for their removal by autophagy. A single case report in human tissue suggests that the absence of fission is likely incompatible with life[18], and is of importance in β-cells whose mitochondria can form elongated and ramified networks[12, 13].

Generating the Substrate for Mitophagy: the Production of Depolarized Mitochondria

Mitochondrial membrane potential ($\Delta\psi_m$) is the driving force for mitochondrial ATP synthesis. Depolarization below a certain $\Delta\psi_m$ may indicate impaired mitochondrial function

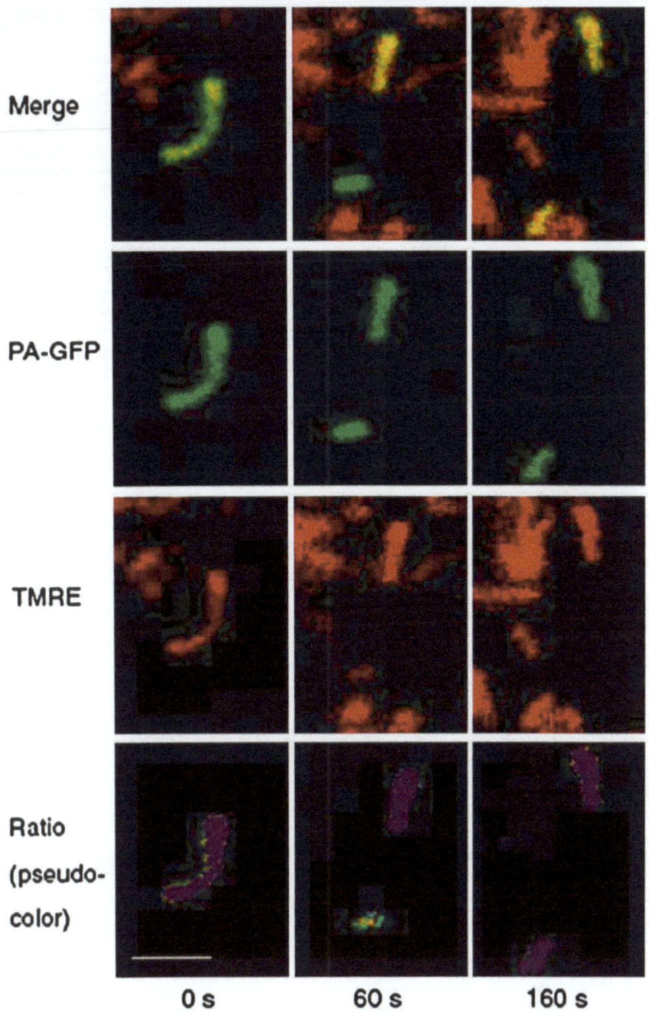

FIGURE 6.1 Mitochondrial fission provides the substrate for mitophagy. An example of a fission event that yields a ~9 mV difference between the two daughter mitochondria in INS1 β-cells. The pseudo-color represents the membrane potential and is derived from a ratio image of TMRE and matrix-targeted photoactivatable-GFP. The pseudo-colored image depicts two time points, 60 sec before fission and 160 sec after fission. Scale bar, 2 μm. Charts of $\Delta\psi_m$ over time for typical fission events under control conditions are shown to the right (Image is modified from [14]).

and is a prerequisite for mitophagy[9, 14]. Mitochondrial depolarization appears to precede the translocation of proteins that tag mitochondria for mitophagy such as Parkin and Pink1 (PTEN-induced Putative Kinase 1)[19, 20]. Yet, while essential for mitophagy, depolarization alone is insufficient to trigger mitophagy.

In primary and INS1 β-cells, prolonged tracking (up to two hours) revealed that mitochondria retain a stable $\Delta\psi_m$ in the absence of fusion or fission events[13, 14]. This suggests that a continuous deterioration in $\Delta\psi_m$ is an unlikely (or infrequent) route for the generation of

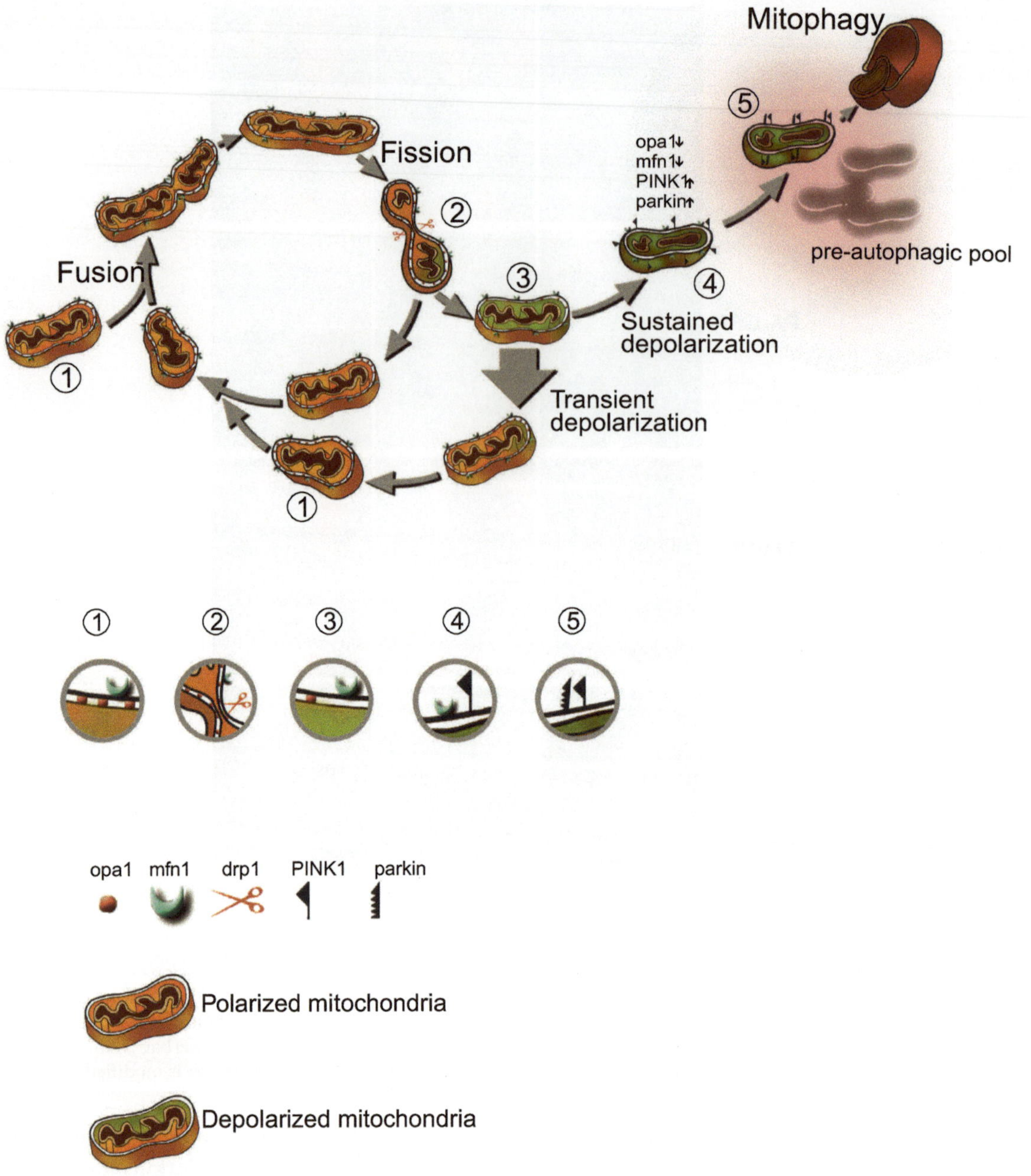
Mitophagy
Fission
Fusion
opa1↓
mfn1↓
PINK1↑
parkin↑
pre-autophagic pool
Sustained depolarization
Transient depolarization
①
②
③
④
⑤
opa1
mfn1
drp1
PINK1
parkin
Polarized mitochondria
Depolarized mitochondria

depolarized mitochondria under normal conditions. However, fission events were associated with large changes in $\Delta\psi_m$. Direct measurements of mitochondrial membrane potential in INS1 cells showed that ~5% of fission events are associated with sustained depolarization and are destined to produce one daughter mitochondrion that will be targeted by the mitophagy machinery[14]. These live imaging data are supported by EM tomography in neurons, showing that fission events yield asymmetric daughter mitochondria[21, 22] and unequal nucleoid distribution[23]. Given the high frequency of mitochondrial fission in INS1 and primary β-cells[12, 14, 24, 25] (a fission event every ~30 and 60 min respectively), mitochondrial fission constitute a principal route for generating (depolarized) mitochondria that are later targeted by the mitophagy machinery. The disruption of this quality control mechanism either by altering fission or inhibiting mitophagy, results in the accumulation of damaged mitochondria with excess of oxidized and nitrosylated mitochondrial proteins with depolarized potential. Importantly, the rate of reactive oxygen production remains unchanged[14]. At the cellular level, oxygen consumption is decreased and glucose-stimulated insulin secretion is impaired[14].

The Regulation of Mitophagy by Mitochondrial-Dynamics-Related Proteins

Table 6.1 summarizes studies that linked mitochondrial dynamics and mitophagy in INS1 β-cells and other cell lines. The view of mitochondrial fission as a prerequisite for mitophagy is supported by genetic manipulation of the pro-fission proteins Fis1 and Drp1 (Table 6.1). Knockdown of FIS1 (by siRNA) or overexpression of a dominant negative isoform of DRP1 ($DRP1^{K38A}$), reduced mitochondrial autophagy exclusively in INS1 cells, as confirmed by the lack of an increase in ER mass inside the autophagosomes[14, 23]. Overexpression of hFis1 reduced the mitochondrial (but not ER) mass in HeLa[26, 27] and INS1 cells[28], consistent with the notion that stimulation of fission facilitates mitochondrial autophagy under some circumstances. Arnoult *et al.* showed that overexpression of Drp1 facilitated mitochondrial elimination under various pro-apoptotic stimuli[23]. Endophilin B1 (also known as Bif-1), a Drp1-dependent mediator of mitochondrial fission[29], interacts with Beclin 1 and co-localizes with autophagosome markers LC3, Atg5, and Atg9 in response to nutrient starvation[30]. Its loss is associated with tubulation of mitochondria[29] and suppression of autophagy[30]. Hailey *et al.* suggested that under starvation conditions, the mitochondrial outer membrane contributes components of the autophagosome in a Mfn2-dependent manner[31]. Collectively, these findings indicate that mitochondrial-dynamics related proteins are principal regulators of mitophagy rate.

FIGURE 6.2 A schematic illustration of the mitochondrion's life cycle and the roles of mitochondrial dynamics and autophagy in the segregation of dysfunctional mitochondria in β-cells. The mitochondrion cyclically shifts between a post-fusion state (networked) that lasts tens of seconds and a post-fission state (solitary) that can last tens of minutes. Following a fission event the mitochondrion can depolarize and restore an intact potential (thick arrow) or remain in a depolarized level. Sustained $\Delta\psi_m$ depolarization triggers cleavage of Opa1, accumulation of PINK1/Parkin and reduction in mitofusin capacity. The mitochondrion may spend several hours in this pre-autophagic state before being targeted by the autophagic machinery. Protein compositions during various steps along these pathways are indicated by a relevant close-up (Scheme is modified from[60]).

TABLE 6.1 The Effect of Manipulations in Mitochondrial Proteins Opa1, Mfn1/2, Drp1 and Fis1 on Autophagy and Mitophagy. Note that in some works mitophagy was assessed directly by co-localization between mitochondrial and AP markers, indirectly as changes in mitochondrial mass, or was not distinguished from autophagy

Manipulation	Cell Type	Effect on Mitophagy	Comments
Fis1 RNAi	INS1 β-cells	Reduction of 70% in APs containing mitochondria[14]	No change in the mass of APs containing ER
Fis1 Overexpression	HeLa cells, INS1 β-cells	Reduce total mitochondrial volume by ~50%[26, 28]	No change in ER morphology or total ER area
	HeLa cells	50% increase in total APs mass. No direct assessment of mitophagy[27]	Qualitative analysis suggest medium-high correlation between LC3 and mitochondrial markers
Drp1^{K38A}	INS1 β-cells	Reduction of 75% in APs containing mitochondria[14]	No change in the mass of APs containing ER
Drp1 overexpression	HeLa cells	70% decrease in mitochondrial mass[23]	No specific tests for mitophagy
Opa1 Overexpression	INS1 β-cells	Reduction of 63% in APs containing mitochondria[14]	No change in the mass of APs containing ER
Opa1 heterozygous mutant mice	Retinal Ganglion cells	An increase in general autophagy[61]	No specific tests for mitophagy
Mfn1 Overexpression	INS1 β-cells	40% decrease in mitochondrial volume[28]	No specific tests for mitophagy
Mfn1 Loss of function (Overexpression of Mfn1 DN)	INS1 β-cells	Unchanged mitochondrial mass[28]	Insignificant slight decrease in mitochondrial volume. No specific tests for mitophagy
Mfn2 KO	MEFs	Inhibition of autophagy[31]	No specific tests for mitophagy

Mitochondrial Fragmentation and an Increase in Mitophagic Load form a Universal Mitochondrial-Stress Response

As shown above, mitochondrial fission is the event that alters the energetic state of the mitochondrion and thereby determines the fate of the organelle. While under some conditions mitochondria hyperfuse as a first line of defense to non-mitochondrial metabolic insults[32], other insults that directly affect mitochondrial metabolism are associated with significantly unbalanced fission mainly due to a decreased fusion rate (hereto referred to as mitochondrial fragmentation). Depletion of Atp either by inhibiting Atp synthase[33, 34], collapsing $\Delta\psi_m$[35-37] or inhibiting the Na^+/K^+ Atpase[34] triggers general fragmentation of the mitochondrial web due to cleavage of the profusion protein Opa1[38]. Oxidative stress induced by hyperglycemia (25-50 mM) fragments the mitochondria of β-cells[12, 39]. Direct application of hydrogen peroxide is associated with a similar architecture phenotype[40]. Common to the above observations is a stress-induced mitochondrial damage, parallel to fragmentation, and a selective increase in mitochondrial localization in autophagosomes. This raises the possibility that mitochondrial fragmentation, which can be induced by various insults, is a common stress response that is principally required to segregate and eliminate dysfunctional mitochondria from the web by mitophagy.

The Effect of Free Fatty Acids and Glucose on Autophagy in β-Cells

Free fatty acids (FFAs) have attracted much attention, particularly in the context of pancreatic β-cells where they both impair insulin secretion[41-44] and induce cell death by apoptosis[12, 45-47]. Lipotoxicity (or glucolipotoxicity when FFAs are combined with high concentrations of glucose) is associated with induction of ER-stress [48], mitochondrial network fragmentation [12], and oxidative damage to proteins[49]. Figure 6.3 shows a representative example of autophagosome-containing mitochondria in INS1 β-cell.

Autophagy is increasingly implicated in lipotoxicity. *In vivo*, the β-cells of mice fed with a high fat diet[3, 50] and the β-cells of ob/ob mice and those of db/db mice were shown to have an increase in autophagosome mass[3, 51]. *In vitro*, during glucotoxicity[6], lipotoxicity and glucolipotoxicity[3, 5] the number of autophagosomes was increased. Treatment with FFAs led to a particularly strong effect. Within four hours of exposure to FFAs, the number of autophagosomes that accumulate in the cytoplasm of a β-cell exceeds that observed during amino acid starvation, the strongest known stimulator of autophagy[5]. Autophagosomes can increase due to:

1. an increase in their formation
2. a decrease in their clearance by the lysosomes
3. a combination of both increased formation and decreased degradation.

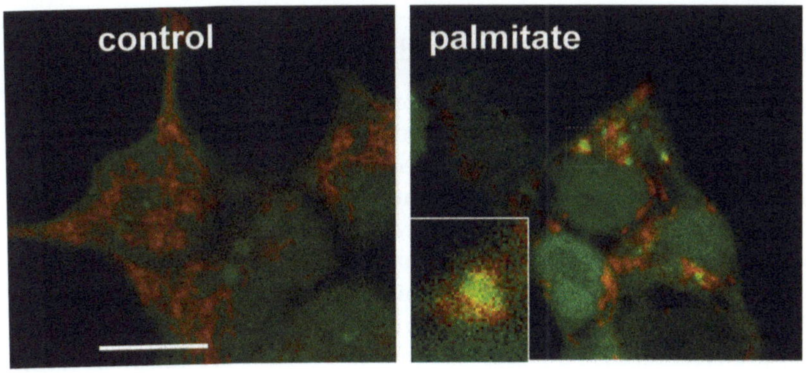

FIGURE 6.3 Mitophagy under palmitate exposure. Confocal images of INS1 cells stably expressing GFP-LC3 and Ds-Red and treated for 14h with either 0.5% BSA alone (control) or with 0.5% BSA bound to 0.4mM palmitate. White square denotes an autophagosome (green) engulfing mitochondria (red) in the presence of palmitate. Bar, 10 μm.

Free fatty acids were shown to induce a bi-phasic pattern on autophagosome mass, with a first peak occurring after a 6 hour treatment and a second peak occurring after 12 hours of treatment[5]. While increased autophagosome formation seems to contribute to the appearance of both waves, the second wave also shows a decrease in long-lived protein degradation, including that of LC3, indicating a decrease in autophagic flux. The impairment in autophagosome clearance is associated with a decrease in hydrolase function induced by alkalization of the lysosomal lumen[5]. The exact cause of lysosomal alkalization remains to be determined, but impairment in mitochondrial function could play a role in this phenomenon. Treatment of β-cells with FFAs suppressed oxygen consumption and decreased cellular Atp content, which could alter the lysosomal proton pump activity[5]. Alternatively, treatment of β-cells with FFA could change the lipid composition of lysosomal membrane so that it would lead to proton leak[52]. Thus, FFAs suppress autophagic flux over time due to decreased lysosome formation, and therefore the apparent increase in autophagosome mass does not play any protective role.

Like with glucolipotoxicity, the long term suppression of autophagic flux by FFAs is amplified when high concentrations of glucose are present[5]. The analogy with glucolipotoxicity does not occur with oleate, a FFA that is not in itself lipotoxic. However, oleate was shown to suppress autophagic flux to the same extent as palmitate[5]. This suggests that the suppression of autophagy by FFAs cannot exclusively explain lipotoxicity. However, mice with a β-cell specific Atg7 deletion show a higher susceptibility to a high fat diet than wild type mice[3]. INS1 cells treated with Atg5 siRNA show a high sensitivity to glucolipotoxicity[53]. It is therefore possible that the alteration of autophagy by FFAs precludes the defense mechanism provided by autophagy against nutrient overload, thus enhancing their toxic effect.

Treatment of INS-1 cells with the antioxidants N-acetyl cysteine or glutathione does not prevent autophagosome accumulation under glucolipotoxicity[53]. On the other hand, treatment with 4-phenylbutyrate (4-PBA), a chemical chaperone, alleviates the accumulation of autophagosomes, suggesting that ER-stress is involved[53]. Against this possibility, however, stands the evidence that oleate, which is reported to be a much weaker inducer of ER-stress than palmitate[48], stimulates autophagosome accumulation to the same extent as palmitate[5]. To further uncover the mechanism underlying the effect of FFAs on autophagy and their effect on the suppression of autophagosome clearance, it will be necessary to distinguish between the stimulatory and inhibitory effects that occur in the long term, and not just rely on autophagosome counting.

The Link Between Insulin Secretory Granules and Autophagy

While glucose regulates the secretion of insulin from β-cells by exocytosis, insulin synthesis is maintained even under non-stimulatory glucose concentrations. Autophagy (originally termed, in this case, *crinophagy*) was suggested as the balance mechanism that maintains the number of granules and total insulin content in β-cells by fusing insulin granules to lysosomes[54-56]. Insulin granules contain several common characteristics with lysosomes including a low pH and several common proteins. Figure 6.4 shows immunostaining for cathepsin, a lysosomal protein, in β-cells and its high degree of co-localization with insulin. In fact, proteomics performed on isolated insulin granules showed that out of 57 intravesicular proteins, 33 are lysosomal[57]. Several membrane proteins are common to the

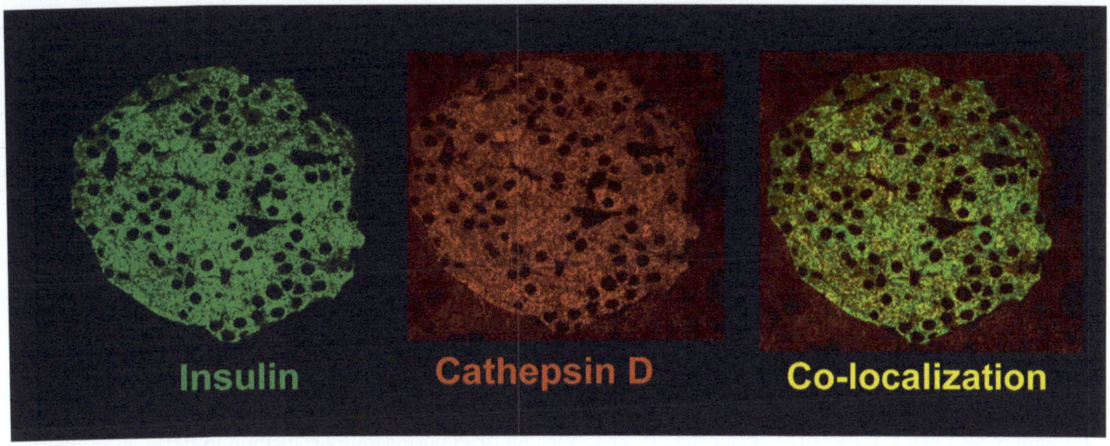

FIGURE 6.4 Expression of cathepsin D in β-cells. Immunofluorescence of pancreas retrieved from C57Bl6 mice. The section was stained against insulin (green) and cathepsin D (red). Note the strong co-localization (yellow) indicating the high expression of cathepsin D inside the insulin granules of β-cells.

two organelles as well[57]. The exact role(s) of the lysosomal proteins in the insulin vesicles is not fully clear. Most of the hydrolases in the insulin vesicles are not mature, and thus are non-functional. This is possibly due to the pH, which in the insulin granules is not acidic enough to allow self-cleavage and activation. Cathepsin L was shown to be secreted from the insulin granules, and could hypothetically have a signaling role outside the cell[58]. Other cathepsins were not shown to be secreted however, and thus their role is still unclear. The most consistent hypothesis explaining the high expression of lysososomal proteins inside the insulin granules is the "sorting by retention hypothesis", according to which, in secreting cells the maturation of lysosomal enzymes go through immature secreting vesicles after they have left the trans-golgi[59]. The immature secreting granules serve as a sorting point for proteins targeted to the lysosomes or to the mature insulin granules.

In vivo evidence for the role of autophagy in degradation of insulin granules emerged from the Rab3A-deficient mice[55]. These mice have impaired insulin secretion, but retain normal insulin biosynthesis and total insulin content due to a marked upregulation in autophagic degradation of insulin granules[55]. Hence, autophagy may play a protective buffering role in β-cells during the signaling disconnection between insulin synthesis and secretion.

To conclude, autophagy is important for proper β-cell function and viability. The removal of damaged mitochondria by mitophagy plays a critical role in the preservation of mitochondrial homeostasis, energetic balance and insulin secretion and is tightly regulated by mitochondrial dynamics. Total insulin content and the number of insulin granules is also regulated by autophagy. Free fatty acids suppress autophagy flux overtime due to a decreased lysosome formation, which contributes to the pathology rather than protects from it.

References

1. Hayashi-Nishino M, Fujita N, Noda T, Yamaguchi A, Yoshimori T, Yamamoto A. A subdomain of the endoplasmic reticulum forms a cradle for autophagosome formation. *Nat Cell Biol* 2009;**11**(12):1433–7.

2. Masini M, Bugliani M, Lupi R, et al. Autophagy in human type 2 diabetes pancreatic beta cells. *Diabetologia* 2009;**52**(6):1083–6.
3. Ebato C, Uchida T, Arakawa M, et al. Autophagy is important in islet homeostasis and compensatory increase of beta cell mass in response to high-fat diet. *Cell Metab* 2008;**8**(4):325–32.
4. Jung HS, Chung KW, Won Kim J, et al. Loss of autophagy diminishes pancreatic beta cell mass and function with resultant hyperglycemia. *Cell Metab* 2008;**8**(4):318–24.
5. Las G, Serada SB, Wikstrom JD, Twig G, Shirihai OS. Fatty acids suppress autophagic turnover in beta-cells. *J Biol Chem* 2011;**286**(49):42534–44.
6. Kaniuk NA, Kiraly M, Bates H, Vranic M, Volchuk A, Brumell JH. Ubiquitinated-protein aggregates form in pancreatic beta-cells during diabetes-induced oxidative stress and are regulated by autophagy. *Diabetes* 2007;**56**(4):930–9.
7. Takano-Ohmuro H, Mukaida M, Kominami E, Morioka K. Autophagy in embryonic erythroid cells: its role in maturation. *Eur J Cell Biol* 2000;**79**(10):759–64.
8. Shitara H, Kaneda H, Sato A, et al. Selective and continuous elimination of mitochondria microinjected into mouse eggs from spermatids, but not from liver cells, occurs throughout embryogenesis. *Genetics* 2000;**156**(3):1277–84.
9. Elmore SP, Qian T, Grissom SF, Lemasters JJ. The mitochondrial permeability transition initiates autophagy in rat hepatocytes. *FASEB J* 2001;**15**(12):2286–7.
10. Crompton M. The mitochondrial permeability transition pore and its role in cell death. *Biochem J* 1999; **341**(Pt 2):233–49.
11. Grivennikova VG, Kareyeva AV, Vinogradov AD. What are the sources of hydrogen peroxide production by heart mitochondria? *Biochim Biophys Acta* 2010. Feb 17.
12. Molina AJ, Wikstrom JD, Stiles L, et al. Mitochondrial Networking Protects Beta Cells from Nutrient Induced Apoptosis. *Diabetes* 2009;**58**(10):2303–15.
13. Wikstrom JD, Katzman SM, Mohamed H, et al. beta-Cell mitochondria exhibit membrane potential heterogeneity that can be altered by stimulatory or toxic fuel levels. *Diabetes* 2007;**56**(10):2569–78.
14. Twig G, Elorza A, Molina AJ, et al. Fission and selective fusion govern mitochondrial segregation and elimination by autophagy. *EMBO J* 2008;**27**(2):433–46.
15. Twig G, Hyde B, Shirihai OS. Mitochondrial fusion, fission and autophagy as a quality control axis: The bioenergetic view. *Biochim Biophys Acta* 2008;**1777**(9):1092–7.
16. Mouli PK, Twig G, Shirihai OS. Frequency and selectivity of mitochondrial fusion are key to its quality maintenance function. *Biophys J* 2009;**96**(9):3509–18.
17. Zhang Z, Wakabayashi N, Wakabayashi J, et al. The dynamin-related GTPase Opa1 is required for glucose-stimulated ATP production in pancreatic beta cells. *Mol Biol Cell* 2011;**22**(13):2235–45.
18. Waterham HR, Koster J, van Roermund CW, Mooyer PA, Wanders RJ, Leonard JV. A lethal defect of mitochondrial and peroxisomal fission. *N Engl J Med* 2007;**356**(17):1736–41.
19. Narendra D, Tanaka A, Suen DF, Youle RJ. Parkin is recruited selectively to impaired mitochondria and promotes their autophagy. *J Cell Biol* 2008;**183**(5):795–803.
20. Narendra DP, Jin SM, Tanaka A, et al. PINK1 is selectively stabilized on impaired mitochondria to activate Parkin. *PLoS Biol* 2010;**8**(1):e1000298.
21. Barsoum MJ, Yuan H, Gerencser AA, et al. Nitric oxide-induced mitochondrial fission is regulated by dynamin-related GTPases in neurons. *EMBO J* 2006;**25**(16):3900–11.
22. Mannella CA. Structure and dynamics of the mitochondrial inner membrane cristae. *Biochim Biophys Acta* 2006;**1763**(5-6):542–8.
23. Arnoult D, Rismanchi N, Grodet A, et al. Bax/Bak-dependent release of DDP/TIMM8a promotes Drp1-mediated mitochondrial fission and mitoptosis during programmed cell death. *Curr Biol* 2005; **15**(23):2112–8.
24. Twig G, Liu X, Liesa M, et al. Biophysical properties of mitochondrial fusion events in pancreatic {beta}-cells and cardiomyocytes unravel potential control mechanisms of its selectivity. *Am J Physiol Cell Physiol* 2010;**299**:C477–87.
25. Karbowski M, Arnoult D, Chen H, Chan DC, Smith CL, Youle RJ. Quantitation of mitochondrial dynamics by photolabeling of individual organelles shows that mitochondrial fusion is blocked during the Bax activation phase of apoptosis. *J Cell Biol* 2004;**164**(4):493–9.

26. Frieden M, James D, Castelbou C, Danckaert A, Martinou JC, Demaurex N. Ca(2+) homeostasis during mitochondrial fragmentation and perinuclear clustering induced by hFis1. *J Biol Chem* 2004;**279**(21):22704–14.
27. Gomes LC, Scorrano L. High levels of Fis1, a pro-fission mitochondrial protein, trigger autophagy. *Biochim Biophys Acta* 2008;**1777**:860–6.
28. Park KS, Wiederkehr A, Kirkpatrick C, et al. Selective actions of mitochondrial fission/fusion genes on metabolism-secretion coupling in insulin-releasing cells. *J Biol Chem* 2008;**283**(48):33347–56.
29. Karbowski M, Jeong SY, Youle RJ. Endophilin B1 is required for the maintenance of mitochondrial morphology. *J Cell Biol* 2004;**166**(7):1027–39.
30. Takahashi Y, Coppola D, Matsushita N, et al. Bif-1 interacts with Beclin 1 through UVRAG and regulates autophagy and tumorigenesis. *Nat Cell Biol* 2007;**9**(10):1142–51.
31. Hailey DW, Rambold AS, Satpute-Krishnan P, et al. Mitochondria supply membranes for autophagosome biogenesis during starvation. *Cell* 2010;**141**(4):656–67.
32. Tondera D, Grandemange S, Jourdain A, et al. SLP-2 is required for stress-induced mitochondrial hyperfusion. *EMBO J* 2009;**28**(11):1589–600.
33. De Vos KJ, Allan VJ, Grierson AJ, Sheetz MP. Mitochondrial function and actin regulate dynamin-related protein 1-dependent mitochondrial fission. *Curr Biol* 2005;**15**(7):678–83.
34. Pletjushkina OY, Lyamzaev KG, Popova EN, et al. Effect of oxidative stress on dynamics of mitochondrial reticulum. *Biochim Biophys Acta* 2006;**1757**(5-6):518–24.
35. Ishihara N, Jofuku A, Eura Y, Mihara K. Regulation of mitochondrial morphology by membrane potential, and DRP1-dependent division and FZO1-dependent fusion reaction in mammalian cells. *Biochem Biophys Res Commun* 2003;**301**(4):891–8.
36. Legros F, Lombes A, Frachon P, Rojo M. Mitochondrial fusion in human cells is efficient, requires the inner membrane potential, and is mediated by mitofusins. *Mol Biol Cell* 2002;**13**(12):4343–54.
37. Malka F, Guillery O, Cifuentes-Diaz C, et al. Separate fusion of outer and inner mitochondrial membranes. *EMBO Rep* 2005;**6**(9):853–9.
38. Song Z, Ghochani M, McCaffery JM, Frey TG, Chan DC. Mitofusins and OPA1 Mediate Sequential Steps in Mitochondrial Membrane Fusion. *Mol Biol Cell*; 2009. May 28.
39. Bindokas VP, Kuznetsov A, Sreenan S, Polonsky KS, Roe MW, Philipson LH. Visualizing superoxide production in normal and diabetic rat islets of Langerhans. *J Biol Chem* 2003;**278**(11):9796–801.
40. Maechler P, Jornot L, Wollheim CB. Hydrogen peroxide alters mitochondrial activation and insulin secretion in pancreatic beta cells. *J Biol Chem* 1999;**274**(39):27905–13.
41. Las G, Mayorek N, Dickstein K, Bar-Tana J. Modulation of insulin secretion by fatty acyl analogs. *Diabetes* 2006;**55**(12):3478–85.
42. Mason TM, Goh T, Tchipashvili V, et al. Prolonged elevation of plasma free fatty acids desensitizes the insulin secretory response to glucose in vivo in rats. *Diabetes* 1999;**48**(3):524–30.
43. Sako Y, Grill VE. A 48-hour lipid infusion in the rat time-dependently inhibits glucose-induced insulin secretion and B cell oxidation through a process likely coupled to fatty acid oxidation. *Endocrinology* 1990;**127**(4):1580–9.
44. Zhou YP, Grill VE. Long-term exposure of rat pancreatic islets to fatty acids inhibits glucose-induced insulin secretion and biosynthesis through a glucose fatty acid cycle. *J Clin Invest* 1994;**93**(2):870–6.
45. El-Assaad W, Buteau J, Peyot ML, et al. Saturated fatty acids synergize with elevated glucose to cause pancreatic beta-cell death. *Endocrinology* 2003;**144**(9):4154–63.
46. El-Assaad W, Joly E, Barbeau A, et al. Glucolipotoxicity alters lipid partitioning and causes mitochondrial dysfunction, cholesterol, and ceramide deposition and reactive oxygen species production in INS832/13 ss-cells. *Endocrinology* 2010;**151**(7):3061–73.
47. Shimabukuro M, Zhou YT, Levi M, Unger RH. Fatty acid-induced beta cell apoptosis: a link between obesity and diabetes. *Proc Natl Acad Sci USA* 1998;**95**(5):2498–502.
48. Cnop M, Igoillo-Esteve M, Cunha DA, Ladriere L, Eizirik DL. An update on lipotoxic endoplasmic reticulum stress in pancreatic beta-cells. *Biochem Soc Trans* 2008;**36**(Pt 5):909–15.
49. Schrauwen P, Hesselink MK. Oxidative capacity, lipotoxicity, and mitochondrial damage in type 2 diabetes. *Diabetes. Jun* 2004;**53**(6):1412–7.
50. Fujimoto K, Hanson PT, Tran H, et al. Autophagy regulates pancreatic beta cell death in response to Pdx1 deficiency and nutrient deprivation. *J Biol Chem* 2009;**284**(40):27664–73. Oct 2.

51. Komiya K, Uchida T, Ueno T, et al. Free fatty acids stimulate autophagy in pancreatic beta-cells via JNK pathway. *Biochem Biophys Res Commun* 2010;**401**(4):561–7.
52. Koga H, Kaushik S, Cuervo AM. Altered lipid content inhibits autophagic vesicular fusion. *Faseb J* 2010;**24**(8):3052–65.
53. Choi SE, Lee SM, Lee YJ, et al. Protective role of autophagy in palmitate-induced INS-1 beta-cell death. *Endocrinology* 2009;**150**(1):126–34.
54. Halban PA. Structural domains and molecular lifestyles of insulin and its precursors in the pancreatic beta cell. *Diabetologia* 1991;**34**(11):767–78.
55. Marsh BJ, Soden C, Alarcon C, et al. Regulated autophagy controls hormone content in secretory-deficient pancreatic endocrine beta-cells. *Mol Endocrinol* 2007;**21**(9):2255–69.
56. Uchizono Y, Alarcon C, Wicksteed BL, Marsh BJ, Rhodes CJ. The balance between proinsulin biosynthesis and insulin secretion: where can imbalance lead? *Diabetes Obes Metab* 2007;**9**(Suppl. 2):56–66.
57. Brunner Y, Coute Y, Iezzi M, et al. Proteomics analysis of insulin secretory granules. *Mol Cell Proteomics* 2007;**6**(6):1007–17.
58. Kuliawat R, Klumperman J, Ludwig T, Arvan P. Differential sorting of lysosomal enzymes out of the regulated secretory pathway in pancreatic beta-cells. *J Cell Biol* 1997;**137**(3):595–608.
59. Tooze SA. Biogenesis of secretory granules in the trans-Golgi network of neuroendocrine and endocrine cells. *Biochim Biophys Acta* 1998;**1404**(1-2):231–44.
60. Twig G, Shirihai OS. The interplay between mitochondrial dynamics and mitophagy. *Antioxid Redox Signal* 2011;**14**(10):1939–51.
61. White KE, Davies VJ, Hogan VE, et al. OPA1 deficiency associated with increased autophagy in retinal ganglion cells in a murine model of dominant optic atrophy. *Invest Ophthalmol Vis Sci* 2009;**50**(6):2567–71.

SECTION III

DISEASES AND PROCESSES

CHAPTER

7

Developmental Autophagy

Valentina Cianfanelli and Francesco Cecconi

Dulbecco Telethon Institute at the Department of Biology, University of Tor Vergata, Rome, 00133 Italy

INTRODUCTION

Autophagy is a powerful catabolic pathway that can be triggered by environmental and hormonal stimuli; thus, it represents a good candidate for driving the rapid cellular remodeling necessary for proper development. Genetic studies carried out on the yeast model allowed the identification of a set of genes, the so-called autophagy-related (*ATG*) genes, which encode components of the autophagy machinery[1]. Because most of these genes are highly conserved among eukaryotes, a thorough analysis of autophagy functions has been carried out by using reverse genetic techniques. These studies have revealed various physiological roles and pathological effects of autophagy[2]; there is also increasing evidence that autophagy has conserved roles in development. Indeed, since the discovery of the *atg* machinery in yeast, it has been noted that autophagy mutants cannot sporulate during starvation; subsequent studies in various model organisms, including fungi, protozoa, worms and insects, confirmed the evolutionarily conserved role of autophagy in development (Table 7.1). The recent creation of defective-autophagy mouse models has made it possible to establish the various phenotypes associated with autophagy defects and consequently the different roles attributable to autophagy in this context (Table 7.2).

In this chapter, we highlight recent advances in our understanding of autophagy-mediated regulation of animal development, focusing mainly on vertebrate embryogenesis. We also draw attention to several autophagy-defective mouse phenotypes suggesting new possible roles for autophagy in development.

AUTOPHAGY OCCURRENCE IN INVERTEBRATE DEVELOPMENT

D. melanogaster

The fruit fly *D. melanogaster* is one of the most commonly used model organisms because it is easy to take care of, has a short life cycle and expresses all *atg* genes. The fruit fly embryo

Autophagy in Health and Disease
http://dx.doi.org/10.1016/B978-0-12-385101-7.00007-3

TABLE 7.1 Role of Autophagy in Development of Some Invertebrate Model Organism

Saccharomyces Cerevisiae	Spore Formation
Dictyostelium discoideum	Fruiting body formation
Caenorhabditis elegans	Larval development; dauer formation; degradation of germline P granules in somatic cells
Drosophila melanogaster	Larval development; degradation of larval tissues (midgut and salivary gland)

goes through three distinct stages during development: larva, pupa and adult insect. Each stage is characterized by its own body plan. Thus, a substantial metamorphosis occurs during embryogenesis; throughout this period, survival and metabolic changes need to be supported by nutrient uptake. Studies carried out on *atg*-defective *D. melanogaster*

TABLE 7.2 Phenotypes of the Principal Knockout Mice of *atg*-Related Genes

Genes	Yeast Ortholog	Function in Autophagy	Embryonic Phenotype
ULK1$^{-/-}$	atg*1*	Autophagy induction	Normal phenotype
FIP200$^{-/-}$	Functional ortholog of atg*17*	Autophagy induction	Defective heart and liver development; embryonic lethality at E13.5-E16.5
Beclin 1$^{-/-}$	atg*6*	Autophagosome nucleation	Cavitation defect; failure in proamniotic canal closure; embryonic lethality at ~E7.5
Ambra1$^{gt/gt}$	-	Autophagosome nucleation	Defects in neural tube development; hyperproliferation of the neuroepithelium; embryonic lethality at ~E14.5
UVRAG*	Functional ortholog of *vps38*	Autophagosome nucleation	Abnormal left-right axis formation; heterotaxy and other malformations
atg3$^{-/-}$, ***atg5***$^{-/-}$, ***atg7***$^{-/-}$, ***atg9***$^{-/-}$, ***atg16L***$^{-/-}$	*atg3, atg5, atg7, atg9, atg16*	Autophagosome expansion, completion, and recycling	Reduced blood amino acid levels; suckling defect; perinatal lethality
atg4C$^{-/-}$	*atg4*	Autophagosome expansion and closure	Normal phenotype
LC3B$^{-/-}$	*atg8*	Autophagosome expansion and closure	Normal phenotype
GABARAP$^{-/-}$	*atg8*	Autophagosome expansion and closure	Normal phenotype

Gt: gene-trapped allele.
**Disrupted in a human case.*

demonstrate that autophagy is important for proper development in all three developmental stages. For example, *atg*6 and *atg*18 mutations are lethal at the larval stage, while larvae devoid of *atg*1 survive the larval stage, albeit with reduced viability, but they cannot develop beyond pupae[3]. *Tor* mutation decreases both cell size and viability in flies, but a more severe phenotype is observed in *atg*1/*tor* double mutants: in this case flies do not survive beyond the larval stage[3]. Thus autophagy is essential for *D. melanogaster* metamorphosis from the larval stage onward. On the other hand, mutation of *atg*7 impairs autophagy but does not result in morphologic defects during metamorphosis, probably due to compensatory mechanisms[4]. *D. melanogaster* metamorphosis is driven by the rise of hormones, in particular the steroid hormone 20-hydroxyecdysone (ecdysone), whose levels increase at the end of the larval stage, so determining the length of the development period and controlling larval molting and metamorphosis[5]. Autophagy occurs as a response to hormone increase and is responsible for the remodeling which precedes morphogenesis of adult tissues. In more detail, it has been demonstrated that ecdysone signaling negatively regulates phosphoinositide (PI) 3-kinase (PI3K) activity. This results in a reduction of phosphatidylinositol-3,4,5-tris-phosphate (PIP3) pool which, in turn, induces autophagy[6]. Other evidence supports the role of ecdysone as an autophagy inducer; increasing mRNA levels of some *atg* genes are observed following the ecdysone wave in the fat body[6], the intestine, and salivary glands[7]. All three structures go on to degenerate at the end of the larval stage. Similarly, it has been demonstrated that autophagy, but not apoptosis, is essential for ecdysone-induced degradation of dying midgut cells. Thus, this degradation does not occur in flies devoid of *atg*1, *atg*2 or *atg*18[8, 9], and is delayed in *atg*7 mutants[4]. During *D. melanogaster* development, autophagy and apoptosis can occur in parallel to mediate the removal of an entire tissue: salivary gland degeneration is a well-characterized example of this phenomenon. Ca. 10 hours after puparium formation (Apf), a massive "type II cell death", as autophagic cell death is sometimes referred to[10], is observed in the salivary gland. Autophagosomes are detected in the same cells and ecdysone levels, probably responsible for autophagy induction, remain elevated up to 16 hours apf[6]. A large body of evidence confirms the importance of autophagy in this process: mutants devoid of *atg*1, *atg*2, *atg*3, *atg*6, *atg*7, *atg*8, *atg*12 or *atg*18 fail to degrade salivary glands. Moreover, overexpression of *agt*1 results in both autophagy induction and caspase-independent cell death[8], suggesting that autophagy is the main factor responsible for salivary gland degradation.

The dynein light chain 1 (*ddlc1*), a component of the dynein motor complex, has recently been implicated in autophagy regulation during fly development. In salivary glands of *ddlc1* mutants, cells undergo apoptosis, whereas autophagy is attenuated, resulting in impaired salivary gland degradation[10]. *atg*1 overexpression is sufficient to rescue the *ddlc1* mutant phenotype, although the molecular mechanism by which ddlc1 regulates autophagy in the fly is not clear. On the other hand, in mammals microtubules are important for autophagosome transport[11]; they also recruit pro-autophagy proteins under basal conditions, and release them upon starvation, thereby initiating autophagy[12]. However, as dynein heavy chain mutation does not affect autophagy in the salivary gland[9], it could be that *ddlc1* regulates autophagy in a motor-independent manner. Halfway through embryogenesis, the embryo is covered by an extraembryonic membrane, the amnioserosa (AS), that is subsequently eliminated in a caspase-dependent manner. The degenerating tissue also shows autophagic hallmarks, suggesting a role for autophagy in this process[13].

A large body of evidence reported in literature implicates autophagy in both cell survival and cell death. However, it is unclear how these two phenomena are distinguished and how the cell can switch from "pro-survival" to "pro-death" autophagy. Recently, Baehrecke's group demonstrated that an engulfment factor, Draper (Drpr), is required for autophagy-mediated cell death of salivary gland cells in *D. melanogaster*, but not for autophagy induced by starvation in the fat body[14]; thus Drpr represents a good candidate for distinguishing autophagy that is linked to cell death from that one that is associated with survival.

C. elegans

Similar to other eukaryotic organisms, different autophagy genes play distinct roles during worm embryogenesis, and some of these gene phenotypes show partial compensation for each other. For example, deficiency in the yeast *atg*6 homolog *bec-1* is responsible for a lethal phenotype and worms die before or during the first larval stage, showing increased vacuolization and molting defects[15]. In addition to the effect on viability, *bec*-1 also seems to be involved in fertility, since mutant worms that reach adulthood are sterile. A lethal phenotype is also observed in worms deficient in the *C. elegans* ortholog of Class III PI3K *vps34 (let-512), atg*8 *(lgg-1)* or *atg*18 *(F41E6,13)*[16, 17]. In contrast to *D. melanogaster*, *C. elegans* devoid of the *atg*1 ortholog *unc-51* are viable, although they exhibit axonal defects that impair motor neuron function[18], similar to what is observed in vertebrates, as will be illustrated later.

It is known that *C. elegans* larvae respond to stressful conditions, such as starvation, high population density or increased temperature by arresting development at the third larval stage (L3)[18]. These dauer larvae undergo a number of metabolic and morphological changes that are distinct from normal L3 larvae, which allow their survival. If external conditions improve, dauer larvae can resume reproductive development, reach the adult stage, and have a normal lifespan[19]. When the food supply is limited, autophagy activation provides a pool of amino acids essential for the synthesis of pro-survival proteins. Although the regulation of dauer development has been characterized, the cellular pathways involved in dauer morphogenesis are only partially known. In dauer larvae with a *daf-2* (the insulin-like receptor) mutation, an increase of autophagy in hypodermal precursor cells is measured[17]. In *daf-2* mutants in which autophagy is inactivated by knockdown of *bec-1, unc-51, atg*7, *atg*8 and *atg*18 dauer formation is defective, with worms dying within a few days. These studies demonstrate that autophagy is required for normal dauer morphogenesis, by acting downstream of the insulin/IGF pathway.

Very recently, a novel role has been proposed for selective autophagy in degrading the P granules in somatic cells during the early stages of embryo segmentation in *C. elegans*[20]. P granules are cytosolic aggregates containing both proteins and RNA. Initially, they are present in every cell of the worm blastula, but they quickly redistribute exclusively into the germ cell, thus determining its fate as a precursor of the germinal lineage.

Two lines of evidence reveal that during the neuronal development of *C. elegans* the clustering of neurotransmitter receptors is mediated by autophagy. First, *unc-51* worm mutants are characterized by impairment of the axonal outgrowth along the antero-posterior axis[21]. Second, GABA receptors that are internalized in the absence of pre-synaptic inputs in muscle cells are specifically degraded by autophagy. Notably, the mammalian $GABA_A$-receptor associated protein Gabarap is an ortholog of the yeast autophagy protein *atg8*. Receptor internalization and degradation are associated with autophagy since receptors co-localize both with the *atg8* ortholog *lgg-1* and with *bec-1* in autophagosomes of non-innervated muscles. By blocking endocytosis, receptors do not aggregate in autophagosomes, suggesting that trafficking of receptors from cell surface to autophagosomes likely occurs via the endocytic pathway[21].

In the last year, two research groups addressed, by means of *C. elegans*, a biological question common to eukaryote development: the cause of the exclusive maternal inheritance of mitochondrial DNA (mtDNA)[22]. After fertilization, sperm mitochondria enter into the ooplasm but, by 16-cell-stage embryo, paternal mtDNA disappears, so justifying the maternal origin of mitochondria. One possible explanation was that the low amount of mitochondria provided by the sperm is being diluted during the following cell divisions. In *C. elegans* this hypothesis is now abandoned: Paternal mitochondria are actively degraded by autophagy in the fertilized oocyte. A few minutes after fertilization, autophagosomes surround and degrade sperm mitochondria and the membranous organelles (MOs; a sperm vesicular structure, which is functionally homologous to the mammalian acrosome). Once in the oocyte, MOs are ubiquitylated. This ubiquitylation probably provides the signal for autophagosome formation and selective degradation of both MOs and the close-associated paternal mitochondria[23, 24]. An open question that has yet to be addressed is whether autophagic degradation of paternal mitochondria is conserved among eukaryotes. Investigating this aspect would help to better understand the pathogenesis of mitochondrial diseases.

DIFFERENT ROLES FOR AUTOPHAGY IN EMBRYOGENESIS OF VERTEBRATES

Preimplantation and Early Development

The specific importance of autophagy in the development of higher eukaryotes is evident from very early in embryogenesis (Figure 7.1). The earliest autophagic event in mammalian development is observed in fertilized oocytes in mouse[25].The oocyte is a highly differentiated cell, whose fate is finally to grow into a fully functioning organism. After fertilization, the oocyte undergoes a sudden "reprogramming", including degradation of maternally supplied mRNAs and proteins after the two-cell stage. The zygotic genome is then activated and new mRNAs and proteins encoded by it are synthesized[26, 27], leading to marked changes in the protein species synthesized after the four-cell to eight-cell stages[28].

The study by Tsukamoto *et al.* shows that autophagy occurs at only low levels in unfertilized oocytes but is massively induced within four hours after fertilization[25]. This induction of autophagy is completely dependent on fertilization and is not due to starvation after ovulation, because autophagy is not induced in the ovulated oocyte unless it has been fertilized. Moreover, parthenogenetic activation is sufficient to induce autophagy in oocytes, suggesting that calcium oscillation could be responsible for triggering this type of autophagy. The phosphorylation of S6 kinase, a substrate of mTOR, is high in unfertilized oocytes and decreases in the zygote, hinting that mTOR inhibition may mediate autophagy induction downstream of calcium signaling[25]. Interestingly, autophagy is transiently suppressed

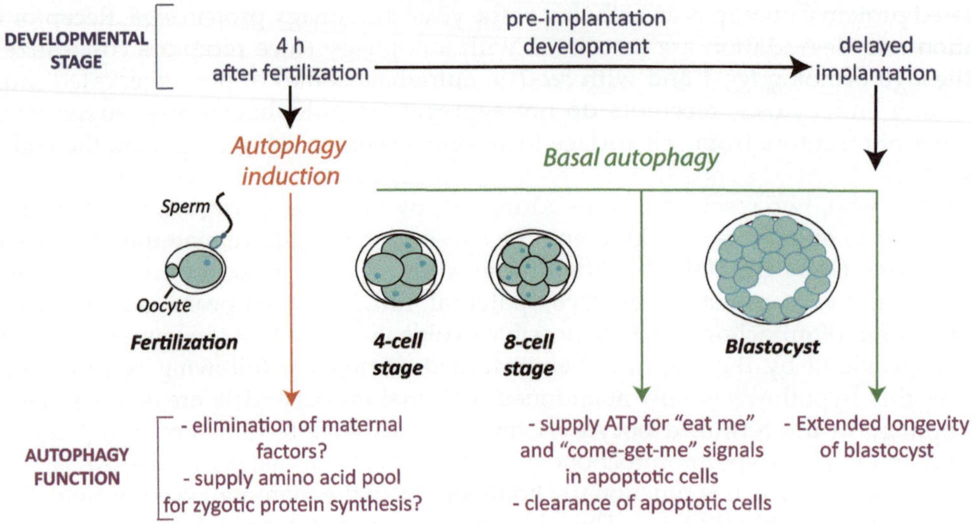

FIGURE 7.1 **Roles of autophagy in pre-implantation development.** Principal stages of the pre-implantation development are shown (in blue). After the fertilization, a burst of autophagy induction occurs, while basal levels of autophagy are measured during the rest of pre-implantation development. In each stage, autophagy functions are reported (in violet).

from the late one-cell to middle two-cell stages, and then reactivated, perhaps to avoid the degradation of nuclear factors temporarily released to the cytoplasm during cell division and essential for later development[29]. The suppression of autophagy recently observed in cultured mammalian cells during the mitotic phase is consistent with this hypothesis[30]. In fact, Cyclin-dependent kinase 1 (CDK1) is essential for both the zygote-to-two-cell stage division[31] and for autophagy inhibition during mitosis, further highlighting the close correlation between these cellular processes[32]. In addition, very recently a role for autophagy has been proved in delayed implantation, a widespread phenomenon in mammals. During delayed implantation, the blastocyst remains dormant for a long period, but resumes implantation competence upon favorable conditions. The molecular mechanisms maintaining extended blastocyst dormancy are poorly characterized, but data collected by Lee *et al.* demonstrated that autophagy is responsible for extended longevity of blastocysts during delayed implantation[33].

The data reported above suggest an important role for autophagy in early embryogenesis, even if mice derived by atg$5^{-/-}$ oocytes overcome these early stages, perhaps because the presence of maternally supplied *atg*5 protein is sufficient to support autophagy induction[34]. In support of this, elimination of the maternal *atg*5 protein in oocyte-specific *atg*5-knockout mice results in embryonic lethality at the four-cell to eight-cell stages[25], thus demonstrating that autophagy is essential for the preimplantation embryonic development. However, the precise role of autophagy during this process is not fully understood. Given that it is well recognized that preimplantation development depends on protein synthesis and the rate of protein translation is reduced in autophagy-defective embryos, normal levels of autophagy may be necessary for the production of sufficient aminoacids for zygotic protein synthesis[25]. Moreover, autophagy is a bulk degradation process, so it represents a good candidate for active elimination of maternally provided proteins and organelles which must be removed after fertilization. In addition, autophagy-mediated degradation of maternal suppressors of the zygotic gene program could enable the radical "remodeling" necessary for developmental progression. While not losing sight of both recycling and degradative functions of autophagy, we should bear in mind that these different possibilities are not mutually exclusive.

The necessity of competent autophagy machinery for completion of embryogenesis can also be observed during later stages of development, as confirmed by the embryonic lethal phenotype of Beclin 1 knockout mice. These mutant embryos exhibit a severe developmental delay, leading to reduced size and lethality at embryonic day seven (E7.5)[35]. These embryos show massive cell death and failure to close the preamniotic canal. A more detailed morphological analysis also revealed a defect in the size and cellular organization of the visceral endoderm (VE), an extra embryonic tissue that participates in developmental processes preceding and during gastrulation. *Beclin* $1^{-/-}$ embryonic stem cells are viable, although the corresponding knockout mice are early embryonic lethal, suggesting that Beclin 1 is dispensable *in vitro* but critically important for development *in vivo*. The role of autophagy in this context has been clarified by using the aggregates of inner-cell-mass-derived embryonic stem cells (embryoid bodies, EB) as an *in vitro* mouse model of early embryogenesis. EBs are made up of undifferentiated cell aggregates which subsequently undergo cavitation, a phase of gastrulation where the solid embryonic ectoderm is transformed in a columnar epithelium surrounding a cavity. During this morphogenetic process, apoptosis is activated in the inner ectodermal cells; dead cells are then removed, leading to the formation of the proamniotic cavity[36]. It has been demonstrated that EBs composed of *atg*5- or Beclin 1-defective cells fail to cavitate due to accumulation of dead cells, specifying a defect in cell clearance[37]. The process of cell removal is regulated by the expression of recognition signals on the dying cell and soluble chemo attractant production, resulting in engulfment of dead corpses by the phagocytic cells. These recognition signals consist of phosphatidylserine (PS) exposure (the so-called "eat me" signal) and lysophosphatidylcholine (LP) release (or "come-get-me" signal), both signals being faulty in autophagy-defective EBs[36, 37].

Cell clearance is thus an active, programmed process that needs to be sustained by providing the cell with energy. The engulfment defects observed in *atg*5- and Beclin 1-deficient models are associated with low cellular ATP levels. Restoration of normal ATP levels rescues PS exposure and LP secretion defects observed in autophagy null EBs, strongly suggesting that autophagy-dependent ATP production is necessary for completion of the apoptotic program during early development[37]. Data collected *in vivo* by Qu and colleagues are consistent with this hypothesis and strongly support the significance of autophagy in this context. A massive accumulation of apoptotic bodies has been observed in the embryonic lungs and retina of *atg*5 null mice[37] as well as during the early stages of retinal development (E5) in a chick model following pharmacological inhibition of autophagy[38]. In addition to morphogenesis, the efficient removal of apoptotic cells is essential in multicellular organisms to prevent unwanted inflammatory responses. As expected, *atg*5 null retina and lungs are characterized by an increased number of lymphocytes and neutrophils[37], indicating that activation of autophagy in apoptotic cells may be necessary to ensure the efficient apoptotic corpse engulfment and, in turn, to prevent damaging inflammatory responses. In light of all the above data, autophagy can therefore be defined as a crucial event in cell clearance during the morphogenesis, participating in embryo cavitation and neural retina formation.

Roles of Autophagy Genes in Embryomorphogenesis

The analysis of mice expressing the autophagy reporter green fluorescent protein (GFP) LC3, made it possible to exclude high levels of autophagy occurrence during later stages of embryogenesis, except in certain tissues such as the developing thymus[39]. Accordingly, the majority of *atg* gene knockout embryos (such as *atg*3$^{-/-}$, *atg*5$^{-/-}$, *atg*7$^{-/-}$, *atg*9$^{-/-}$ and *atg*16L1$^{-/-}$) survive the entire embryonic period and are born at Mendelian frequency[40-43]. This suggests that all these genes are not essential for embryogenesis (unless during the early stages when maternally inherited proteins allow survival in conventional knockout mice). However, further studies are required for the specification of possible roles of *ATG* genes in more subtle processes occurring during embryo development or in order to rule out this hypothesis.

Alternatively, mild phenotypic abnormalities of *atg*5$^{-/-}$ embryos (see above and below) and severe phenotypes observed in Beclin 1 (see above), *Ambra1* and *FIP200* deficient mice indicate that impairment of basal level of autophagy may be sufficient to interfere with normal embryogenesis[35, 44, 45]. For instance, it has been recently observed that disruption of an autophagy machinery component could be responsible for aberrant left-right axis formation[46]. The morphogen Sonic hedgehog (Shh) plays a central role in this process in vertebrates. Shh is asymmetrically expressed in the node; during gastrulation, the unidirectional movement of cilia on the node cells causes a flow of this morphogen[47], directed from the right to the left side, thereby inducing a specific gene expression pattern in the left cells of the primitive streak. This event marks one of the earliest indications of the establishment of left-right asymmetry in vertebrates. A mutation in the *UVRAG* (UV radiation Resistance-Associated Gene protein) gene, which encodes for a component of the Beclin 1/Vps34 complex[48] is associated with a human case of abnormal left-right axis formation, resulting in heterotaxy and other malformations[46]. Interestingly, a similar left-right axis aberration has been observed in *D. melanogaster* mutated in the UVRAG ortholog gene[49]. A displaced expression of Shh is also detected in *Ambra1* (Activating molecule in Beclin 1-regulated autophagy)—deficient embryos produced by gene-trap mutagenesis (*Ambra1*$^{gt/gt}$)[44]. As a result, a common mechanism by which autophagy regulates Shh distribution could be hypothesized.

Ambra1 protein directly interacts with Beclin 1, thus positively regulating the activity of the entire Vps34 complex and in turn the early steps of autophagosome formation, both in basal state and upon autophagy induction[44]. *Ambra1*$^{gt/gt}$ mice are embryonic lethal at day E16.5 and show evident defects in the development of the neural tube, a tissue in which *Ambra1* is highly expressed in wt embryos. Homozygous mutation of this gene causes hyperproliferation of the neuroepithelium, associated with an excess of apoptotic cells[44], consistent with the prosurvival role of Ambra1 recently demonstrated by Strappazzon *et al*[50]. In this case, the strict relationship between autophagy, apoptosis and cell proliferation is also evident[51]. As a result of both apoptosis and proliferation alteration, the morphogenesis of the central nervous system (CNS) turns out to be profoundly altered: *Ambra1*$^{gt/gt}$ embryos are affected by midbrain and hindbrain exencephaly and defective closure of the neural tube[44], constituting a striking example of the crucial role of autophagy in CNS development. A function in neurodevelopment has also been attributed to Ulk-1 (Unc-51-like kinase 1), an *atg*1 homolog in human. A link between this kinase and Ambra1 has recently been identified: Di Bartolomeo *et al* demonstrated that Ulk1 promotes the dissociation of the Ambra1/Beclin 1 complex from the dynein motor complex, thus enabling execution of autophagy[52]. *Ulk1* is highly expressed in developing cerebellum. Immature granule cells infected with a dominant negative form of *Ulk1* show impairment of axon outgrowth and neuron differentiation[53]; Ulk1 activity at the growth cones is also shown to be crucial for neurite extension and the formation of fibers, thus allowing the progression of cerebellar development[53]. Because the mechanism by which Ulk1 controls the phenomenon is still unknown, it remains to be clarified whether or not the role of Ulk1 in neurodevelopment depends on its function in regulating autophagy. The phenotype of an Ulk1-interacting protein, FIP200 (Focal adhesion kinase family Interacting Protein of M_r 200K), has been recently characterized, showing embryonic lethality between E13.5 and E16.5 as a result of defective heart and liver development[45]. FIP200 has a molecular function similar to that of yeast *atg*17, although it shows no homology with any yeast *atg* protein[54].

It is not completely understood why the phenotypes of distinct *ATG* gene knockout mouse models are so different. Because Beclin 1/Vps34 complexes have several functions[55] and FIP200 has other interacting partners (such as focal adhesion kinase, tuberous sclerosis complex 1, p53, etc.) in addition to ULK1, some of the abnormalities that affect Beclin $1^{-/-}$, Ambra$1^{gt/gt}$ and FIP$200^{-/-}$ embryos may be caused by defects in other pathways as well as autophagy deficiency. Alternatively, the different phenotypes of *atg* gene knockout mice may depend on the steps of autophagy in which each factor is involved. Beclin 1/Vps34 and ULK1-FIP200 complexes function at the autophagosome nucleation step, while *atg*3, *atg*5, *atg*7 and *atg*16L function in autophagosome elongation. Thus, upstream factors may show more severe phenotypes, or, as recently reported, downstream factors could be dispensable for a particular type of macroautophagy[56]. Also, different phenotypes of *atg* gene knockout mice could be explained by different levels of functional redundancy between *atg* proteins. Further studies are needed to determine whether the distinct stages of the embryonic development after the oocyte-to-embryo transition are regulated by the autophagy pathway itself or by autophagy-independent functions of certain components.

Different possible roles for autophagy in development are hypothesized and our incomplete knowledge about them emerges from the data reported above. Many interesting findings in the literature merit additional in-depth study. Striking functions of autophagy in this context could be unraveled, such as the severe phenotype of FIP$200^{-/-}$embryonic heart[57], consistent with the high expression level of *Ambra1*gene in the same organ during embryogenesis (Figure 7.2a and b). Moreover, the first study of *atg*5 (originally referred to as *paddy*) expression pattern showed high *atg*5 mRNA levels in the zone of polarizing activity (ZPA), the area of limb bud responsible for antero-posterior axis formation[58]. This evidence suggests a not well-characterized role for autophagy in limb development (Figure 7.2 c, d and e).

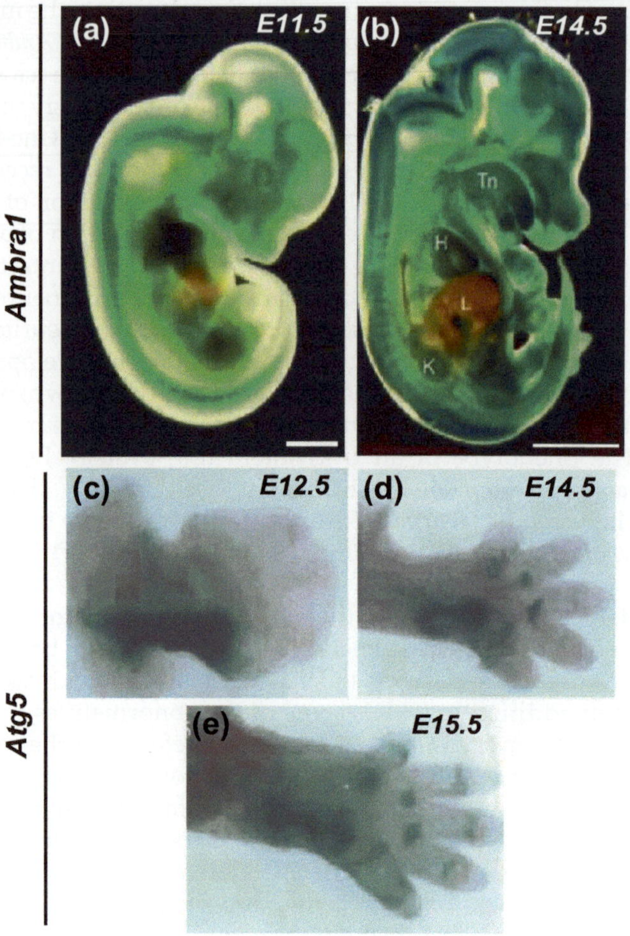

FIGURE 7.2 **Expression pattern of *Ambra1* and *atg5* during embryogenesis. a-b.** Whole-mount β-Gal staining of *Ambra1*$^{+/gt}$mouse embryo at E11.5 and E14.5 is shown. The blue areas are expressing *Ambra1* gene; (**a**) at E11.5, stained regions correspond to the encephalic vesicles, the neural retina, the limbs and the dorsal root ganglia. (**b**) At E14.5 *Ambra1* expression extends to the developing nervous system, to the body mesenchyme, to the tongue and to the heart. **c-e** β-Gal staining of whole-mount *Paddy*$^{+/gt}$ (the original name for *Atg5*) mouse embryo at E12.5-E15.5. *Atg5* expression is detected in the limb at different stages. gt: gene-trapped allele; Tn: tongue; H: heart; L: liver; K: kidney.

Autophagy to Overcome Early Neonatal Starvation

Throughout mammalian embryogenesis, necessary nutrients are maternally provided to the embryo. This supply is suddenly interrupted at birth, when newborns have to face severe starvation. Accordingly, massive autophagy induction is observed during the early neonatal period in all tissues, except the brain, until one or two days after birth[39]. Despite normal appearance of the newborns, *atg*3, *atg*5, *atg*7, *atg*9 and *atg*16 L knockout mice die within one day after birth[40-43] as a result of autophagy defects, demonstrating that autophagy is essential for survival during the neonatal period. Amino acid levels in plasma are decreased by 20% in both *atg*5$^{-/-}$ and *atg*7$^{-/-}$ neonates with respect to wild-type ones, whereas blood glucose and fatty acid concentrations are unchanged. This suggests that autophagy is necessary to maintain the amino acid pool[40], even if it is not known precisely how these amino acids are used. One possibility is that the amino acid pool supplied by autophagy could be necessary to satisfy the high energy demand of certain neonatal tissues after birth, this hypothesis being supported by the activation of low-energy sensing kinase AMPK (AMP-activated protein Kinase) in *atg*5-deficient hearts[40].

Nevertheless, it is yet to be established whether the decrease in amino acid levels alone can explain the perinatal lethality of *atg* knockout newborns, given that they are affected by other abnormalities. First, the impairment of basal autophagy in neurons may be responsible for a suckling defect in these mice, which in turn exacerbates their malnutritional state[40, 59]. Second, autophagy may also be important for degradation of macromolecules beyond proteins. At birth, glycogen stores are catabolized by delivery to lysosomes, mediated by autophagosomes. In this way autophagy plays a role in glucose production, thereby contributing to the energy supply of skeletal muscles which start working actively at birth, such as the diaphragm and muscles involved in suckling[59-61]. In addition to amino acid starvation, insufficient energy production during the perinatal period could be caused by less adipose tissue in *ATG* knockout mice as a result of defects in adipogenesis[62]. Finally, given that both environmental status and metabolism change at birth[63, 64], autophagy could be involved in degradation of embryonic factors and cellular re-modeling to allow newborn adaptation.

CONCLUSION

The extensive analysis carried out on *atg*-deficient animal models allowed us to establish the widespread role of autophagy in the development of both invertebrates and vertebrates.

During fly and worm embryogenesis, autophagy is induced by hormones or nutrient stress, operating principally in parallel with apoptosis to mediate deep reorganization of the embryo. In this context, entire tissues are disrupted by autophagy, so as to allow proper embryo development or adaptation to unfavorable environments. On the other hand, innovative and subtle roles of autophagy in embryogenesis are also provided by the study of invertebrate models. For example, the only role to have been found for selective autophagy in this respect is in worms, where the degradation of GABA receptors is specifically mediated by autophagy. This finding could represent the first of a series of similar roles introducing an autophagy-mediated regulation of cell signaling during embryogenesis, thereby explaining the widely differing-phenotypes associated with autophagy deficiency in mammals.

In mammals, a burst of autophagy occurs immediately after fertilization and again during neonatal starvation; in both cases the great importance of this catabolic process has been demonstrated. Nevertheless, despite the absence of high levels of autophagy during the rest of embryogenesis, the basal autophagy is relevant for proper embryo morphogenesis. Additionally, most of the molecular mechanisms by which autophagy plays these roles are not fully understood. Given autophagy's importance in development, such mechanisms need to be better characterized. By shedding light on the molecular mechanisms responsible for all the phenomena noted above, we could therefore lay the foundation for development of clinical tools for both diagnosis and treatment of abnormalities in embryogenesis.

References

1. Klionsky DJ, Cregg JM, Dunn Jr WA, Emr SD, Sakai Y, Sandoval IV, et al. A unified nomenclature for yeast autophagy-related genes. *Dev Cell* 2003;**5**:539–45.
2. Yang Z, Klionsky DJ. Eaten alive: a history of macroautophagy. *Nat Cell Biol* 2010;**12**:814–22.
3. Scott RC, Schuldiner O, Neufeld TP. Role and regulation of starvation-induced autophagy in the Drosophila fat body. *Dev Cell* 2004;**7**:167–78.
4. Juhasz G, Erdi B, Sass M, Neufeld TP. *Atg7*-dependent autophagy promotes neuronal health, stress tolerance, and longevity but is dispensable for metamorphosis in Drosophila. *Genes Dev* 2007;**21**:3061–6.
5. King-Jones K, Charles JP, Lam G, Thummel CS. The ecdysone-induced DHR4 orphan nuclear receptor coordinates growth and maturation in Drosophila. *Cell* 2005;**121**:773–84.
6. Rusten TE, Lindmo K, Juhasz G, Sass M, Seglen PO, Brech A, et al. Programmed autophagy in the Drosophila fat body is induced by ecdysone through regulation of the PI3K pathway. *Dev Cell* 2004;**7**:179–92.
7. Lee CY, Baehrecke EH. Steroid regulation of autophagic programmed cell death during development. *Development* 2001;**128**:1443–55.
8. Denton D, Shravage B, Simin R, Baehrecke EH, Kumar S. Larval midgut destruction in Drosophila: not dependent on caspases but suppressed by the loss of autophagy. *Autophagy* 2010;**6**:163–5.
9. Denton D, Shravage B, Simin R, Mills K, Berry DL, Baehrecke EH, et al. Autophagy, not apoptosis, is essential for midgut cell death in Drosophila. *Curr Biol* 2009;**19**:1741–6.
10. Berry DL, Baehrecke EH. Growth arrest and autophagy are required for salivary gland cell degradation in Drosophila. *Cell* 2007;**131**:1137–48.
11. Longatti A, Tooze SA. Vesicular trafficking and autophagosome formation. *Cell Death Differ* 2009;**16**:956–65.
12. Di Bartolomeo S, Corazzari M, Nazio F, Oliverio S, Lisi G, Antonioli M, et al. The dynamic interaction of AMBRA1 with the dynein motor complex regulates mammalian autophagy. *J Cell Biol* 2010;**191**:155–68.
13. McCall K. Eggs over easy: cell death in the Drosophila ovary. *Dev Biol* 2004;**274**:3–14.
14. McPhee CK, Baehrecke EH. The engulfment receptor Draper is required for autophagy during cell death. *Autophagy* 2010;**6**:1192–3.
15. Takacs-Vellai K, Vellai T, Puoti A, Passannante M, Wicky C, Streit A, et al. Inactivation of the autophagy gene bec-1 triggers apoptotic cell death in C. elegans. *Curr Biol* 2005;**15**:1513–7.
16. Roggo L, Bernard V, Kovacs AL, Rose AM, Savoy F, Zetka MP, et al. Membrane transport in Caenorhabditis elegans: an essential role for VPS34 at the nuclear membrane. *EMBO J* 2002;**21**:1673–83.
17. Melendez A, Talloczy Z, Seaman M, Eskelinen EL, Hall DH, Levine B. Autophagy genes are essential for dauer development and life-span extension in C. elegans. *Science.* **301**:1387-91.
18. McIntire SL, Garriga G, White J, Jacobson D, Horvitz HR. Genes necessary for directed axonal elongation or fasciculation in C. elegans. *Neuron* 1992;**8**:307–22.
19. Riddle DL. *Genetic and environmental regulation of dauer larva development*. Cold Spring Harbor, NY: Cold Spring Harbor Laboratory Press; 1997.
20. Zhang Y, Yan L, Zhou Z, Yang P, Tian E, Zhang K, et al. SEPA-1 mediates the specific recognition and degradation of P granule components by autophagy in C. elegans. *Cell* 2009;**136**:308–21.
21. Rowland AM, Richmond JE, Olsen JG, Hall DH, Bamber BA. Presynaptic terminals independently regulate synaptic clustering and autophagy of GABAA receptors in Caenorhabditis elegans. *J Neurosci* 2006;**26**:1711–20.
22. Ankel-Simons F, Cummin JM. Misconceptions about mitochondria and mammalian fertilization implications for theories on human evolution. *Proc Natl Acad Sci USA* 1996;**93**:13859–63.
23. Al Rawi S, Louvet-Vallée S, Djeddi A, Sachse M, Culetto E, Hajjar C, et al. Postfertilization autophagy of sperm organelles prevents paternal mitochondrial DNA transmission. *Science* 2011;**334**:1144–7.
24. Sato M, Sato K. Degradation of paternal mitochondria by fertilization-triggered autophagy in C. elegans embryos. *Science* 2011;**334**:1141–51.

25. Tsukamoto S, Kuma A, Murakami M, Kishi C, Yamamoto A, Mizushima N. Autophagy is essential for preimplantation development of mouse embryos. *Science* 2008;**321**:117–20.
26. Merz EA, Brinster RL, Brunner S, Chen HY. Protein degradation during preimplantation development of the mouse. *J Reprod Fertil* 1981;**61**:415–8.
27. Stitzel ML, Seydoux G. Regulation of the oocyte-to-zygote transition. *Science* 2007;**316**:407–8.
28. Van Blerkom J, Brockway GO. Qualitative patterns of protein synthesis in the preimplantation mouse embryo. I. Normal pregnancy. *Dev Biol* 1975;**44**:148–57.
29. Egli D, Rosains J, Birkhoff G, Eggan K. Developmental reprogramming after chromosome transfer into mitotic mouse zygotes. *Nature* 2007;**447**:679–85.
30. Eskelinen EL, Prescott AR, Cooper J, Brachmann SM, Wang L, Tang X, et al. Inhibition of autophagy in mitotic animal cells. *Traffic* 2002;**3**:878–93.
31. Santamaría D, Barrière C, Cerqueira A, Hunt S, Tardy C, Newton K, et al. Cdk1 is sufficient to drive the mammalian cell cycle. *Nature* 2007;**448**:811–5.
32. Furuya T, Kim M, Lipinski M, Li J, Kim D, Lu T, et al. Negative regulation of Vps34 by Cdk mediated phosphorylation. *Mol Cell* 2010;**38**:500–11.
33. Lee JE, Oh HA, Song H, Jun JH, Roh CR, Xie H, et al. Autophagy Regulates Embryonic Survival During Delayed Implantation. *Endocrinology* 2011.
34. Hosokawa N, Hara Y, Mizushima N. Generation of cell lines with tetracycline-regulated autophagy and a role for autophagy in controlling cell size. *FEBS Lett* 2006;**580**:2623–9.
35. Yue Z, Jin S, Yang C, Levine AJ, Heintz N. Beclin 1, an autophagy gene essential for early embryonic development, is a haploinsufficient tumor suppressor. *Proc Natl Acad Sci USA* 2003;**100**:15077–82.
36. Coucouvanis E, Martin GR. Signals for death and survival: a two-step mechanism for cavitation in the vertebrate embryo. *Cell* 1995;**83**:279–87.
37. Qu X, Zou Z, Sun Q, Luby-Phelps K, Cheng P, Hogan RN, et al. Autophagy gene-dependent clearance of apoptotic cells during embryonic development. *Cell* 2007;**128**:931–46.
38. Mellen MA, de la Rosa EJ, Boya P. The autophagic machinery is necessary for removal of cell corpses from the developing retinal neuroepithelium. *Cell Death Differ* 2008;**15**:1279–90.
39. Mizushima N, Yamamoto A, Matsui M, Yoshimori T, Ohsumi Y. In vivo analysis of autophagy in response to nutrient starvation using transgenic mice expressing a fluorescent autophagosome marker. *Mol Biol Cell* 2004;**15**:1101–11.
40. Kuma A, Hatano M, Matsui M, Yamamoto A, Nakaya H, Yoshimori T, et al. The role of autophagy during the early neonatal starvation period. *Nature* 2004;**432**:1032–6.
41. Sou YS, Waguri S, Iwata J, Ueno T, Fujimura T, Hara T, et al. The *atg*8 conjugation system is indispensable for proper development of autophagic isolation membranes in mice. *Mol Biol Cell* 2008;**19**:4762–75.
42. Saitoh T, Fujita N, Hayashi T, Takahara K, Satoh T, Lee H, et al. *Atg*9a controls dsDNA-driven dynamic translocation of STING and the innate immune response. *Proc Natl Acad Sci USA* 2009;**106**:20842–6.
43. Saitoh T, Fujita N, Jang MH, Uematsu S, Yang BG, Satoh T, et al. Loss of the autophagy protein *atg*16L1 enhances endotoxin-induced IL-1beta production. *Nature* 2008;**456**:264–8.
44. Fimia GM, Stoykova A, Romagnoli A, Giunta L, Di Bartolomeo S, Nardacci R, et al. Ambra1 regulates autophagy and development of the nervous system. *Nature* 2007;**447**:1121–5.
45. Gan B, Peng X, Nagy T, Alcaraz A, Gu H, Guan JL. Role of FIP200 in cardiac and liver development and its regulation of TNF alpha and TSC-mTOR signaling pathways. *J Cell Biol* 2006;**175**:121–33.
46. Iida A, Emi M, Matsuoka R, Hiratsuka E, Okui K, Ohashi H, et al. Identification of a gene disrupted by inv(11)(q13.5;q25) in a patient with left-right axis malformation. *Hum Genet* 2000;**106**:277–87.
47. Hirokawa N, Tanaka Y, Okada Y, Takeda S. Nodal flow and the generation of left-right asymmetry. *Cell* 2006;**125**:33–45.
48. Liang C, Feng P, Ku B, Dotan I, Canaani D, Oh BH, et al. Autophagic and tumour suppressor activity of a novel Beclin1-binding protein UVRAG. *Nat Cell Biol* 2006;**8**:688–99.
49. He C, Orvedahl A. Keystone Symposium on Autophagy in Health and Disease. *Autophagy* 2007;**3**:527–36.
50. Strappazzon F, Vietri-Rudan M, Campello S, Nazio F, Florenzano F, Fimia GM, et al. Mitochondrial BCL-2 inhibits AMBRA1-induced autophagy. *EMBO J* 2011.
51. Cecconi F, Di Bartolomeo S, Nardacci R, Fuoco C, Corazzari M, Giunta L, et al. A novel role for autophagy in neurodevelopment. *Autophagy* 2007;**3**:506–8.

52. Di Bartolomeo S, Corazzari M, Nazio F, Oliverio S, Lisi G, Antonioli M, et al. The dynamic interaction of Ambra1 with the dynein motor complex regulates mammalian autophagy. *J Cell Biol* 2010;**191**:155–68.
53. Tomoda T, Bhatt RS, Kuroyanagi H, Shirasawa T, Hatten ME. A mouse serine/threonine kinase homologous to C. elegans UNC51 functions in parallel fiber formation of cerebellar granule neurons. *Neuron* 1999;**24**:833–46.
54. Hara T, Takamura A, Kishi C, Iemura S, Natsume T, Guan JL, et al. FIP200, a ULK-interacting protein, is required for autophagosome formation in mammalian cells. *J Cell Biol* 2008;**181**:497–510.
55. Funderburk SF, Wang QJ, Yue Z. The Beclin 1-VPS34 complex–at the crossroads of autophagy and beyond. *Trends Cell Biol* 2010;**20**:355–62.
56. Longatti A, Tooze SA. Vesicular trafficking and autophagosome formation. *Cell Death Differ* 2009;**16**:956–65.
57. Gan B, Peng X, Nagy T, Alcaraz A, Gu H, Guan JL. Role of FIP200 in cardiac and liver development and its regulation of TNF alpha and TSC-mTOR signaling pathways. *J Cell Biol* 2006;**175**:121–33.
58. Pires-DaSilva A, Gruss P. Gene trap insertion into a novel gene expressed during mouse limb development. *Dev Dyn* 1998;**212**:318–25.
59. Komatsu M, Waguri S, Ueno T, Iwata J, Murata S, Tanida I, et al. Impairment of starvation-induced and constitutive autophagy in *atg7*-deficient mice. *J Cell Biol* 2005;**169**:425–34.
60. Kotoulas OB, Kalamidas SA, Kondomerkos DJ. Glycogen autophagy in glucose homeostasis. *Pathol Res Pract* 2006;**202**:631–8.
61. Schiaffino S, Mammucari C, Sandri M. The role of autophagy in neonatal tissues: just a response to amino acid starvation? *Autophagy* 2008;**4**:727–30.
62. Baerga R, Zhang Y, Chen PH, Goldman S, Jin S. Targeted deletion of *autophagy related 5 (atg5)* impairs adipogenesis in a cellular model and in mice. *Autophagy* 2009;**5**:1118–30.
63. Cifuentes D, Martinez Pons C, Garcia Rocha M, Galina A, de Pouplana LR, Guinovart JJ. Hepatic glycogen synthesis in the absence of glucokinase: the case of embryonic liver. *J Biol Chem* 2008;**283**:5642–9.
64. Pallardo FV, Sastre J, Asensi M, Rodrigo F, Estrela JM, Vina J. Physiological changes in glutathione metabolism in foetal and newborn rat liver. *Biochem J* 1991;**274**:891–3.

CHAPTER

8

Autophagy in Diabetes and the Metabolic Syndrome

Timothy E. Graham and E. Dale Abel

Division of Endocrinology Metabolism and Diabetes and Program in Molecular Medicine University of Utah School of Medicine, 15 N. 2030E, Bldg. 533 Rm 3110, Salt Lake City, UT 84112

DIABETES AND RELATED SYNDROMES

Diagnosis of Diabetes

Diabetes is diagnosed on the basis of hyperglycemia; in particular, non-fasting plasma glucose of ≥200 mg/dl or fasting plasma glucose of ≥126 mg/dl. Pre-diabetic hyperglycemic states of impaired fasting glucose (IFG; fasting glucose 100-125 mg/dl) or impaired glucose tolerance (IGT; glucose 149-199 mg/dl after an oral glucose challenge) indicate increased risk for diabetes. Table 8.1 summarizes the clinical spectrum of diabetes.

Type 1 or Autoimmune Diabetes

Diabetes occurs through one of two distinct disease processes. In a minority of diabetics (<10% of cases worldwide), it is caused by autoimmune, T-cell mediated destruction of the pancreatic beta islet cells leading to progressive loss of insulin, termed Type 1 diabetes[1, 3]. A preceding viral infection may trigger autoimmunity in some cases, and auto-antibodies directed against pancreatic islet β-cells are present. Onset is typically rapid and severe, and characterized by symptoms of polyuria and polydipsia, weight loss, dehydration, and metabolic acidosis. Reduced plasma concentrations of insulin enables release of large quantities of free fatty acids from adipose tissue (lipolysis); the fatty acids are converted to circulating ketones and ketoacids by the liver, resulting in a severe and life-threatening condition known as diabetic ketoacidosis. The course of Type 1 diabetes is usually unrelenting, and lifetime insulin replacement therapy is nearly always required.

Autophagy in Health and Disease
http://dx.doi.org/10.1016/B978-0-12-385101-7.00008-5

TABLE 8.1 Common Forms of Diabetes

	Type 1 Diabetes	Type 2 Diabetes	Metabolic Syndrome
Clinical Features	• Normal insulin sensitivity • Absolute insulin deficiency • Ketosis causing metabolic acidosis • Rapid onset, usually in childhood or early adulthood • Polyuria/polydipsia • Weight loss	• Insulin resistance in multiple tissues • Persistent hyperglycemia (≥ 200 mg/dL) • Hyperinsulinemia initially • Progressive β cell failure • Polyuria/polydipsia • Features of Metabolic Syndrome	• Insulin resistance without diabetic-range hyperglycemia • IFG and/or IGT • Abdominal adiposity • Hypertension • Dyslipidemia: ↑ triglycerides ↓ HDL cholesterol • Elevated uric acid • Albuminuria
Contributing factors	• Autoimmune T-cell mediated destruction of pancreatic β islet cells • Antibodies to β islet antigens • HLA type • Preceding viral illness	• Obesity • Nutrition • Genetics • Inactivity • Aging • Sleep apnea	

Type 2 Diabetes

Type 2 diabetes, the most common cause of diabetes (>90% of cases worldwide), is characterized by hyperglycemia in the setting of impaired insulin action (or "insulin-resistance") in multiple tissues[1]. Risk factors for Type 2 diabetes are also risk factors for insulin-resistance, and include: overweight and obesity, physical inactivity, aging, and family history of Type 2 diabetes[4, 5]. Increased plasma insulin concentrations (hyperinsulinemia) may be present for months or years preceding frank diabetes, and may be sustained for years after onset. Hyperinsulinemia results from an increase of insulin production by the pancreatic islet β-cell necessary to compensate for the insulin-resistant state. Post-mortem studies of pancreata of subjects with recent onset Type 2 diabetes or insulin-resistance without diabetes reveal islet hypertrophy with increased numbers and size of insulin producing cells, and abnormal islet organization and architecture. The excessive demands placed on the β-cell to compensate for insulin resistance may eventually cause a transition from enhanced insulin secretion to a progressive loss of insulin secretion and cell death. When insulin secretion begins to wane in the face of insulin resistance, hyperglycemia and clinical diabetes ensue. When individuals cross the glycemic threshold defining Type 2 diabetes, plasma insulin concentrations are usually sufficiently high to suppress adipose tissue lipolysis, and so diabetic ketoacidosis occurs much less frequently than in Type 1 diabetes. Lifestyle modifications such as diet and exercise and weight loss can improve insulin sensitivity, normalize plasma glucose, and ultimately prevent diabetes in susceptible individuals. Insulin-sensitizing medications are not as effective as lifestyle modifications, and co-treatment with insulin secretagogue medications, which enhance β-cell insulin production, may be needed to control glycemia. Exogenous insulin therapy is typically started when the combined medications cannot maintain glycemia.

Metabolic Syndrome

Individuals destined to develop Type 2 diabetes exhibit characteristic metabolic abnormalities well in advance of frank diabetes. The clustering of abnormalities include: Obesity with a central distribution of adipose tissue, hypertension, dyslipidemia characterized by elevated triglycerides and low HDL cholesterol, increased uric acid, and increased urinary albumin excretion; in addition to, impaired fasting glucose or impaired glucose tolerance[6]. These features are collectively referred to as the Metabolic Syndrome and are hallmarks of insulin resistance, itself. Because clinical criteria for Metabolic Syndrome includes several conditions which, by themselves, are predictive of diabetes and cardiovascular disease, the "added value" of the diagnosis of Metabolic Syndrome has been questioned[6]. Even so, it is evident that individuals with Metabolic Syndrome are at increased risk of developing premature cardiovascular disease, diabetic-like microvascular complications, and neuropathy, in the absence of diabetic-range hyperglycemia[7-9].

AUTOPHAGY AND TYPE 1 DIABETES

Insulin Deficiency, β-Cell Survival, and the Systemic Catabolic State

Type 1 diabetes occurs through T-cell mediated inflammation of pancreatic β islets (insulitis) that progresses to frank destruction of β islet cells by both apoptotic and necrotic mechanisms. Though not yet studied in detail, it is probable that autophagy is involved in the pathogenesis of Type 1 diabetes, given its known roles in MHC antigen peptide processing, immune cell function, and cell survival[10]. β-cell survival in the setting of the immune cell-mediated injury of Type 1 diabetes is most likely governed by factors different from than those involved in Type 2 diabetes. A decade of experience in patients with Type 1 diabetes receiving pancreatic islet heterologous transplants revealed that rapamycin (Sirolimus), or glucocorticoids that induce autophagy in vitro may hasten failure of the grafted β-cells[11-15]. While it is not known whether the deleterious effects in vivo are exerted directly or indirectly on the grafted β-cells, the observations suggest that pharmacologic activation of autophagy could hasten β-cell death in the autoimmune milieu. Autophagy also regulates systemic metabolism in Type 1 diabetes. Insulin is a critical hormonal inhibitor of autophagy, and systemic insulin deficiency leads to activation (or de-repression) of autophagy in multiple tissues, leading to a state of unrestrained whole body catabolism ("starvation in the midst of plenty") characteristic of untreated Type 1 diabetes[16-19]. The suppressive effect of insulin or insulin-like signaling on autophagy and total body catabolism is highly conserved from vertebrates through lower species such as Drosophila and C. elegans[20]. Reports that autophagy mediates lipolysis in adipocytes[21] (discussed below) further suggest that loss of insulin suppression of autophagy may drive ketogenesis and subsequent systemic ketoacidosis in the setting of Type 1 diabetes.

AUTOPHAGY AND OBESITY, INSULIN RESISTANCE, METABOLIC SYNDROME, AND TYPE 2 DIABETES

Autophagy: Fundamental Cellular Response to Nutrients

Obesity, insulin resistance, Metabolic Syndrome, and diabetes result from impaired responses to nutrients in individual tissues and impaired coordination of responses via endocrine factors such as insulin and leptin[22]. Autophagy itself is a fundamental cellular response to nutrients conserved evolutionarily from yeast to humans[23]. Cells utilize multiple mechanisms to sense environmental nutrients and activate adaptive responses. Some mechanisms respond directly to the presence or absence of specific nutrients or metabolites whereas others respond indirectly to the nutrient-dependent cellular energy status (e.g., the AMP activated protein kinase (AMPK) or Sirtuin family NAD^+-dependent deacetylases). Different cell types display different hierarchies of regulation that may reflect their specialized functions. Nutrient-dependent signaling by the mTOR regulatory complex 1 (mTORc1) is an important node at which multiple nutrient-sensing mechanisms converge to regulate autophagy[24]. Activation of mTORc1 by nutrients inhibits activity of Unc51-like kinases, Ulk1/2 (mammalian Atg1 proteins), via direct phosphorylation of the kinases and associated regulatory proteins, Atg13, FIP200, and Atg101[25].

mTORc1 is primarily responsive to amino acids, some being more potent activators than others. However, other nutrients have been shown to regulate autophagy. However, mTORc1-independent pathways of autophagy regulation exist, and requirement for mTORc1 signaling may be cell-type specific. Indeed, some non-nutrient regulators of autophagy are insensitive to rapamycin and may utilize mTOR regulatory complex 2 (mTORC2)[26]. Glucose deprivation alone may induce autophagy in some cell types or act synergistically with amino acid deprivation. 2-deoxyglucose, a specific inhibitor of glucose uptake and glycolysis, can activate autophagy even in the presence of nutrients[27, 28]. Cholesterol deprivation induces autophagy in several cell types[29], and statin drugs that inhibit the rate-limiting enzyme of cholesterol biosynthesis, HMG-Co-A reductase, activate autophagy[30, 31]. Deprivation of oxygen (hypoxemia), the critical substrate for cellular respiration, activates transcription factor HIF1α to induce autophagy via Bnip3/Nix-dependent and -independent pathways[32-43]. In contrast to derivational nutrient responses, addition of free fatty acids to culture media induces autophagy in pancreatic islet β-cells and other tissues[35, 36], possibly through activation of Toll-like 4 receptor signaling[37, 38].

The AMP-activated protein kinase (AMPK) is a master regulator of energy metabolism that shifts cell machinery from energy consuming to energy-producing pathways in the setting of cellular energy depletion (increased AMP/ATP ratio). AMPK directly phosphorylates tuberin (TSC2) to activate the TSC1/TSC2 Rheb-GTPase activating protein and thereby inhibits mTORC1 to activate autophagy. AMPK can also bypass mTORC1 by directly phosphorylating and activating Ulk1/2 to activate autophagy[39-41]. Direct action of AMPK on Ulk1/2 is subject to an immediate short loop feedback whereby activated Ulk1/2 can in turn phosphorylate AMPK and terminate its activity[42]. Induction of autophagy by AMPK may be sustained by phosphorylation and activation of the transcription factor, FoxO3, which regulates autophagy through transcriptional induction of LC3, Bnip3, Nix, and other components of the autophagy machinery[43]. Sirt1 and Sirt2, nutrient-dependent members of the sirtuindeacetylase family, also regulate autophagy in response to cellular energy depletion (increased NAD^+/NADH ratio)[44]. Sirt1 also directly interacts with and deacetylates Atg5, Atg7, and Atg8 in a NAD^+-dependent manner[45], and may additionally sustain autophagy by deacetylation and activation of FoxO3 and other FoxO forkhead family transcription factors. Sirt2 deacetylates FoxO1 and FoxO3 in a nutrient-dependent manner[46, 47]. However, under conditions of serum deprivation or oxidative stress, FoxO1 does not interact with Sirt2, and becomes predominately cytosolic and transcriptionally inactive[48]. Cytosolic acetylated FoxO1 can interact directly with Atg7 to induce autophagy under these conditions[48].

Autophagy and Cellular Mechanisms of Insulin Resistance

Numerous monogenic forms of obesity and Type 2 diabetes have been identified. However, these conditions are rare, and for the majority of affected individuals, obesity and diabetes result from interactions between complex, polygenic traits and environmental factors such as diet and physical inactivity. Genome-wide association studies identified a handful of alleles that increase risk for Type 2 diabetes; however, most of these studies used hyperglycemia as an endpoint, and identified genes related more to beta cell function and survival than to insulin resistance[49]. Nevertheless, recent years have seen major advances in understanding the molecular pathophysiology of obesity and insulin resistance.

Insulin resistance in humans and rodents is characterized by post-receptor defects in insulin signaling occurring at insulin receptor substrates 1 and 2 (IRS-1/2), large adapter molecules that undergo tyrosine phosphorylation by the insulin receptor to initiate recruitment and activation of class I PI3K and several other proteins required for insulin signaling. Serine/threonine (Ser/Thr) phosphorylation of IRS1/2 at several sites produces uncoupling of insulin signaling from specific downstream pathways, and in some cases, degradation of the IRS1/2 proteins themselves and global attenuation of insulin signaling. Several intracellular processes lead to Ser/Thr phosphorylation of IRS1/2 and insulin resistance, including nutrient-dependent activation of the ribosomal S6 kinase by mTOR complex 1[50, 51].

Autophagy either regulates or is regulated by several processes known to affect insulin action at the cellular level, including lipid accumulation/lipotoxicity, reactive oxygen species (ROS) production, inflammation, endoplasmic reticulum stress, adipokine secretion, and mitochondrial turnover[52, 53]. The role of autophagy in these processes is discussed in detail in other chapters. Basic information regarding their relationship to autophagy and insulin resistance is summarized in Figure 8.1 and Table 8.2.

Autophagy and Pancreatic β-Cells: Insulin Secretion and β-Cell Survival

Ultrastructural studies of the human and rodent pancreas many years ago revealed that lysosomes play a role in maintaining the health of the pancreatic β-cell. Pancreatic beta cells constantly synthesize proinsulin, which is packaged in granules and processed, and then either secreted or recycled along with other granule contents. The majority of insulin granules that are generated undergo recycling rather than exocytosis, and lysosomes play a key role in the

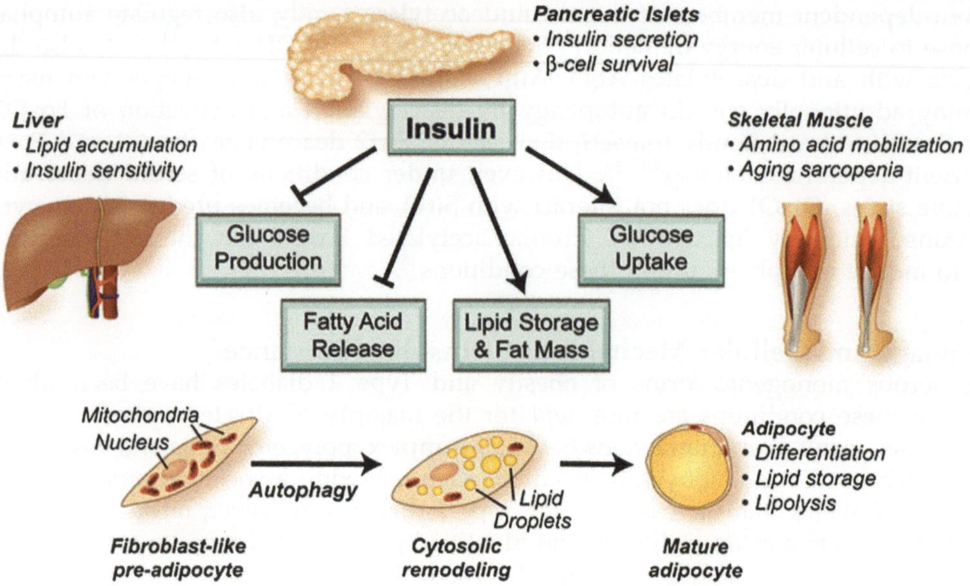

FIGURE 8.1 Physiology of insulin-glucose homeostasis and processes potentially regulated by autophagy in specific target tissues of insulin action. Insulin acts to suppress endogenous glucose production in the liver and to stimulate glucose uptake in skeletal muscle, cardiac muscle (not shown), and adipocytes. In adipocytes, insulin suppresses fatty acid release (lipolysis) and stimulates adipogenesis and triacylglyceride synthesis and storage; in addition, autophagy is required for cytosolic remodeling and reduction in mitochondrial content during adipocyte differentiation. Autophagy-dependent processes affecting insulin production or action are listed under each target tissue.

recycling process. Catabolism and recycling of insulin granules and their contents occurs by direct docking of granules with lysosomes (crinophagy) or by engulfment of granules and trafficking to lysosomes via macroautophagy[54, 55]. In the setting of insulin resistance there is increased demand for insulin, and β-cells expand in size and number, and upregulate the expression and processing of proinsulin[56]. As a result, there is increased protein passing through the endoplasmic reticulum, which must be properly folded and targeted to the insulin granules, and the half-life of insulin granules within the β-cells decreases due to increased rates of both exocytosis and recycling. These extraordinary synthesis, trafficking, and recycling demands place enormous stress on the β-cell, and ultimately contribute to the insulin secretory dysfunction and loss of β-cell mass which mark the onset of Type 2 diabetes. Given the high demand for protein synthesis, it is not surprising that β-cells express high levels of ER stress/unfolded protein response (UPR) transducer proteins such as IRE1α, PERK, XBP1, GRP94 and BiP.23[57]. Activation of the UPR stimulates autophagy, which plays a critical role in alleviating ER stress by clearing poorly soluble polyubiquitinyl protein aggregates and through direct engulfment of impaired endoplasmic reticulum in some cases. Islets of subjects with Type 2 diabetes show increased ER stress[58], and impairment of UPR signaling leads to loss of insulin secretion and beta-cell death, even in the absence of insulin resistance or Type 2 diabetes[59].

TABLE 8.2

Process	Mechanism Leading to Insulin Resistance	Potential Role of Autophagy
Lipid Accumulation/Lipotoxicity	• Activation of toll-like receptor-4 by free fatty acids (FFA) • Activation of novel protein kinase-C family by free fatty acids	• Mobilizes small lipid droplets – lipophagy • Mediates lipolysis
Oxidative Stress/ROS	• Activation of c-Jun N-terminal kinase (Jnk)	• Activated by ROS and decreases susceptibility to ROS • Degrades dysfunctional mitochondria producing ROS
ER Stress/Unfolded Protein Response	• Activation of protein kinase RNA-like endoplasmic reticulum kinase (PERK)	• Activated by UPR and alleviates ER stress • Degrades insoluble protein aggregates
Inflammation	• Activation of inhibitor of kappa kinase alpha (iκKα) • FoxO1 activation	• Cytokine receptor Jak/Stat signaling • Processing of NFkB • Inflammasome activation in adipose macrophages
Adipose tissue dysfunction	• Release of FFA contributes to lipid accumulation in other tissues • Decreased production of beneficial adiponectin • Increased production of deleterious adipokines (e.g., RBP4 and resistin)	• Adipocyte differentiation: mediates cytosolic remodeling to reduce mitochondria, enhance lipid storage • Processes high-molecular weight form of adiponectin

β islets from pancreata of live human multi-organ transplant donors (Type 2 diabetics and non-diabetic age-matched controls) have been studied with regard to autophagy. Ultrastructural TEM reveals increased numbers of enlarged autophagic vesicles (i.e., autophagic vacuoles) and increased cell death among β-cells from Type 2 diabetics vs. control patients (2.2% for diabetic islets vs. 0.7%, for controls)[35]. Distinct hallmarks of either apoptotic cell death, characterized by chromatin condensation, or autophagic death, characterized by massive vacuolization, can be identified, and apoptotic and autophagic mechanisms appear to contribute equally to increased β-cell death in diabetic subjects (1-1.1% for diabetics vs. 0.3-0.4% for controls for each mechanism). Autophagic vacuole formation in diabetic β-cells was associated with decreased mRNA expression of LAMP1 and the lysosomal proteases, Cathepsin B and D, but normal expression of autophagy regulator genes Atg1 and Beclin-1/Atg 6[35]. These transcriptional changes and the observation of large, overloaded autophagic vacuoles may indicate a primary disorder in lysosomal function or impaired formation of the autolysosome at terminal stages of autophagy. Importantly, these autophagic changes are not observed in glucagon-secreting α-cells or somatostatin-secreting δ-cells[35]. Remarkably, treatment of isolated live islets of control subjects with free fatty acids in vitro was found to increase autophagic vacuoles and percentage of cells displaying autophagic cell death, and co-treatment with metformin, which enhances β oxidation of lipids, prevented these effects[35]. Treatment of live islets from diabetic subjects with metformin also reduced β-cell autophagic death to the rate observed in untreated islets from control subjects[35]. It has also been reported that treatment of cultured INS-1 β-cells in vitro with free fatty acids induces autophagy but hyperglycemia does not[36]. These findings implicate alterations in autophagy and/or lysosome function in the pathogenesis of decreased insulin secretion and reduced β-cell survival which cause the transition to hyperglycemia and frank diabetes in states of insulin resistance; in addition, these findings suggest that suppression of autophagy may be a specific mechanism by which free fatty acids induce lipotoxic injury to β-cells.

Genetic studies in mice confirm that autophagy plays an important role in β-cell function. β-cell-specific Cre/loxP-mediated KO of Atg7 impairs basal β-cell autophagy, as expected, and leads to degenerative changes in β-cells, reduced insulin secretion, and glucose intolerance[60]. β-cell histological changes of extensive vacuolization similar to those observed in β-cells isolated from human subjects with Type 2 diabetes are present, and large cytoplasmic inclusions containing polyubiquitinyl proteins and p62/SQSTM are also present[60]. Atg7 KO in β-cells do not undergo expansion and enhanced insulin section with high fat diet feeding in vivo, resulting in the early onset of frank diabetes[61]. Hence, autophagy appears to be absolutely required for β-cells to meet the increased metabolic demands of the insulin resistant state induced by high fat diet in vivo. While still poorly understood, the requirement for autophagy probably reflects its critically important housekeeping role, including counteracting ER stress, an important source of injury for β-cells; however, it may also reflect involvement of autophagy in unique aspects of β-cell regulation, such as control of membrane levels of the GLUT2 glucose transporter, a critical mediator of glucose-stimulated insulin secretion.

At face value, the observation that autophagy must be intact for β-cell expansion and survival is potentially at odds with observations of increased autophagic cell death in diabetic human β-cells treated with free fatty acids ex vivo (discussed above); however, autophagy in mammalian cells has evolved a "double duty" as both housekeeper and mediator of cell death under some conditions[62, 63]. Furthermore, it is likely that diabetic β-cells already exist in a state of sustained decompensation with regard to cellular autophagy and lysosomal function such that further stimulation by free fatty acids only serves to hasten autophagic cell death. Consistent with this, metformin, which induces clearance of free fatty acids by β-oxidation, reduces autophagic cell death in diabetic β-cells. Even though autophagy seems to be induced selectively by free fatty acids, it may play a critical role in β-cell compensatory responses to hyperglycemia. Cultured INS-1 β-cells exposed to hyperglycemic conditions in vitro form polyubiquitinyl protein/p62 cytoplasmic inclusions similar to those observed in diabetic and autophagy-deficient β-cells. When normoglycemic conditions are restored, the cells are capable of clearing the inclusions; however, inhibition of autophagy by treatment with 3-MA completely inhibits this process of clearance and leads to increased cell death. Taken together, these findings indicate that autophagy plays an important role in normal β-cell function, and impairment of β-cell autophagy may be a critical factor contributing to the pathogenesis of diabetes in insulin resistant states.

Autophagy in Liver: Hepatocyte Lipotoxicity ER Stress, and Insulin Resistance

Inhibition of autophagy in immortalized rat hepatocytes by treatment with 3-MA or Atg5 RNAi causes accumulation of intracellular lipids (TGs and cholesterol) in the form of small lipid droplets in the cytoplasm[64]. Culturing hepatocytes in methionine and choline deficient (MCD) media (analogous to MCD rodent diets known to cause hepatic steatosis in vivo) also produces TG accumulation. Under these conditions, TG accumulation is due to decreased lipid beta-oxidation rather than increased lipogenesis[64]. TEM studies of hepatocyte ultrastructure reveal direct interactions between lipid droplets and autophagic vesicles and engulfment of portions of lipid droplets within LC3-positive double membranes (autophagosome), and similar findings are observed in liver of mice after 72 hour starvation; in addition, lipid droplet associated proteins are seen to accumulate in isolated autophagosomes and lysosomes under these conditions[64]. These findings indicate that small lipid droplets can undergo direct autophagic lysosomal degradation in a fashion similar to other cytoplasmic organelles. Consistent with this, hepatocyte-specific knockdown of Atg7 exhibit increased hepatic lipid (triglyceride and cholesterol) and increased amounts of TIP47, indicating that autophagy may be the principal pathway for degradation of this lipid droplet-associated protein[64].

Chronic high fat diet feeding of mice, which induces insulin resistance and hepatic steatosis, also impairs lipid autophagy and expression of several autophagy proteins in liver[65, 66]. Autophagy proteins are similarly decreased in livers of leptin-deficient ob/ob mice[66]. This effect may result in part from insulin action to inhibit FoxO1-induced transcription of multiple genes in the autophagy pathway in liver[65]. Treatment of mice with the pancreatic β-cell toxin streptozotocin to cause insulin deficiency (Type 1 diabetes) alleviates suppression of FoxO1, resulting in induction of autophagy gene expression in liver[65]. Remarkably, chronic insulin treatment that causes insulin resistance at the level of IRS1 and Akt phosphorylation of IRS1 does not impair insulin action to suppress autophagy[65]. Hence, chronic hyperinsulinemia present in insulin resistant states could potentially exert tonic inhibition of autophagy in hepatocytes and thereby contribute to ectopic lipid accumulation and lipotoxicity. However, hyperinsulinemia is not likely the sole cause of reduced liver autophagy, since normalization of plasma insulin levels in vivo is not sufficient to restore Atg7 expression in liver of severely hyperinsulinemic ob/ob and db/db mouse models.

Suppression of liver autophagy in insulin resistant states may itself be an important cause of insulin resistance. Atg5 null and Atg7 null MEFs exhibit reduced insulin-stimulated phosphorylation of insulin receptor and Akt[66]. Remarkably, even short term knockdowns of Atg7 in cultured hepatocytes or in liver via adenoviral shRNAi in vivo are sufficient to induce insulin resistance[66]. It is likely that mechanisms other than lipotoxicity are involved since insulin resistance develops in a timeframe that is too rapid to be accounted for by lipid accumulation. The effect of short-term impairment of autophagy may be explained by acute activation of the unfolded protein response by subsequent ER stress[66]. Autophagy has been shown to play an important role in relieving ER stress by enhancing the clearance of protein aggregates[67]. Restoration of autophagy in liver of ob/ob mice, which are severely insulin resistant and exhibit chronic ER stress in liver and other tissues, is sufficient to improve systemic insulin-glucose homeostasis[66]. These findings suggest that impaired autophagy in liver is both a consequence of and pathogenic factor in systemic insulin resistance and glucose intolerance. Strategies to enhance liver autophagy may have potential as therapies to prevent or treat Type 2 diabetes.

Autophagy in Adipose Tissue: Adipogenesis and Obesity

Two studies of adipose tissue autophagy using fat biopsies from human subjects have reported that autophagasome numbers and autophagic flux are increased in subjects with obesity or Type 2 diabetes[68, 69]. In addition, expression of Atg5 and Atg8 (LC3a and LC3b) is increased in adipocytes of obese subjects, especially in biopsies from the omental fat depot, and in biopsies from obese subjects with insulin resistance vs. obese with normal insulin sensitivity[69]. Adipocyte size, which is highly correlated with insulin resistance in most cohorts, was also highly correlated with Atg5 and Atg8 expression[69]. These findings are consistent with reports that Atg5 and Atg8, along with several other autophagy genes, are regulated by FoxO transcription factors whose activity is known to be unregulated in adipocytes of insulin resistant subjects; in addition FoxO1 binds Atg7 and may directly regulate its activity[48]. A 40% reduction in mTOR complex 1 (mTORC1)-mediated phosphorylation of the ribosomal S6 kinase was observed in adipocytes of Type 2 diabetics[68]; however, this finding is consistent with an expected reduction in insulin/Akt signaling in adipocytes of these subjects, and whether mTORC1 signaling regulates adipocyte autophagy generally has yet to be determined. Together these studies provide initial data indicating that adipocyte autophagy is increased in the setting of obesity and Type 2 diabetes. Given a multitude of observations that ER stress, oxidative stress, tissue hypoxia, inflammation, and insulin resistance are all present in adipocytes of obese and Type 2 diabetic subjects, it is possible that the observed increase in autophagy represents a normal response to the state of sustained cellular stress in adipocytes. More work is needed to understand whether adipocyte autophagy plays a beneficial or a pathogenic role per se in human adipose tissue function and whole body insulin-glucose homeostasis.

Animal models can provide insight into pathogenesis; however, animal studies of adipocyte autophagy had been limited by the early mortality of mice with total body KO of autophagy genes, in which death occurred before formation of mature adipose depots[70-72]. Recent studies employing mice with Cre/loxP-mediated adipocyte-specific knockout of Atg7 (adipocyte Atg7 KO mice) have revealed a critical role for autophagy in adipogenesis and obesity, which play important roles in insulin resistance and Type 2 diabetes. Two groups independently reported that mice with adipose tissue loss of autophagy genes are lean and have enhanced insulin sensitivity and glucose tolerance[21, 73]. These mice have normal birth weight but gain fat mass less quickly than wild-type mice after adulthood and are protected from obesity on high fat diet. The same flox-Atg7 allele and the aP2 (fatty acid binding protein 4)-promoter Cre (aP2-Cre) transgenic mouse lines were used in each study, and the phenotypes of the two adipocyte Atg7 KO lines were essentially identical. The aP2 (fatty acid binding protein 4) promoter-driven Cre transgene used in these studies is active in both white and brown adipose tissue (WAT and BAT). Adipocyte Atg7 KO WAT depots are reduced in size and grossly abnormal in appearance and function: they have the appearance of BAT, including the presence of large numbers of mitochondria and multilocular small lipid droplets rather than a single large droplet. The suprascapular BAT depot of adipocyte Atg7 KO mice is increased in total size and contains increased numbers of mitochondria[21, 73].

Beta oxidation of lipids is greatly increased in both WAT and BAT of adipocyte Atg7 KO mice indicating that the expanded mitochondrial population is functional and can contribute to whole body lipid utilization[21, 73]. In addition to the increased utilization of fatty acids, WAT of adipocyte Atg7 KO mice also exhibits decreased catecholamine-stimulated lipolysis in vitro and enhanced sensitivity to suppression of lipolysis by insulin in vivo[21]. Secretion of leptin by WAT and plasma leptin levels are lower in adipocyte Atg7 KO mice, consistent with reduced WAT mass, but adiponectin secretion and plasma levels are unchanged[73]. Studies of in vitro adipocyte differentiation of Atg5 null mouse embryonic fibroblasts (MEFs) also reveals impaired adipogenesis characterized by normal early stages of adipocyte differentiation followed by failure to accumulate lipid droplets[74]. There are increased rates of apoptosis in later stages of differentiation, and reduced numbers of WAT-like adipocytes in Atg5 total body KO mouse embryos[74]. Knockdown of Atg5 or Atg7 or treatment with the lysosome inhibitor chloroquine also impairs differentiation of 3T3L1 cells[74], a widely used model for studying in vitro adipogenesis.

UCP1 and PGC1α protein expression are increased in WAT and BAT of adipocyte Atg7 KO mice, consistent with increased mitochondrial content; however, expression of BAT lineage-specific genes such as *Elovl3*, *Cidea*, and *Prdm16*, is not increased. Furthermore, the number and proliferative capacity of WAT lineage-specific precursors in the adipose tissue stroma of adipocyte Atg7 KO mice are not altered[73]. Therefore, adipocyte Atg7 KO appears to cause transdifferentiation of committed WAT lineage precursor cells to a BAT-like phenotype rather than changing the number or commitment of WAT/BAT lineage-specific precursors. Indeed, expansion of the number of mitochondria normally occurs during early WAT differentiation and mitochondrial content is subsequently reduced during terminal differentiation to produce the mature WAT adipocyte adapted to lipid storage rather than utilization. The phenotype of Atg7 KO adipocytes suggests that autophagy plays a crucial role in this remodeling process, such that a BAT-like phenotype results when mitochondria cannot be cleared during terminal differentiation.

There are several important caveats for interpreting these data with regard to the pathophysiology of obesity and diabetes. First, the aP2-Cre promoter used in each study is active in development as early as embryonic day six. Hence, adipocyte autophagy KO mice generated with this promoter do not develop normal white adipose tissue at any point in their lives and are best viewed as models of impaired primary adipogenesis. Furthermore, the leanness of these mice is a key determinant of their enhanced insulin sensitivity and glucose tolerance, such that their observed metabolic phenotypes may be irrelevant to the in vivo role of autophagy in mature white adipose tissue of adult rodents or humans with obesity, insulin resistance, or Type 2 diabetes. Nevertheless, the metabolic phenotype of the adipocyte autophagy KO mice is informative because it differs strikingly from other models of impaired adipogenesis, which typically present with insulin resistant and diabetic phenotypes arising from obligate ectopic lipid storage producing lipotoxicity in liver, muscle, and pancreas. In contrast, adipocyte autophagy KO mice do not exhibit increased lipid accumulation in non-adipose organs, likely owing to the expansion of BAT-like adipose tissue and increased lipid beta-oxidation providing increased body lipid clearance. While the lean and insulin-sensitive phenotype of adipocyte Atg7 KO mice is obesity resistant and appears to be metabolically "healthy", the observation that the mice begin to die from unknown causes at eight weeks of life and exhibit ~40% mortality by 12 weeks of age cannot be overlooked, as it indicates that autophagy in adipose tissue is likely necessary for life. Therefore, therapies for obesity that would target adipocyte autophagy potentially pose inherent risks that are not fully understood at this time. More work in this area will be needed to understand both the physiological and pathophysiological roles of autophagy in adipocytes in obesity, insulin resistance, and Type 2 diabetes.

Autophagy and Skeletal Muscle

Fasting metabolism, muscle mass, mitochondrial dysfunction, age-related sarcopenia, the progressive reduction in skeletal muscle mass with aging, are major factors contributing to insulin resistance, Type 2 diabetes risk, and morbidity/mortality due to falls and injuries in older individuals[75-78]. Skeletal muscle mass is increased by anabolic endocrine factors such as insulin, IGF-1, and testosterone. Autophagy is induced in skeletal muscle during fasting[79], which serves to maintain circulating amino acids and glucose (by supplying gluconeogenic amino acids). Insulin and IGF-1 levels are reduced during fasting, which may be permissive for induction of autophagy. Consistent with this, activation of inducible Akt to mimic insulin/IGF1 signaling in skeletal muscle suppresses the normal autophagic response during fasting[80, 81]. Akt-dependent inhibition of the transcription factor FoxO3 negatively regulates expression of multiple autophagy genes, including LC3, GAPARAPL1, Bnip3, Nix, and Atg4b[80, 81]. Regulation of skeletal muscle autophagy by FoxO3 is rapamycin-insensitive and requires the mTOR adapter molecule RICTOR, indicating involvement of mTOR complex 2[80, 81].

While autophagy plays a critical catabolic role in skeletal muscle in the setting of fasting, it nevertheless is required to maintain normal muscle mass and strength in aging. Mice with skeletal muscle-specific knockout of Atg7[26, 82] exhibit normal birth weight and growth until 40 days of age, at which point they exhibit reduced weight gain. Autophagy is impaired, as expected, and histological examination reveals degenerative changes of vacuolization and central nucleation, with glycolytic and oxidative fibers equally affected. Some fibers show necrotic changes without generalized myolysis[26, 82]. Force generation in skeletal muscle of older Atg7 KO mice is severely impaired, and both histological and functional changes are accelerated by denervation in vivo[26, 82]. Ultrastructural exam by TEM reveals accumulation of enlarged and dysmorphic mitochondria in muscle fibers of aging Atg7 KO muscle, suggesting impaired mitophagy[26, 82]. A massive increase in protein carbonylation is present in Atg7 KO muscle without a change in AMP-activated protein kinase activity, suggesting that increased mitochondrial ROS production rather than impaired ATP production may drive the degenerative process[26, 82]. The metabolic phenotype of these mice has not yet been reported. However, recent studies of mitochondrial changes in skeletal muscle biopsies of subjects with Type 2 diabetes have revealed a similar phenotype of increased mitochondrial ROS production without impairment of ATP production[83]. Taken together, these findings implicate skeletal muscle autophagy in normal fed/fasting nutrient homeostasis and maintenance of skeletal muscle mass during aging, processes known to be important in the pathogenesis of insulin resistance and Type 2 diabetes. Further work is needed to determine how skeletal muscle autophagy may influence insulin-glucose homeostasis in rodents and humans.

AUTOPHAGY AND DIABETES COMPLICATIONS

Complications of diabetes represent the major causes of morbidity and mortality that are associated with this chronic metabolic disorder[84]. Specifically, cardiovascular disease is the leading cause of mortality in subjects with diabetes, which represents the major cause of end stage renal disease, blindness and limb amputations in developed and many developing societies. Broadly speaking, diabetes complications can be categorized as macrovascular, which are those affecting large and medium size blood vessels and microvascular complications, which involve small blood vessels such as small arterioles. Macrovascular complications include accelerated atherosclerosis, leading to increased coronary artery disease, increased cerebrovascular disease and stroke, and increased peripheral arterial disease that contributes to critical limb ischemia. There is also strong evidence that diabetes leads to direct adverse effects in the heart that leads to diabetic cardiomyopathy[85]. Microvascular impairment leads to the complications of diabetic retinopathy, diabetic nephropathy and diabetic neuropathy. The pathogenesis of these diverse complications of diabetes is complex, is only partially understood and has been the subject of many reviews. Only recently has attention turned to the potential contribution of changes in autophagy due to the complications of diabetes.

As reviewed in other chapters, autophagy may serve a protective role in cells where it promotes cellular survival by removing damaged organelles such as mitochondria. Therefore in these contexts, a decrease in autophagy could contribute to cellular dysfunction that occurs on the basis of accumulation of dysfunctional organelles or irreversibly damaged cellular substructures. Conversely, excessive levels of autophagy can contribute to cell death. Thus a challenge in elucidating the contribution of autophagy to diabetes complications rests in the difficulty in discerning if any observed change in autophagic flux represents an adaptation or a maladaptation that is contributing to tissue injury or to its preservation in the context of the abnormal metabolic milieu that characterizes diabetes. Thus, although descriptions of changes in autophagy in diabetes complications are of interest, interpretation of the significance of the direction of the changes observed is more challenging. Ultimately, the most robust interpretations will be made when autophagy is independently manipulated in the context of a diabetes complication and a determination made of the impact of this manipulation on the progression of the particular complication in question.

Prior to summarizing what is known thus far regarding potential contributions of autophagy to diabetes complications, it will be of value to summarize mechanisms that may regulate autophagy in susceptible tissues in the diabetic milieu. Diabetes is characterized by hyperglycemia, hyperlipidemia, hypoinsulinemia (in the case of poorly controlled Type 1 diabetes) and hyperinsulinemia and insulin resistance in the case of type 2 diabetes. In addition, there might be increased concentrations of circulating cytokines, advanced glycation end products and systemic oxidative stress. Each of the components of the altered metabolic milieu could potentially influence autophagic signaling. Glucose enters cells via facilitative glucose transporters[86]. In most tissues, hyperglycemia will therefore be associated with increased cellular glucose uptake, increased glycolysis and the activation of metabolic pathways that originate from glycolytic intermediates such as the hexosamine biosynthesis pathway, the pentose phosphate shunt and the polyol pathway. Byproducts of these metabolic pathways can activate signaling mechanisms that may contribute to tissue injury in the context of diabetes[87]. A variety of studies that will be described later, in the context of specific complications, clearly indicate that exposure of cells to increased glucose concentrations can modulate autophagy in various ways. On the one hand it is clear that glucose deprivation is a potent activator of autophagy acting via mechanisms that include activation of AMPK, and activation of Sirt1[88, 89]. However, increased delivery of glucose to cells does not necessarily result in a suppression of autophagy but may activate autophagy via mechanisms that involve increased oxidative stress that arise from mitochondrial ROS overproduction, ER stress or the repression of inhibitors of autophagy such as GATA4[90-93]. Advanced glycation end products have also been shown to promote autophagy via mechanisms that are incompletely understood[94].

In many cell types, activation of Class 1A PI3K leading to activation of mTOR is a potent repressor of autophagy[95]. Thus it is plausible that reduced or defective insulin signaling could lead to increased autophagy in the context of diabetes. Whether or not this represents a dominant mechanism or is superseded by other metabolic regulators of autophagy in respective tissues is incompletely understood. Other potential regulators of autophagy that could be at play in the context of diabetes include consequences of increased FA uptake and delivery, and signaling via inflammatory cytokines. Additional key signaling regulators of autophagy such AMPK, Sirt1, FOXO are altered by the diabetic milieu and as such could potentially influence autophagic flux in susceptible tissues and organs.

SPECIFIC COMPLICATIONS—MICROVASCULAR

Neuropathy

In neural tissue it is widely accepted that constitutive levels of autophagy play an important "housekeeping" role to recycle degenerating or damaged organelles. Deletion of autophagy genes in drosophila and in the nervous system of mice is associated with accelerated neurodegeneration[96-99]. Thus an increase in autophagy in the context of diabetes could represent an adaptive response to diabetes-related neuronal damage. The abnormal metabolic milieu that characterizes diabetes drives the pathogenesis of diabetic neuropathy. There is strong evidence that increased metabolism of glucose and lipids promote oxidative stress from mitochondrial and non-mitochondrial sources. There is compelling evidence that mitochondrial dysfunction, characterized by proteome remodeling, defects in the electron transport chain and altered mitochondrial dynamics and biogenesis, develop in experimental models of diabetic neuropathy[100-104]. Dysfunctional mitochondria may trigger an autophagic response. Therefore it is not surprising that increased autophagy has been described in dorsal root ganglia of STZ-diabetic rats[105]. In this study, the autophagy mediator LC3 was found to co-localize with the FAS (CD 95)-cascade mediator FADD. FADD is involved in the membrane-death receptor extrinsic apoptosis pathway. It has been speculated that the activation of autophagy in this model leads to the sequestration of and degradation of FADD, which in turn leads to reduced activation of the extrinsic apoptotic pathway. Thus these data suggest that the activation of autophagy in this model could represent an adaptive mechanism that mitigates neuronal damage. Interestingly, it was recently reported that sera from patients with diabetes and neuropathy harbored antibodies that interact with FAS. Moreover, incubation of the neuroblastoma cell line (SH-SY5Y) with these sera resulted in an activation of autophagy[105]. These data would suggest that activation of neuronal autophagy in the context of diabetic neuropathy could be mediated via intrinsic and extrinsic mechanisms. In contrast, insulin withdrawal promotes autophagic cell death in cultured hippocampal neurons[106]. Thus in this context, it is clear that excessive autophagy is deleterious. While there is no doubt that autophagy is increased in neurons in diabetes, the significance of these changes to the pathophysiology of diabetic neuropathy remains to be completely clarified. If the increase in autophagy is an adaptive response that mitigates neuronal injury, then the induction of diabetes in appropriate animal models in which autophagic capacity is decreased would be predicted to accelerate neuronal injury. In contrast, if enhanced autophagy is pathogenic, then reducing autophagy should be beneficial.

Nephropathy

Like neuropathy, the pathogenesis of diabetic nephropathy is complex and multifactorial. Critical pathophysiological mediators include hyperglycemia, lipotoxicity and oxidative stress[107]. There is also strong evidence that increased podocyte apoptosis contributes importantly to the pathogenesis of glomerular injury early in the course of diabetic nephropathy[108]. There are relatively few studies that have directly examined the impact of autophagy on the pathogenesis of diabetic nephropathy. In a series of morphological studies, Pfeifer and colleagues described inhibition of autophagy within five days of onset of STZ-induced diabetes, which was reversed following insulin administration[109, 110]. The relevance of these observations to the pathogenesis of diabetic nephropathy is not clear, given the relatively long duration of hyperglycemia, which precedes the development of diabetic nephropathy in rodent models. Moreover, no published studies to date have examined the impact of autophagy in glomeruli or podocytes, which play a critical role in the pathogenesis of diabetic nephropathy. There is growing evidence that activation of mTOR signaling may play a role in the pathogenesis of diabetic nephropathy and that mTOR inhibition with rapamycin or genetic strategies might protect against diabetic nephropathy[111-114]. Conversely inhibition of AMPK has been observed in experimental models of diabetic nephropathy and may contribute to the activation of mTOR[115, 116]. Indeed, activation of AMPK using pharmacological strategies reduces renal hypertrophy and decreases protein synthesis in part via inhibition of mTOR. Activation of AMPK and repression of mTOR activity would promote autophagy, whereas inhibition of AMPK and activation of mTOR would repress autophagy. These associations between mTOR and AMPK signaling with diabetic nephropathy supports the hypothesis that inhibition of autophagy as a consequence of mTOR activation or AMPK inhibition early in the course of diabetic nephropathy may impair cell survival and accelerate the progression of diabetic nephropathy.

An increase in autophagy and apoptosis occurs in concert with renal tubular necrosis following ischemia/reperfusion injury in the kidney[117]. These changes are associated with ATP depletion, and manipulations such as PARP inhibition that preserve tubular ATP content mitigates tubular necrosis that occurs in this context[118]. It is probable that the severe stress of ischemia activates multiple cell death programs. Ischemia-associated ATP depletion likely activates autophagy via mechanisms such as AMPK activation. If ischemia becomes more protracted, then apoptosis is likely to be triggered as a consequence of increased cytochrome C release and activation of mitochondrial programmed cell death pathways. The relevance of changes associated with ischemia/reperfusion injury to the pathogenesis of diabetic nephropathy is not clear. However, mitochondrial dysfunction has been implicated in the pathogenesis of podocyte loss in diabetic nephropathy[119, 120]. Mitochondrial dysfunction is a trigger of mitophagy and can also stimulate activation of bulk autophagy secondary to reduced energetics and AMPK activation. Thus it is plausible to hypothesize that activation of autophagy may occur early in the course of diabetic nephropathy. This might represent an initial adaptation to diabetes-associated mitochondrial dysfunction early in the course of diabetic nephropathy, and that suppression of autophagy might therefore accelerate glomerular injury. Therefore, there is a critical need for studies that seek to dissect the role of autophagy in glomerular and tubular injury in the course of diabetic nephropathy.

SPECIFIC COMPLICATIONS—MACROVASCULAR

Atherosclerosis

Diabetes is characterized by accelerated and more advanced atherosclerosis[121]. The mechanisms for this are incompletely understood, but likely represent the combined effects of endothelial dysfunction that is secondary to the effects of exposure to increased circulating concentrations of lipids and glucose, which induce oxidative stress and reduces the availability of nitric oxide[121]. Increased autophagy has been described in atherosclerotic plaques[122, 123]. This occurs in macrophages and is believed to play a protective role by reducing macrophage number and stabilizing atheromatous plaques. Multiple mechanisms have been implicated in the activation of these macrophages, and include hypoxia, exposure to inflammatory cytokines, oxidized LDL and ER stress[124]. These mechanisms would be expected to increase with diabetes. Whereas some cytokines such as TNFα might increase autophagy by inducing Beclin 1 content, others such as IL-4 and IL-13 may suppress autophagy via PI3-K mediated activation of mTOR[125, 126]. Regulation of autophagy in vascular smooth muscle cells (VSMC), isolated from atherosclerotic plaques has also been described, with TNFα promoting autophagy by inducing microtubule-associated protein 1 light chain 3 (MAPLC-3) and Beclin 1[127]. This induction of autophagy is mediated via c-Jun terminal kinase signaling pathways and is associated with increased cell death. By contrast, exposure of these VSMCs to IGF-1 reduced autophagic cell death, which would be predicted to increase plaque stability[125]. The relative contribution of pro-autophagic versus anti-autophagic signaling in the arterial wall in the context of diabetes is incompletely understood. Thus, the activation of autophagy in macrophages in atherosclerotic plaques might represent a protective mechanism, although activation of autophagy in VSMCs might contribute to plaque instability. Future studies are therefore required to determine if the increased susceptibility to atherosclerosis, which develops in diabetes, occurs in part as a result of the imbalance in adaptive versus maladaptive effects of autophagy in specific cell types in the vessel wall.

Cardiomyopathy

Autophagy plays an important role in the pathophysiology of cardiac disease. Genetic deletion of the autophagy gene Atg5 in the adult heart leads to heart failure, confirming a critical role for basal rates of autophagy in maintaining cardiac contractile homeostasis[128]. Conversely, whereas the increase in autophagy that accompanies ischemic injury is a protective adaptation, the increase in autophagy that occurs following reperfusion injury is maladaptive[129, 139]. The same might be true for the myocardial adaptation to pressure overload hypertrophy wherein an increase in autophagy accelerates the transition to heart failure[131-133]. The role of autophagy in diabetic cardiomyopathy is relatively understudied but a number of observations provide strong support for the hypothesis that autophagy may contribute importantly to its pathogenesis. For example, fasting induces a striking increase in rates of autophagy in the heart[88]. Fasting is associated with hypoinsulinemia and increased circulating concentrations of fatty acids. These metabolic changes also occur in diabetes. It is therefore plausible that these two metabolic changes could contribute to the increase in autophagic flux observed. It was recently reported that in the fructose-fed model of insulin resistance, increased myocardial autophagy and increased fibrosis was associated with reduced Akt and mTOR signaling and oxidative stress. These findings provide preliminary evidence for increased autophagy in the hearts of an animal model of insulin resistance.

Fasting is also associated with de-phosphorylation of FOXO1/3 in the heart and nuclear FOXO promotes the expression of a family of genes that encode critical regulators of autophagy[134]. Therefore, in states in which insulin signaling is deficient it is reasonable to hypothesize that autophagy could be increased in this context. Moreover, starvation activates Sirt1, which de-acetylates and further activates FOXO. Indeed expression of a mutant FOXO, which fails to interact with Sirt1 leads to a starvation-induced cardiac dysfunction. In these contexts, autophagy clearly plays an adaptive role[88]. Diabetes is associated with activation of Sirt1 in the heart[135], and therefore it is reasonable to hypothesize that a similar mechanism may lead to an increase in autophagy in the heart in diabetes. This hypothesis remains to be directly proven.

Glucose deprivation of cardiomyocytes *in vitro* induces autophagy, whereas repletion of glucose in this context suppresses autophagy[136]. In beating hearts (*in vivo* and *in vitro*) diabetes is associated with decreased glucose utilization[137]. The glucose metabolite G-6-P has been suggested to activate mTOR, which would be expected to suppress autophagy[138]. Thus reduced glucose utilization in the diabetic heart could contribute to an increase in autophagy. This hypothesis needs to be directly evaluated in appropriate experimental models, and whether or not such an increase in autophagy is adaptive or maladaptive also needs to be established. Conversely, exposure of isolated cardiomyocytes to increased glucose *in vitro* promotes cell death. Incubation of H9C2 cells and neonatal cardiomyocytes with 28mM glucose promoted cell death that was associated with evidence of ER stress, oxidative stress, induction of MCP1 and induction of autophagy. ROS scavengers and inhibitors of ER stress attenuated cell death and autophagy. Moreover, inhibition of autophagy promoted cellular survival in cells cultured under high glucose conditions[90-91]. These observations would suggest that glucose toxicity might promote autophagy in cultured cardiomyocytes *in vitro*, and in this context the excessive autophagy would be maladaptive. It was recently reported that the transcription factor GATA4 inhibits doxorubicin-induced autophagy and cardiomyocyte death[92]. The same group had earlier reported that streptozotocin-induced diabetes reduced myocardial levels of GATA4, and that overexpression of GATA4 prevented the death of cardiomyocytes cultured under high glucose conditions[93]. Taken together, the available evidence provides strong evidence of multiple mechanisms that may modulate autophagic flux in the heart in context of diabetes. Future studies will be needed to determine which changes represent adaptive responses to the diabetic milieu that maintain cardiac contractility versus those that contribute to cardiac dysfunction.

In summary, we have reviewed evidence that suggests a role for autophagy in the pathogenesis of various diabetes complications. The study of autophagy as it relates to diabetes is in its nascent stages and we anticipate that in the near future, there will be a flood of new information that will further inform the field and will clarify the role of autophagy in the pathogenesis of diabetes complications where it may serve as a friend or foe.

References

1. Diagnosis and classification of diabetes mellitus. *Diabetes Care* 2006;**29**(Suppl. 1):S43–8.
2. Aroda VR, Ratner R. Approach to the patient with prediabetes. *J Clin Endocrinol Metab* 2008;**93**:3259–65.
3. van Belle TL, Coppieters KT, von Herrath MG. Type 1 diabetes: etiology, immunology, and therapeutic strategies. *Physiol Rev* 2011;**91**:79–118.
4. Chudyk A, Petrella RJ. Effects of Exercise on Cardiovascular Risk Factors in Type 2 Diabetes: A meta-analysis. *Diabetes Care* 2011;**34**:1228–37.
5. Goodyear LJ, Kahn BB. Exercise, glucose transport, and insulin sensitivity. *Ann Rev Med* 1998;**49**:235–61.
6 Reaven GM. The metabolic syndrome: time to get off the merry-go-round? *J Intern Med* 2011;**269**:127–36.
7. Cuspidi C, Valerio C, Giudici V, Negri F, Sala C, Zanchetti A, et al. Metabolic syndrome and multiple organ damage in essential hypertension. *Blood Press.* 2008;**17**:195–203.
8. Kawasaki R, Tielsch JM, Wang JJ, Wong TY, Mitchell P, Tano Y, et al. The metabolic syndrome and retinal microvascular signs in a Japanese population: the Funagata study. *Br J Ophthalmol* 2008;**92**:161–6.
9. Smith AG, Rose K, Singleton JR. Idiopathic neuropathy patients are at high risk for metabolic syndrome. *J Neurol Sci* 2008;**273**:25–8.
10. Levine B, Mizushima N, Virgin HW. Autophagy in immunity and inflammation. *Nature* 2011;**469**:323–35.
11. Hirshberg B, Rother KI, Digon 3rd BJ, Lee J, Gaglia JL, Hines K, et al. Benefits and risks of solitary islet transplantation for type 1 diabetes using steroid-sparing immunosuppression: the National Institutes of Health experience. *Diabetes Care* 2003;**26**:3288–95.
12. Rilo HL, Carroll PB, Zeng YJ, Fontes P, Demetris J, Ricordi C. Acceleration of chronic failure of intrahepatic canine islet autografts by a short course of prednisone. *Transplantation* 1994;**57**:181–7.
13. Zhang N, Su D, Qu S, Tse T, Bottino R, Balamurugan AN, et al. Sirolimus is associated with reduced islet engraftment and impaired beta-cell function. *Diabetes* 2006;**55**:2429–36.
14. Tanemura M, Saga A, Kawamoto K, Machida T, Deguchi T, Nishida T, et al. Rapamycin induces autophagy in islets: relevance in islet transplantation. *Transplant Proc* 2009;**41**:334–8.
15. Johnson JD, Ao Z, Ao P, Li H, Dai LJ, He Z, et al. Different effects of FK506, rapamycin, and mycophenolate mofetil on glucose-stimulated insulin release and apoptosis in human islets. *Cell Transplant* 2009;**18**:833–45.
16. Amherdt M, Harris V, Renold AE, Orci L, Unger RH. Hepatic autography in uncontrolled experimental diabetes and its relationships to insulin and glucagon. *J Clin Invest* 1974;**54**:188–93.
17. Lenk SE, Bhat D, Blakeney W, Dunn Jr WA. Effects of streptozotocin-induced diabetes on rough endoplasmic reticulum and lysosomes of rat liver. *Am J Physiol* 1992;**263**:E856–62.
18. Han K, Lehringer-Polzin M, Zhou H, Pfeifer U. Cellular autophagy in proximal tubules of early diabetic rats following insulin treatment and islet transplantation. *Virchows Arch B-Cell Pathol Incl Mol Pathol* 1992;**61**:367–73.
19. Zhou Z, Wu S, Li X, Xue Z, Tong J. Rapamycin induces autophagy and exacerbates metabolism associated complications in a mouse model of type 1 diabetes. *Indian J Exp Biol* 2010;**48**:31–8.
20. Jia K, Levine B. Autophagy and longevity: lessons from C. elegans. *Adv Exp Med Biol* 2010;**694**:47–60.
21. Zhang Y, Goldman S, Baerga R, Zhao Y, Komatsu M, Jin S. Adipose-specific deletion of autophagy-related gene 7 (atg7) in mice reveals a role in adipogenesis. *Proc Natl Acad Sci USA* 2009;**106**:19860–5.
22. Herman MA, Kahn BB. Glucose transport and sensing in the maintenance of glucose homeostasis and metabolic harmony. *J Clin Invest* 2006;**116**:1767–75.

23 Kroemer G, Marino G, Levine B. Autophagy and the integrated stress response. *Mol Cell* 2010;**40**:280–93.
24. Kapahi P, Chen D, Rogers AN, Katewa SD, Li PW, Thomas EL, et al. With TOR, less is more: a key role for the conserved nutrient-sensing TOR pathway in aging. *Cell Metab* 2010;**11**:453–65.
25. Kundu M. ULK1, Mammalian Target of Rapamycin, and Mitochondria: Linking Nutrient Availability and Autophagy. *Antioxid Redox Signal* 2011;**14**:1953–8.
26. Masiero E, Agatea L, Mammucari C, Blaauw B, Loro E, Komatsu M, et al. Autophagy is required to maintain muscle mass. *Cell Metab* 2009;**10**:507–15.
27. Matsuda F, Fujii J, Yoshida S. Autophagy induced by 2-deoxy-D-glucose suppresses intracellular multiplication of Legionella pneumophila in A/J mouse macrophages. *Autophagy* 2009;**5**:484–93.
28. Wang Q, Liang B, Shirwany NA, Zou MH. 2-Deoxy-D-glucose treatment of endothelial cells induces autophagy by reactive oxygen species-mediated activation of the AMP-activated protein kinase. *PLoS One* 2011;**6**:e17234.
29. Cheng J, Ohsaki Y, Tauchi-Sato K, Fujita A, Fujimoto T. Cholesterol depletion induces autophagy. *Biochem Biophys Res Commun* 2006;**351**:246–52.
30. Araki M, Motojima K. Hydrophobic statins induce autophagy in cultured human rhabdomyosarcoma cells. *Biochem Biophys Res Commun* 2008;**367**:462–7.
31. Parikh A, Childress C, Deitrick K, Lin Q, Rukstalis D, Yang W. Statin-induced autophagy by inhibition of geranylgeranyl biosynthesis in prostate cancer PC3 cells. *Prostate* 2010;**70**:971–81.
32. Papandreou I, Lim AL, Laderoute K, Denko NC. Hypoxia signals autophagy in tumor cells via AMPK activity, independent of HIF-1, BNIP3, and BNIP3L. *Cell Death Differ* 2008;**15**:1572–81.
33. Bellot G, Garcia-Medina R, Gounon P, Chiche J, Roux D, Pouyssegur J, et al. Hypoxia-induced autophagy is mediated through hypoxia-inducible factor induction of BNIP3 and BNIP3L via their BH3 domains. *Mol Cell Biol* 2009;**29**:2570–81.
34. Chiavarina B, Whitaker-Menezes D, Migneco G, Martinez-Outschoorn UE, Pavlides S, Howell A, et al. HIF1-alpha functions as a tumor promoter in cancer associated fibroblasts, and as a tumor suppressor in breast cancer cells: Autophagy drives compartment-specific oncogenesis. *Cell Cycle* 2010;**9**:3534–51.
35. Masini M, Bugliani M, Lupi R, del Guerra S, Boggi U, Filipponi F, et al. Autophagy in human type 2 diabetes pancreatic beta cells. *Diabetologia* 2009;**52**:1083–6.
36. Komiya K, Uchida T, Ueno T, Koike M, Abe H, Hirose T, et al. Free fatty acids stimulate autophagy in pancreatic beta-cells via JNK pathway. *Biochem Biophys Res Commun* 2010;**401**:561–7.
37. Sims K, Haynes CA, Kelly S, Allegood JC, Wang E, Momin A, et al. Kdo2-lipid A, a TLR4-specific agonist, induces de novo sphingolipid biosynthesis in RAW264.7 macrophages, which is essential for induction of autophagy. *J Biol Chem* 2010;**285**:38568–79.
38. Doyle A, Zhang G, Abdel Fattah EA, Eissa NT, Li YP. Toll-like receptor 4 mediates lipopolysaccharide-induced muscle catabolism via coordinate activation of ubiquitin-proteasome and autophagy-lysosome pathways. *FASEB J* 2011;**25**:99–110.
39. Egan DF, Shackelford DB, Mihaylova MM, Gelino S, Kohnz RA, Mair W, et al. Phosphorylation of ULK1 (hATG1) by AMP-activated protein kinase connects energy sensing to mitophagy. *Science* 2011;**331**:456–61.
40. Lee JW, Park S, Takahashi Y, Wang HG. The association of AMPK with ULK1 regulates autophagy. *PLoS One* 2010;**5**:e15394.
41. Kim J, Kundu M, Viollet B, Guan KL. AMPK and mTOR regulate autophagy through direct phosphorylation of Ulk1. *Nat Cell Biol* 2011;**13**:132–41.
42. Loffler AS, Alers S, Dieterle AM, Keppeler H, Franz-Wachtel M, Kundu M, et al. Ulk1-mediated phosphorylation of AMPK constitutes a negative regulatory feedback loop. *Autophagy* 2011;**7**.
43. Chiacchiera F, Matrone A, Ferrari E, Ingravallo G, Lo Sasso G, Murzilli S, et al. p38alpha blockade inhibits colorectal cancer growth in vivo by inducing a switch from HIF1alpha- to FoxO-dependent transcription. *Cell Death Differ* 2009;**16**:1203–14.
44. Marino G, Morselli E, Bennetzen MV, Eisenberg T, Megalou E, Schroeder S, et al. Longevity-relevant regulation of autophagy at the level of the acetylproteome. *Autophagy* 2011;**7**.
45. Lee IH, Cao L, Mostoslavsky R, Lombard DB, Liu J, Bruns NE, et al. A role for the NAD-dependent deacetylase Sirt1 in the regulation of autophagy. *Proc Natl Acad Sci USA* 2008;**105**:3374–9.
46. Wang F, Nguyen M, Qin FX, Tong Q. SIRT2 deacetylates FOXO3a in response to oxidative stress and caloric restriction. *Aging Cell* 2007;**6**:505–14.

47. Jing E, Gesta S, Kahn CR. SIRT2 regulates adipocyte differentiation through FoxO1 acetylation/deacetylation. *Cell Metab* 2007;**6**:105–14.
48. Zhao Y, Yang J, Liao W, Liu X, Zhang H, Wang S, et al. Cytosolic FoxO1 is essential for the induction of autophagy and tumour suppressor activity. *Nat Cell Biol* 2010;**12**:665–75.
49. Billings LK, Florez JC. The genetics of type 2 diabetes: what have we learned from GWAS? *Ann NY Acad Sci* 2010;**1212**:59–77.
50. Boura-Halfon S, Zick Y. Phosphorylation of IRS proteins, insulin action, and insulin resistance. *Am J Physiol Endocrinol Metab* 2009;**296**:E581–91.
51. Yecies JL, Manning BD. Transcriptional Control of Cellular Metabolism by mTOR Signaling. *Cancer Res* 2011;**71**:2815–20.
52. Shoelson SE, Lee J, Goldfine AB. Inflammation and insulin resistance. *J Clin Invest* 2006;**116**:1793–801.
53. Donath MY, Shoelson SE. Type 2 diabetes as an inflammatory disease. *Nat Rev Immunol* 2011;**11**:98–107.
54. Schnell AH, Swenne I, Borg LA. Lysosomes and pancreatic islet function. A quantitative estimation of crinophagy in the mouse pancreatic B-cell. *Cell Tissue Res* 1988;**252**:9–15.
55. Schnell Landstrom AH, Andersson A, Borg LA. Lysosomes and pancreatic islet function: adaptation of beta-cell lysosomes to various metabolic demands. *Metabolism* 1991;**40**:399–405.
56. Wang J, Chen Y, Yuan Q, Tang W, Zhang X, Osei K. Control of Precursor Maturation and Disposal Is an Early Regulative Mechanism in the Normal Insulin Production of Pancreatic beta-Cells. *PLoS One* 2011;**6**:e19446.
57. Oyadomari S, Araki E, Mori M. Endoplasmic reticulum stress-mediated apoptosis in pancreatic beta-cells. *Apoptosis* 2002;**7**:335–45.
58. Marchetti P, Bugliani M, Lupi R, Marselli L, Masini M, Boggi U, et al. The endoplasmic reticulum in pancreatic beta cells of type 2 diabetes patients. *Diabetologia* 2007;**50**:2486–94.
59. Harding HP, Zeng H, Zhang Y, Jungries R, Chung P, Plesken H, et al. Diabetes mellitus and exocrine pancreatic dysfunction in perk-/- mice reveals a role for translational control in secretory cell survival. *Mol Cell* 2001;**7**:1153–63.
60. Jung HS, Chung KW, Won Kim J, Kim J, Komatsu M, Tanaka K, et al. Loss of autophagy diminishes pancreatic beta cell mass and function with resultant hyperglycemia. *Cell Metab* 2008;**8**:318–24.
61. Ebato C, Uchida T, Arakawa M, Komatsu M, Ueno T, Komiya K, et al. Autophagy is important in islet homeostasis and compensatory increase of beta cell mass in response to high-fat diet. *Cell Metab* 2008;**8**:325–32.
62. Shintani T, Klionsky DJ. Autophagy in health and disease: a double-edged sword. *Science* 2004;**306**:990–5.
63. White E, DiPaola RS. The double-edged sword of autophagy modulation in cancer. *Clin Cancer Res* 2009;**15**:5308–16.
64. Singh R, Kaushik S, Wang Y, Xiang Y, Novak I, Komatsu M, et al. Autophagy regulates lipid metabolism. *Nature* 2009;**458**:1131–5.
65. Liu HY, Han J, Cao SY, Hong T, Zhuo D, Shi J, et al. Hepatic autophagy is suppressed in the presence of insulin resistance and hyperinsulinemia: inhibition of FoxO1-dependent expression of key autophagy genes by insulin. *J Biol Chem* 2009;**284**:31484–92.
66. Yang L, Li P, Fu S, Calay ES, Hotamisligil GS. Defective hepatic autophagy in obesity promotes ER stress and causes insulin resistance. *Cell Metab* 2010;**11**:467–78.
67. Momoi T. Conformational diseases and ER stress-mediated cell death: apoptotic cell death and autophagic cell death. *Curr Mol Med* 2006;**6**:111–8.
68. Ost A, Svensson K, Ruishalme I, Brannmark C, Franck N, Krook H, et al. Attenuated mTOR signaling and enhanced autophagy in adipocytes from obese patients with type 2 diabetes. *Mol Med* 2010;**16**:235–46.
69. Kovsan J, Bluher M, Tarnovscki T, Kloting N, Kirshtein B, Madar L, et al. Altered autophagy in human adipose tissues in obesity. *J Clin Endocrinol Metab* 2011;**96**:E268–77.
70. Kuma A, Hatano M, Matsui M, Yamamoto A, Nakaya H, Yoshimori T, et al. The role of autophagy during the early neonatal starvation period. *Nature* 2004;**432**:1032–6.
71. Komatsu M, Waguri S, Ueno T, Iwata J, Murata S, Tanida I, et al. Impairment of starvation-induced and constitutive autophagy in Atg7-deficient mice. *J Cell Biol* 2005;**169**:425–34.
72. Sou YS, Waguri S, Iwata J, Ueno T, Fujimura T, Hara T, et al. The Atg8 conjugation system is indispensable for proper development of autophagic isolation membranes in mice. *Mol Biol Cell* 2008;**19**:4762–75.

73. Singh R, Xiang Y, Wang Y, Baikati K, Cuervo AM, Luu YK, et al. Autophagy regulates adipose mass and differentiation in mice. *J Clin Invest* 2009;**119**:3329–39.
74. Baerga R, Zhang Y, Chen PH, Goldman S, Jin S. Targeted deletion of autophagy-related 5 (atg5) impairs adipogenesis in a cellular model and in mice. *Autophagy* 2009;**5**:1118–30.
75. Narici MV, Maffulli N. Sarcopenia: characteristics, mechanisms and functional significance. *Br Med Bull* 2010;**95**:139–59.
76. Lim S, Kim JH, Yoon JW, Kang SM, Choi SH, Park YJ, et al. Sarcopenic obesity: prevalence and association with metabolic syndrome in the Korean Longitudinal Study on Health and Aging (KLoSHA). *Diabetes Care* 2010;**33**:1652–4.
77. Fry CS, Rasmussen BB. Skeletal Muscle Protein Balance and Metabolism in the Elderly. *Curr Aging Sci*; 2011.
78. Koopman R. Dietary protein and exercise training in ageing. *Proc Nutr Soc* 2011;**70**:104–13.
79. Ogata T, Oishi Y, Higuchi M, Muraoka I. Fasting-related autophagic response in slow- and fast-twitch skeletal muscle. *Biochem Biophys Res Commun* 2010;**394**:136–40.
80. Zhao J, Brault JJ, Schild A, Cao P, Sandri M, Schiaffino S, et al. FoxO3 coordinately activates protein degradation by the autophagic/lysosomal and proteasomal pathways in atrophying muscle cells. *Cell Metab* 2007;**6**:472–83.
81. Mammucari C, Schiaffino S, Sandri M. Downstream of Akt: FoxO3 and mTOR in the regulation of autophagy in skeletal muscle. *Autophagy* 2008;**4**:524–6.
82. Masiero E, Sandri M. Autophagy inhibition induces atrophy and myopathy in adult skeletal muscles. *Autophagy* 2010;**6**:307–9.
83. Lefort N, Glancy B, Bowen B, Willis WT, Bailowitz Z, De Filippis EA, et al. Increased reactive oxygen species production and lower abundance of complex I subunits and carnitine palmitoyltransferase 1B protein despite normal mitochondrial respiration in insulin-resistant human skeletal muscle. *Diabetes* 2010;**59**:2444–52.
84. Krolewski AS, Warram JH. Epidemiology of late complications of diabetes: A basis for the development and evaluation of preventive program, *Lippincott*. New York: Williams & Wilkins; 2005.
85. Boudina S, Abel ED. Diabetic cardiomyopathy, causes and effects. *Rev Endocr Metab Disord* 2010;**11**:31–9.
86. Thorens B, Mueckler M. Glucose transporters in the 21st Century. *Am J Physiol. Endocrinol Metab* 2010;**298**:E141–5.
87. Brownlee M. Biochemistry and molecular cell biology of diabetic complications. *Nature* 2001;**414**:813–20.
88. Hariharan N, Maejima Y, Nakae J, Paik J, Depinho RA, Sadoshima J. Deacetylation of FoxO by Sirt1 Plays an Essential Role in Mediating Starvation-Induced Autophagy in Cardiac Myocytes. *Circ Res* 2010;**107**:1470–82.
89. Takagi H, Matsui Y, Hirotani S, Sakoda H, Asano T, Sadoshima J. AMPK mediates autophagy during myocardial ischemia in vivo. *Autophagy* 2007;**3**:405–7.
90. Younce CW, Kolattukudy PE. MCP-1 causes cardiomyoblast death via autophagy resulting from ER stress caused by oxidative stress generated by inducing a novel zinc-finger protein, MCPIP. *Biochem J* 2010;**426**:43–53.
91. Younce CW, Wang K, Kolattukudy PE. Hyperglycaemia-induced cardiomyocyte death is mediated via MCP-1 production and induction of a novel zinc-finger protein MCPIP. *Cardiovasc Res* 2010;**87**:665–74.
92. Kobayashi S, Volden P, Timm D, Mao K, Xu X, Liang Q. Transcription factor GATA4 inhibits doxorubicin-induced autophagy and cardiomyocyte death. *J Biol Chem* 2010;**285**:793–804.
93. Kobayashi S, Mao K, Zheng H, Wang X, Patterson C, O'Connell TD, et al. Diminished GATA4 protein levels contribute to hyperglycemia-induced cardiomyocyte injury. *J Biol Chem* 2007;**282**:21945–52.
94. Lo MC, Lu CI, Chen MH, Chen CD, Lee HM, Kao SH. Glycoxidative stress-induced mitophagy modulates mitochondrial fates. *Ann NY Acad Sci* 2010;**1201**:1–7.
95. Yang Z, Klionsky DJ. Mammalian autophagy: core molecular machinery and signaling regulation. *Curr Opin Cell Biol* 2010;**22**:124–31.
96. Juhasz G, Neufeld TP. Drosophila Atg7: required for stress resistance, longevity and neuronal homeostasis, but not for metamorphosis. *Autophagy* 2008;**4**:357–8.
97. Komatsu M, Kominami E, Tanaka K. Autophagy and neurodegeneration. *Autophagy* 2006;**2**:315–7.
98. Komatsu M, Waguri S, Chiba T, Murata S, Iwata J, Tanida I, et al. Loss of autophagy in the central nervous system causes neurodegeneration in mice. *Nature* 2006;**441**:880–4.
99. Komatsu M, Wang QJ, Holstein GR, Friedrich Jr VL, Iwata J, Kominami E, et al. Essential role for autophagy protein Atg7 in the maintenance of axonal homeostasis and the prevention of axonal degeneration. *Proc Natl Acad Sci USA* 2007;**104**:14489–94.

100. Akude E, Zherebitskaya E, Chowdhury SK, Smith DR, Dobrowsky RT, Fernyhough P. Diminished superoxide generation is associated with respiratory chain dysfunction and changes in the mitochondrial proteome of sensory neurons from diabetic rats. *Diabetes* 2011;**60**:288–97.
101. Chowdhury SK, Zherebitskaya E, Smith DR, Akude E, Chattopadhyay S, Jolivalt CG, et al. Mitochondrial respiratory chain dysfunction in dorsal root ganglia of streptozotocin-induced diabetic rats and its correction by insulin treatment. *Diabetes* 2010;**59**:1082–91.
102. Fernyhough P, Roy Chowdhury SK, Schmidt RE. Mitochondrial stress and the pathogenesis of diabetic neuropathy. *Expert Rev. Endocrinol Metab* 2010;**5**:39–49.
103. Vincent AM, Edwards JL, McLean LL, Hong Y, Cerri F, Lopez I, et al. Mitochondrial biogenesis and fission in axons in cell culture and animal models of diabetic neuropathy. *Acta Neuropathol* 2010;**120**:477–89.
104. Edwards JL, Quattrini A, Lentz SI, Figueroa-Romero C, Cerri F, Backus C, et al. Diabetes regulates mitochondrial biogenesis and fission in mouse neurons. *Diabetologia* 2010;**53**:160–9.
105. Towns R, Kabeya Y, Yoshimori T, Guo C, Shangguan Y, Hong S, et al. Sera from patients with type 2 diabetes and neuropathy induce autophagy and colocalization with mitochondria in SY5Y cells. *Autophagy* 2005;**1**:163–70.
106. Baek SH, Kim EK, Goudreau JL, Lookingland KJ, Kim SW, Yu SW. Insulin withdrawal-induced cell death in adult hippocampal neural stem cells as a model of autophagic cell death. *Autophagy* 2009;**5**:277–9.
107. Brosius 3rd FC, Alpers CE, Bottinger EP, Breyer MD, Coffman TM, Gurley SB, et al. Mouse models of diabetic nephropathy. *J Am Soc Nephrol* 2009;**20**:2503–12.
108. Brownlee M. Preventing kidney cell suicide. *Nat Med* 2007;**13**:1284–5.
109. Barbosa Ade Junior A, Zhou Hultenschmidt H, Hultenschmidt D, Totovic V, Jurilj N, Pfeifer U. Inhibition of cellular autophagy in proximal tubular cells of the kidney in streptozotocin-diabetic and uninephrectomized rats. *Virchows Arch B-Cell Pathol Incl Mol Pathol* 1992;**61**:359–66.
110. Pfeifer U, Warmuth-Metz M. Inhibition by insulin of cellular autophagy in proximal tubular cells of rat kidney. *Am J Physiol* 1983;**244**:E109–14.
111. Chen JK, Chen J, Thomas G, Kozma SC, Harris RC. S6 kinase 1 knockout inhibits uninephrectomy- or diabetes-induced renal hypertrophy. *Am J Physiol Renal Physiol* 2009;**297**:F585–93.
112. Lieberthal W, Levine JS. The role of the mammalian target of rapamycin (mTOR) in renal disease. *J Am Soc Nephrol* 2009;**20**:2493–502.
113. Lloberas N, Cruzado JM, Franquesa M, Herrero-Fresneda I, Torras J, Alperovich G, et al. Mammalian target of rapamycin pathway blockade slows progression of diabetic kidney disease in rats. *J Am Soc Nephrol* 2006;**17**:1395–404.
114. Yang Y, Wang J, Qin L, Shou Z, Zhao J, Wang H, et al. Rapamycin prevents early steps of the development of diabetic nephropathy in rats. *Am J Nephrol* 2007;**27**:495–502.
115. Takiyama Y, Harumi T, Watanabe J, Fujita Y, Honjo J, Shimizu N, et al. Tubular Injury in a Rat Model of Type 2 Diabetes Is Prevented by Metformin: A Possible Role of HIF-1α Expression and Oxygen Metabolism. *Diabetes* 2011;**60**:981–92.
116. Lee MJ, Feliers D, Mariappan MM, Sataranatarajan K, Mahimainathan L, Musi N, et al. A role for AMP-activated protein kinase in diabetes-induced renal hypertrophy. *Am J Physiol Renal Physiol* 2007;**292**:F617–27.
117. Isaka Y, Suzuki C, Abe T, Okumi M, Ichimaru N, Imamura R, et al. Bcl-2 protects tubular epithelial cells from ischemia/reperfusion injury by dual mechanisms. *Transplant Proc* 2009;**41**:52–4.
118. Huang Q, Wu YT, Tan HL, Ong CN, Shen HM. A novel function of poly(ADP-ribose) polymerase-1 in modulation of autophagy and necrosis under oxidative stress. *Cell Death Differ* 2009;**16**:264–77.
119. Abe Y, Sakairi T, Kajiyama H, Shrivastav S, Beeson C, Kopp JB. Bioenergetic characterization of mouse podocytes. *Am J Physiol Cell Physiol* 2010;**299**:C464–76.
120. Chacko BK, Reily C, Srivastava A, Johnson MS, Ye Y, Ulasova E, et al. Prevention of diabetic nephropathy in Ins2(+/)(AkitaJ) mice by the mitochondria-targeted therapy MitoQ. *Biochem J* 2010;**432**:9–19.
121. Goldberg IJ, Dansky HM. Diabetic vascular disease: an experimental objective. *Arterioscler Thromb Vasc Biol* 2006;**26**:1693–701.
122. Tabas I. Consequences and therapeutic implications of macrophage apoptosis in atherosclerosis: the importance of lesion stage and phagocytic efficiency. *Arterioscler Thromb Vasc Biol* 2005;**25**:2255–64.
123. Schrijvers DM, De Meyer GR, Herman AG, Martinet W. Phagocytosis in atherosclerosis: Molecular mechanisms and implications for plaque progression and stability. *Cardiovasc Res* 2007;**73**:470–80.

124. Gonzalez CD, Lee MS, Marchetti P, Pietropaolo M, Towns R, Vaccaro MI, et al. The emerging role of autophagy in the pathophysiology of diabetes mellitus. *Autophagy* 2011;**7**:2–11.
125. Jia G, Cheng G, Gangahar DM, Agrawal DK. Insulin-like growth factor-1 and TNF-alpha regulate autophagy through c-jun N-terminal kinase and Akt pathways in human atherosclerotic vascular smooth cells. *Immunol Cell Biol* 2006;**84**:448–54.
126. Deretic V. Multiple regulatory and effector roles of autophagy in immunity. *Curr Opin Immunol* 2009;**21**:53–62.
127. Jia G, Cheng G, Agrawal DK. Autophagy of vascular smooth muscle cells in atherosclerotic lesions. *Autophagy* 2007;**3**:63–4.
128. Nakai A, Yamaguchi O, Takeda T, Higuchi Y, Hikoso S, Taniike M, et al. The role of autophagy in cardiomyocytes in the basal state and in response to hemodynamic stress. *Nat Med* 2007;**13**:619–24.
129. Matsui Y, Takagi H, Qu X, Abdellatif M, Sakoda H, Asano T, et al. Distinct roles of autophagy in the heart during ischemia and reperfusion: roles of AMP-activated protein kinase and Beclin 1 in mediating autophagy. *Circ Res* 2007;**100**:914–22.
130. Takagi H, Matsui Y, Sadoshima J. The role of autophagy in mediating cell survival and death during ischemia and reperfusion in the heart. *Antioxid Redox Signal* 2007;**9**:1373–81.
131. Rothermel BA, Hill JA. Myocyte autophagy in heart disease: friend or foe? *Autophagy* 2007;**3**:632–4.
132. Rothermel BA, Hill JA. Autophagy in load-induced heart disease. *Circ Res* 2008;**103**:1363–9.
133. Zhu H, Tannous P, Johnstone JL, Kong Y, Shelton JM, Richardson JA, et al. Cardiac autophagy is a maladaptive response to hemodynamic stress. *J Clin Invest* 2007;**117**:1782–93.
134. Sengupta A, Molkentin JD, Yutzey KE. FoxO transcription factors promote autophagy in cardiomyocytes. *J Biol Chem* 2009;**284**:28319–31.
135. Vahtola E, Louhelainen M, Merasto S, Martonen E, Penttinen S, Aahos I, et al. Forkhead class O transcription factor 3a activation and Sirtuin1 overexpression in the hypertrophied myocardium of the diabetic Goto-Kakizaki rat. *J Hypertens* 2008;**26**:334–44.
136. Marambio P, Toro B, Sanhueza C, Troncoso R, Parra V, Verdejo H, et al. Glucose deprivation causes oxidative stress and stimulates aggresome formation and autophagy in cultured cardiac myocytes. *Biochim Biophys Acta* 2010;**1802**:509–18.
137. Boudina S, Abel ED. Diabetic cardiomyopathy revisited. *Circulation* 2007;**115**:3213–23.
138. Sharma S, Guthrie PH, Chan SS, Haq S, Taegtmeyer H. Glucose phosphorylation is required for insulin-dependent mTOR signalling in the heart. *Cardiovasc Res* 2007;**76**:71–80.

CHAPTER

9

Autophagy: An Endogenous Survival Mechanism and Cardioprotective Response to Ischemic Stress

Åsa B. Gustafsson[1] and Robert M. Mentzer, Jr.[2,3,4]

[1]Skaggs School of Pharmacy and Pharmaceutical Sciences, University of California, San Diego, La Jolla, CA, [2]Donald P. Shiley BioScience Center, San Diego State University, San Diego, CA, [3]WSU Cardiovascular Research Institute, [4]Departments of Physiology, Surgery, Wayne State University School of Medicine, Detroit MI

INTRODUCTION

Autophagy is an evolutionarily conserved catabolic process that involves degradation of cytoplasmic components[1]. In the setting of ischemia/reperfusion, it involves multiple proteins and signaling pathways and the machinery can be divided into four steps:

1. induction
2. assembly and formation of autophagosomes
3. docking and fusion with lysosomal membranes
4. degradation.

It is upregulated in response to nutrient deprivation to degrade cytoplasmic material in the cytosol and provides the cells with amino acids and fatty acids during the starvation period. Autophagy also plays an important role in cellular quality control where it degrades protein aggregates and damaged or dysfunctional organelles such as leaky mitochondria that can be harmful to the cell[2]. Conditions that activate the autophagy machinery and the autophagic steps that are involved in ischemia/reperfusion are depicted in Figure 9.1.

Likewise, the deregulation of autophagy has been associated with a wide variety of cardiovascular pathologies. For instance, Danon's disease is a fatal cardiomyopathy caused by a primary genetic deficiency in LAMP2 (lysosome-associated membrane protein 2)[3]. LAMP-2 is a major lysosomal membrane glycoprotein that is important for the fusion between autophagosomes and lysosomes. Thus, LAMP-2 deficiency causes an accumulation

Autophagy in Health and Disease
http://dx.doi.org/10.1016/B978-0-12-385101-7.00009-7

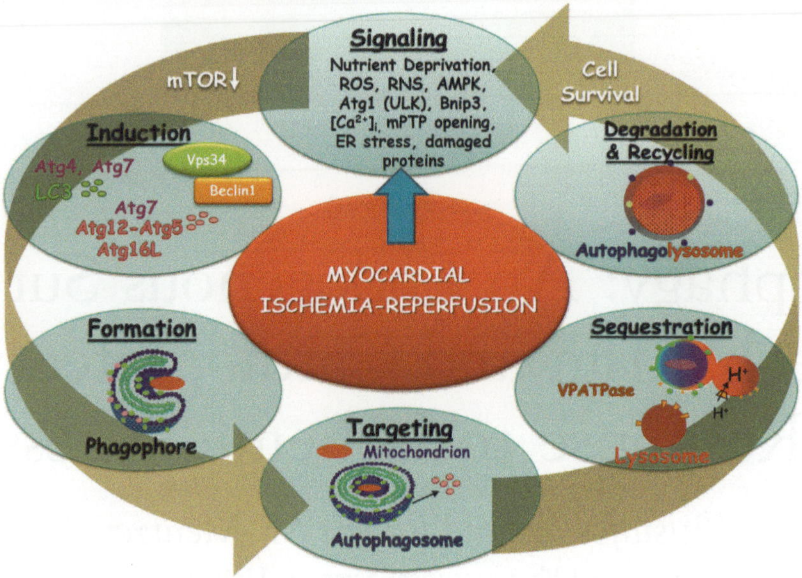

FIGURE 9.1 Cardiac Autophagy and Ischemia-Reperfusion—Adobe PhotoShop EPS fileTORC1=Target of rapamycin (TOR) complex-1; Vps34=The Class III phosphatidylinositol-3 kinase (P13K); mTOR=mammalian target of rapamycin; Bnip3=sub-family of Bcl-2 of proteins induced by hypoxia and acidosis; ROS=reactive oxygen species; RNS=reactive nitrogen species; mPTP=mitochondrial permeability transition pore; $[Ca2+]_i$ = intracellular calcium concentration; ER=endoplasmic reticulum. Induction of autophagy involves three key protein complexes: Atg1 (ULK), TORC1, and Class III PI3K (Vps34). Nutrient deprivation activates AMP-activated protein kinase (AMP) which phosphorylates and inhibits mTOR. Dephosphorylation of Atg13 increases the binding ability of Atg13 to Atg1. The association of Atg13 and Atg1 activates Atg1, a precursor of autophagosome formation. The Atg5 complex and microtubule-associated protein 1 light chain 3 (LC3) work together to initiate elongation and expansion of the isolation membrane (phagophore) that becomes the mature autophagosome. LC3 is first converted to LC3-I through cleavage by Atg4 and is then activated by Atg7 and converted to LC3-II. The phagophore encloses injured organelles and dysfunctional proteins with the ends fusing to form the double membrane autophagosome. LC3-II remains on the autophagosomal membrane. The Atg5 complex dissociates after membrane fusion. The class III PI3K complex including Beclin recruits other proteins to the autophagosomal membrane. Target engulfment and sequestration depend upon specific adaptor proteins including Parkin and p62. In the heart mitochondria are common targets. Degradation of autophagosomal contents by lysosomal hydrolases and export to the cytosol restores energy, reactivates mTOR, and downregulates autophagy.

of autophagosomes in the heart and is associated with the development of a hypertrophic cardiomyopathy[4]. Conditional deletion of Atg5, an essential autophagy protein, in the adult heart leads to disruption of autophagy with subsequent accumulation of mitochondria and development of cardiac dysfunction[5]. This suggests that autophagy is important in the normal turnover of mitochondria in the myocardium. Autophagy is also required in the neonatal period of nutrient deprivation after birth and Atg5-/- mice die during the neonatal period before feeding[6]. Increased autophagy is commonly observed in the heart with acute and chronic ischemia, heart failure and dilated cardiomyopathy[7-12]. Thus, while the functional role of autophagy is still controversial, most studies suggest that cardiac autophagy is a protective response activated by the cell as summarized in Table 9.1. This has given

TABLE 9.1 Upregulation of Autophagy, Ischemia/Reperfusion Injury (I/R), and Cardioprotection*

Reference (Year)	Model	Protocol	Markers/Methods Used to Show the Upregulation of Autophagy	Effect of Upregulation of Autophagy (Outcome)
CULTURED CELLS				
Hamacher-Brady et al (2006)[16]	HL-1 cells	Simulated ischemia / reperfusion (sI/R) with overexpression of Beclin-1 or Atg5; flux measured ± bafilomycin and pepstatin	% cells with numerous GFP-LC3 labeled puncta	↓ apoptosis; ↑autophagic flux
Dosenko et al (2006)[21]	Rat neonatal cardiomyocytes	Anoxia / reoxygenation (simulated IPC and IPoC) ± N-3-methyladenine	Monodansylcadaverine staining; electron microscopy	↓necrosis
Hamacher-Brady et al (2006)[27]	HL-1 cells	sI/R ± overexpression of Bnip3	GFP-LC3 vacuoles; green fluorescing nuclear dye (% YoPro1-positive cells)	↓cell death
Hamacher-Brady et al (2007)[17]	HL-1 cells	sI/R ± Bnip3	GFP-LC3 fluorescence or punctate GFP-LC3 pattern	↓cell death
Yitzhaki et al (2009)[78]	HL-1 cells; rat adult cardiomyocytes	sI/R; simulated ischemia ± CCPA	% cells with GFP-LC3 puncta	↓cell death and LDH release
Gurusamy et al (2009)[84]	H9c2 cardiac myoblast cell line	Hypoxia / reoxygenation (H/R) ± Wortmanin; BAG-1 siRNA	LC3-II; Beclin-1	↓apoptosis and LDH release
Gurusamy et al (2010)[74]	H9c2 cardiac myoblast cell line	H/R ± resveratrol	Monodansylcadaverine staining; LC3-II / LC3-I	Increased cell survival; ↓ apoptosis; ↓LDH release

(*Continued*)

TABLE 9.1 Upregulation of Autophagy, Ischemia/Reperfusion Injury (I/R), and Cardioprotection*—cont'd

Reference (Year)	Model	Protocol	Markers/Methods Used to Show the Upregulation of Autophagy	Effect of Upregulation of Autophagy (Outcome)
ISOLATED HEARTS				
Sybers et al (1976)[88]	Fetal mouse heart	Oxygen + glucose vs. deprivation / resupply	Electron microscopy	Damaged organelles contained in vacuoles—plays role in repair of sublethal injury
Decker et al (1980)[89]	Rabbit	Hypoxia vs. hypoxia and recovery	Electron microscopy	↓cell injury and death
Decker et al (1980)[90]	Rabbit	Hypoxia vs. hypoxia and recovery	Electron microscopy	↓cell injury and death; ↑ cathepsin D release
Gurusamy et al (2009)[84]	Rat	Global I/ R ± Wortmanin; BAG-1 siRNA	LC3-II and Beclin-1	↓infarct size; improved cardiac function; decreased LDH release
Qian et al (2009)[75]	Mouse	Global I/R ± $Hsp20^{S16A}$ overexpression	LC3-II / LC3-I; Beclin-1	Improved cardiac function; decreased LDH release; less apoptosis
Lekli et al (2010)[91]	Rat	Global I/R ± resveratrol and gamma-tocotrienol	Beclin-1; LC3-II; mTOR signaling	Reduced infarct size; improved cardiac function; less apoptosis
Gurusamy et al (2010)[74]	Rat	Global I/R ± resveratrol	LC3-II / LC3-I; Beclin-1; electron microscopy	Improved cardiac function; decreased apoptosis
Huang et al (2010)[77]	Rat	Global I/R ± sulfaphenazole and chelerythrine	LC3-II / LC3-I; % cells with GFP-LC3 puncta	Improved cardiac function; ↓infarct side; ↓CK release
Huang et al (2010)[82]	Rat	Global I/R ± IPC and Tat $Atg5^{K130R}$	% cells with GFP-LC3 puncta; p62 / COXIII	Inhibition of autophagy reduces cardioprotection
Petrovski et al (2011)[92]	Rat	Global I/R ± tunicamycin; thapsigargin	Electron microscopy; LC3-II/LC3-I and Beclin-1	Activation of autophagy by mild ER stress reduced I/R injury

IN-VIVO				
Yan et al (2005)[9]	Pig	Repetitive ischemia / reperfusion; various episodes	Cathepsin D; Hsc73; Beclin-1; LC3-II / LC3-I; electron microscopy	Improved cardiac function; ↓cell death; ↓apoptosis
Matsui et al (2007)[11]	Mouse	Coronary artery occlusion (CAO) (ischemia only) ± 3-methyladenine	LC3-II / LC3-I; % cells with GFP-LC3 puncta; microscopy	↓infarct size (ischemic phase)
May et al (2008)[93]	Mouse (transgenic)	Chronic ischemia and hibernation ± VEGF blockade	Electron microscopy	↑ cell survival
Buss et al (2009)[59]	Rat	CAO ± everolimus	Monodansylcadaverine staining; LC3-II / LC3-I	↓infarct size and remodeling; improved cardiac function
Yan et al (2009)[83]	Pig	Hibernating myocardium (CAO/R) ± IPC	Cathepsin B; Beclin-1; LC3-II /LC3-I	↓infarct size
Sala-Mercado et al (2010)[31]	Pig	I/R ± chloramphenicol	LC3-II; Beclin-1	↓infarct size
Kanamori et al (2011)[54]	Mouse	CAO ± bafilomycin A1 Starvation prior to CAO	Electron microscopy, LC3-II / LC3-I & p62 levels; GFP-LC3 puncta; microscopy	↑cell death and infarct size with bafilomycin A1 ↓infarct size with starvation prior to CAO
Kanamori et al (2011)[52]	Mouse	CAO ± bafilomycin A1 CAO ± rapamycin	Electron microscopy, LC3-II, p62, cathepsin D	Increased cardiac dysfunction and remodeling with bafilomycin; rapamycin reduced remodeling
Zhai et al (2011)[94]	Mouse	CAO and CAO/R in GSK-3β DN TG mice	% cells with GFP-LC3 puncta; p62	Inhibition of autophagy in GSK-3β DN TG mice ↑ischemic injury

**Adapted from Dong et al.*[95] *CCPA=2-chloro-N(6)-cyclopentyladenosine; H/R=hypoxia/reoxygenation; CAO/R=coronary artery occlusion/reperfusion; IPC=ischemic preconditioning; IPoC=ischemic postconditioning.*

rise to the concept that interventions that target autophagy may represent new therapeutic modalities for the treatment of patients with heart disease.

UPREGULATION OF AUTOPHAGY IN MYOCARDIAL ISCHEMIA/ REPERFUSION (I/R)

Studies have shown that autophagy is rapidly upregulated in cardiac cells both *in vitro* and *in vivo* in response to stress such as starvation[10, 11, 13-15], oxygen deprivation[11, 12, 16, 17] and oxidative stress[18, 19]. Most of these studies found that autophagy is upregulated as an adaptive response to prevent death of the cells. For instance, Dosenko *et al* found that anoxia/ reoxygenation caused an upregulation in autophagy in isolated neonatal myocytes[20] and that inhibition of autophagy with 3-methyladenine caused an increase in cell death[21]. Similarly, simulated I/R caused a rapid activation of autophagy in HL-1 cells[16, 17], an atrial-derived cardiac cell line, and enhancing autophagy constituted protection against sI/R-mediated cell death[16].

Recently the Bcl-2 family of proteins has been shown to play a role in regulating autophagy. The Bcl-2 family proteins are important regulators of apoptosis[22]. Anti-apoptotic Bcl-2 and Bcl-X_L have been reported to inhibit autophagy by sequestering Beclin-1,[23] whereas certain pro-apoptotic members can activate autophagy by disrupting the interaction between Beclin-1 and Bcl-2/Bcl-X_L[24, 25]. Bnip3 is a pro-apoptotic Bcl-2 protein that contributes to I/R injury by perturbing mitochondrial function which causes upregulation of autophagy in cardiac cells[17, 26]. Upregulation of autophagy in cardiac myocytes was shown to serve as a protective response against Bnip3-mediated injury by promoting removal of damaged mitochondria[17, 27, 28]. Interestingly, the activation of autophagy by Bnip3 and its homologue occurs independently of activation of the apoptotic pathway[29, 30].

In addition, several studies have demonstrated a protective role for autophagy in vivo. It has been shown that autophagy is induced during ischemia and is further increased during reperfusion in the mouse heart[11, 12]. In vivo studies have confirmed that inhibition or reduction of autophagy by genetic or pharmacologic means result in increased susceptibility to I/R injury and that enhancing autophagy can protect against I/R injury. For instance, Sala-Mercado *et al* recently reported that administration of chloramphenicol prior to or during coronary ligation resulted in rapid upregulation of autophagy and reduced I/R injury[31]. Studies showing a relationship between the upregulation of autophagy and protection against I/R injury in isolated cells, explanted hearts, and in vivo are summarized in Table 9.1. Interestingly, there is evidence that the endoplasmic reticulum (ER) is involved. The ER is responsible for proper folding of newly synthesized proteins and calcium storage[32]. Disturbances of calcium homeostasis in the ER lumen or disturbance of ER function can cause activation of the ER stress pathway and induction of autophagy[33]. Petrovski *et al.* found that treatment of hearts with low doses of tunicamycin or thapsigargin, two different inducers of ER stress, resulted in induction of autophagy which correlated with reduction in apoptosis and infarct size[34].

The mammalian target of rapamycin (mTOR) is a central regulator of cell growth and plays a key role at the interface of the pathways that coordinately regulate the balance between cell growth and autophagy in response to nutritional status, growth factor and stress signals[35].

Moreover, the mTOR pathway is a negative regulator of autophagy in mammalian cells[36, 37] and an important regulator of essential cellular functions during ischemia[38]. For instance, Khan *et al* reported that perfusion of isolated hearts with rapamycin, an inhibitor of mTOR and activator of autophagy, prior to I/R reduced infarct size[39]. The mTOR pathway is regulated by the 5′-AMP-activated protein kinase (AMPK). AMPK is activated during stress such as nutrient deprivation when there are changes in the ATP/AMP ratio. AMPK inhibits mTOR through phosphorylation of TRC2[40]. The AMPK-mTOR pathway is an important regulator of autophagy in response to glucose deprivation in neonatal myocytes and inhibition of AMPK reduced autophagy and increased cell death in neonatal myocytes[11]. Also, the same study found that transgenic mice with cardiac specific expression of a dominant negative AMPK had attenuated induction of autophagy in response to ischemia in vivo. This suggests that AMPK regulation of autophagy via mTOR plays an important role in the adaptation to ischemia.

Another regulator of the mTOR pathway is glycogen synthase kinase-3 (GSK-3). GSK-3 is ubiquitously expressed and functions as an important signaling molecule in several pathways. It is involved in gene transcription, protein translation, and apoptosis, as well as hexose metabolism[41]. GSK-3β is known to function downstream of phosphatidylinositol 3-kinase (PI3K) and Akt[42] and recently the TSC/mTOR signaling pathway has been shown to be downstream of GSK-3β[43]. Similar to AMPK, GSK-3β has been reported to inhibit mTOR signaling via phosphorylation of TSC2[44]. In the heart, inhibition of GSK-3β has been reported to be cardioprotective[12, 42, 45, 46] and recently, Zhai *et al* discovered that GSK-3β is a critical upstream regulator of mTOR during both ischemia and reperfusion in the heart[12]. They reported that suppression of GSK-3β resulted in enhanced injury in response to prolonged ischemia and that the increased injury was prevented in the presence of the mTOR inhibitor rapamycin[12]. In addition, they found that TSC2 heterozygous mice had increased mTOR signaling and reduced autophagy. Infarct size was also greater after prolonged ischemia. These findings suggest that inhibition of mTOR is a protective response and is associated with less injury.

MALADAPTIVE ROLE OF AUTOPHAGY IN MYOCARDIAL I/R

Although most studies suggest that autophagy is a protective response activated during I/R, a few studies have reported that autophagy can also be a detrimental process. For instance, Matsui *et al* found that activation of autophagy during reperfusion after ischemia was detrimental to myocytes[11]. They also found that Beclin-1 expression was dramatically increased during reperfusion and that Beclin-1+/- mice had reduced levels of autophagy and significantly smaller infarcts compared to wild type mice after I/R. This suggests that Beclin-1-mediated autophagy during reperfusion plays a detrimental role. Recently, Zhai *et al* reported that activation of GSK3β decreased mTOR activity during I/R[12] and that inhibition of GSK-3β during reperfusion reduced myocardial injury. This protective effect was reversed in the presence of the mTOR inhibitor rapamycin suggesting that enhanced autophagy during reperfusion contributes to injury[12]. Moreover, TSC2 heterozygous mice have increased mTOR activation and reduced autophagy and these mice are protected against I/R injury[12]. Thus, these results suggest that activation of autophagy via mTOR inhibition or Beclin-1 is detrimental during reperfusion. It is possible that the duration and levels of autophagy plays an important role in determining whether autophagy will be protective or detrimental to the cell in the heart. Certainly it is possible that excessive autophagy could lead to the removal of too many essential organelles and therefore contribute to the development of pathological conditions.

INDUCTION OF AUTOPHAGY BY FOXO TRANSCRIPTION FACTORS IN MYOCARDIAL I/R

The FoxO (Forkhead box-containing protein, O subfamily) transcription factors regulate many diverse cellular functions such as differentiation, metabolism and proliferation by modulating gene expression in cells[47]. FoxO1 and FoxO3 are highly expressed in the heart[48]. Recently, they have been identified as important regulators of autophagy in the heart by activating genes that are involved in autophagosome formation[49]. For instance, cardiac-specific transgenic mice overexpressing FoxO3 resulted in enhanced autophagy and cardiac atrophy[50], whereas FoxO3 deficiency results in cardiac hypertrophy[51]. Sengupta *et al* found that glucose deprivation in cardiac myocytes caused dephosphorylation and translocation of FoxO1 and FoxO3 to the nucleus where they activated expression of genes involved in autophagy[49]. Interestingly, the same group also found that FoxO3, but not FoxO1, was involved in activation of autophagy during I/R[49]. In contrast, Harihanan *et al* found that FoxO1 is important for glucose deprivation-induced autophagy[13]. Clearly, FoxO1 and FoxO3 regulate autophagy and cell size in cardiac myocytes in response to stress. To what degree and how these transcription factors interact to protect the stressed heart remain to be elucidated.

ROLE OF AUTOPHAGY IN ACUTE MYOCARDIAL INFARCTION

During a myocardial infarction (MI), the heart is subjected to sudden ischemia due to interruption of blood flow. The lack of oxygen in the infarcted area leads to rapid loss of cardiac myocytes. There is now evidence that autophagy plays a critical role in limiting the extent of infarct size by salvaging myocytes after a myocardial infarction in vivo. Several studies have found that autophagy is rapidly upregulated in the region bordering the infarction[52-54]. For instance, Kanamori *et al* found that autophagy was upregulated as early as 30 minutes after coronary ligation in the ischemic region of the heart with maximal autophagic activity occurring at four hours. Also, inhibition of autophagy with Bafilomycin A1 immediately prior to coronary ligation resulted in increased serum levels of troponin I and enhanced infarct. Interestingly, starvation of mice for 24 hours prior to MI caused a further enhancement of autophagy after MI and reduced infarct size compared to normally fed mice[54]. The study also found that amino acid and ATP levels were significantly reduced four hours post-infarction but that hearts of mice subjected to starvation prior to MI had preserved amino acid and ATP levels four hours post-MI. Bafilomycin A1 abolished the protective effect of starvation and augmented the reduction in ATP levels four hours post-MI. This suggests that autophagy plays an important role in salvaging myocytes by preserving amino acid and ATP levels during acute MI. Thus, activation of autophagy during MI preserves ATP levels and helps maintain an adequate amino acid pool which can be directly used as an energy source. Autophagy also plays an important role in removing damaged or dysfunctional organelles such as mitochondria[2]. Since MI is associated with increased damage to mitochondria[55], it is also very likely that the enhanced autophagy plays an important role in removing damaged and potentially harmful mitochondria in myocytes in the border zone.

ROLE OF AUTOPHAGY IN CHRONIC MYOCARDIAL INFARCTION

One of the most important factors for improving prognosis after MI is the attenuation of adverse myocardial remodeling. After MI, cardiac fibroblast hyperplasia and interstitial fibrosis in the infarcted area can have a negative impact on the heart. Increased ventricular wall thickness due to interstitial fibrosis and a reduced number of myocytes supporting contraction contribute to impaired cardiac function[56, 57]. Also, fibroblasts and excess extracellular matrix proteins cause morphologic separation of myocytes and may impair the electrical coupling of myocytes.[58] Clearly, the remodeling can have profound impact on myocyte function. Several recent studies suggest that autophagy plays a role in chronic MI to limit post-infarct remodeling. Autophagy was found to be enhanced in surviving myocytes in the area bordering the infarct at one and three weeks post-infarct[52]. This study also found that inhibition of autophagy with bafilomycin A1, two weeks after the infarction, significantly increased post-infarct remodeling and cardiac dysfunction, whereas enhancement of autophagy with rapamycin had the opposite effect[52]. In addition, Buss *et al* reported that targeting the mTOR pathway prevented adverse remodeling and reduced infarct size after MI. They found that treatment of rats with the mTOR inhibitor everolimus (RAD) starting three days post-MI enhanced autophagy in the border zone of the infarct, attenuated post-MI remodeling and improved LV function 28 days post-MI[59]. Interestingly, improvement of LV function was observed three months after MI even when RAD treatment was discontinued after one month.

Expansion of the infarct in a permanent occlusion is usually complete by ~three hours in rats and ~six hours in humans. At this time, reperfusion therapy is ineffective, albeit it may still have a salutary effect on remodeling. Fibroblasts play a key role in remodeling and are responsible for the production of extracellular matrix proteins after MI[60]. There is evidence that autophagy is also important in the regulation of fibroblast activation and collagen production. Therefore, it is very possible that the effect of mTOR is on both myocytes and fibroblasts. Dalla Costa *et al* reported that inhibition of mTOR with rapamycin inhibited stretch-induced proliferation and activation of cardiac fibroblasts[61]. Aranguiz-Urroz *et al* found that increased autophagy in cardiac fibroblasts correlated with increased degradation of intracellular collagen[62]. This suggests that autophagy might also be involved in reducing fibrosis by regulation of the extracellular matrix production in cardiac fibroblasts.

AUTOPHAGY, ISCHEMIC CONDITIONING, AND MYOCARDIAL PROTECTION

In patients, the term myocardial protection refers to mechanical and pharmacological strategies used to increase the heart's tolerance to I/R injury. In patients undergoing heart surgery, the goal is to prevent or ameliorate reversible postischemic ventricular dysfunction (stunning) and irreversible perioperative necrosis (infarction)[63, 64]. With the patient on cardiopulmonary bypass, this usually involves cross clamping the aorta and arresting the heart with a cold hyperkalemic solution. The latter is designed to increase the heart's tolerance to ischemia and simultaneously allows the surgeon to operate in a quiet field. The heart, however, is at risk of ischemic damage while the aorta is clamped and reperfusion injury when the aorta is unclamped and blood flow to the heart restored. In patients with an acute myocardial infarction (AMI) the objective is somewhat different. The goal here is to reduce infarct size after coronary artery blood flow has been restored by way of percutaneous coronary interventions (PCI) or thrombolysis[65]. Reperfusion itself, whether it is at the end of a heart operation or after successful PCI for an AMI ironically contributes to reversible or irreversible injury with resultant cellular damage, loss of contractile function, accelerated apoptosis and necrosis[66, 67].

In an effort to develop effective adjuvant therapies to ameliorate ischemia/reperfusion injury, considerable investigative effort has been undertaken to elucidate the mechanisms underlying the protective effects of ischemic preconditioning (IPC) and postconditioning (IPoC)[66, 68, 69]. Ischemic preconditioning is a phenomenon whereby brief periods of ischemia and reperfusion prior to a prolonged ischemic event results in a striking reduction in infarct size. IPoC confers protection when short cycles of ischemia and reperfusion are applied immediately after the sustained ischemic event. These two powerful interventions have been shown to consistently and markedly reduce infarct size in a variety of species and animal preparations including mouse, rat, rabbit, dog, pigs, and primates[70, 71]. The anti-infarct effectiveness in humans, however, has been difficult to measure. For these reasons, many mechanisms and signaling pathways have been studied and implicated including: formation and release of several autacoids and cytokines; maintenance of acidosis during early reperfusion; activation of protein kinases; preservation of mitochondrial function, and the attenuation of opening of the mitochondrial permeability transition pore (MPTP)[65,72].

Unfortunately, while we now have a much better understanding of the heart's endogenous mechanisms of protection, we still lack an effective therapy to prevent or ameliorate ischemia/reperfusion injury in humans and/or its consequences, namely adverse postinfarction remodeling and heart failure. This is due, in part, to preclinical studies that have relied on findings conducted in disease-free animals, and we have yet to identify all the factors that play a role in mediating IPC and IPoC[63, 73]. One potentially important component of conditioning that has recently come under intense investigation is autophagy. There is now evidence in several isolated cells, ex vivo studies, and intact animals that a number of pharmacologic agents that mimic IPC also induce autophagy, and under unique conditions, inhibition of autophagy may result in loss of the protection normally conferred by IPC[5, 39, 74-79].

One of the earliest reports to implicate a relationship between autophagy, IPC and IPoC was by Dosenko *et al*[21]. The rationale for these experiments was based on an earlier finding that proteasome inhibitors abolished the effects of preconditioning and postconditioning in isolated cardiomyocytes[80]. In this study, isolated neonatal cardiomyocytes were exposed to three cycles of three minutes of anoxia followed by three minutes of reoxygenation before a sustained 30-minute period of anoxia. Postconditioning was performed by exposing the cells to three cycles of one minute of reoxygenation followed by one minute of anoxia after the sustained anoxia. Pharmacologic conditioning was studied using inhibitors of autophagy and apoptosis. Autophagy was assessed by monodansylcadaverine staining of vacuolar structures and electron microscopy. The salient finding was that inhibition of autophagy was associated with an increase in the necrotic cell population during anoxia-reoxygenation and by implication may play a role in IPC and IPoC.

To more directly address the role of autophagy in IPC, Yitzhaki *et al* studied the effects of the A_1 adenosine receptor agonist 2-chloro-N(6)-cyclopentyladenosine (CCPA) on autophagy and cell survival following sI/R in GFP-LC3-infected HL-1 cells, neonatal rat cardiomyocytes, and intact αMHC-mCherry-LC3 transgenic mice[78]. The rationale was based, in part, on the recognition that activation of specific adenosine receptors are involved in triggering IPC[81]. Normoxic HL-1 cells were pretreated with CCPA and found to induce autophagy in a dose-dependent manner; it caused a sustained increase in the number of autophagosomes within 10 minutes of treatment. Similar observations were made in neonatal rat cardiomyocytes, adult cardiomyocytes, and in the hearts of intact mCherry-LC3 transgenic mice. When CCPA HL-1 cells were subjected to simulated I/R, CCPA and the A_1 adenosine receptor antagonist, DPCPX, both cytoprotection and the extent of autophagy were blocked. Likewise the cytoprotective effect of CCPA was blocked in HL-1 cells transfected with the dominant negative inhibitor of autophagy, $Atg5^{K130R}$. Similar results were obtained in adult cardiomyocytes when a cell-permeable recombinant Tat-$Atg5^{K130R}$ was used. These findings provided, for the first time, direct evidence that the cardioprotective properties of CCPA were associated with the induction of autophagy and autophagy might represent an important component of conditioning.

To further evaluate the relationship between autophagy and IPC, Huang *et al* examined the effect of sulfaphenazole on protein kinase C (PKC) translocation. Sulfaphenazole, cimetidine, and chloramphenicol are known inhibitors of cytochrome-P450 enzymes whose mechanisms of cardioprotection are largely unknown[77]. PKC is a family of related isoenzymes with differences in activity, localization, and substrate specificity. Although there is controversy as to which isozyme is responsible for the protective signal, considerable data exist which indicate PKC translocation plays an important role in IPC. In this study, the investigators first examined whether the protection afforded by sulfaphenazole was associated with cardiac autophagy. Isolated perfused rat hearts were exposed to the drug and the distribution of autophagosomes assessed by immunostaining. Western blot was used to detect LC3-I, LC3-II and quantify the LC3II/LC3-I ratio. Autophagosomes were also quantified (number of mCherry-LC3 puncta) in intact mice hearts exposed to sulfaphenazole. In all three preparations—adult rat cardiomyocytes, isolated perfused rat hearts, and the intact mouse—sulfaphenazole treatment was associated with an upregulation of autophagy. In the Langendorff preparation, protection was manifest as enhanced recovery of function, reduced creatine kinase release, and a decrease in infarct size. When the cell-permeable dominant negative inhibitor of autophagy, Tat-$Atg5^{K130R}$ was administered, autophagy and the infarct sparing effect of sulfaphenazole were abolished, indicating that the cardioprotective effect of the drug was mediated, in part, by upregulation of autophagy. The investigators then examined the effect of drug treatment on the redistribution of PKC-ε and PKC-δ. Specifically, they wanted to determine whether the protection afforded by the drug was associated with PKC translocation. While PKC-ε failed to show a consistent pattern of change, PKC-δ decreased in the cytosol and increased in the particulate fraction in isolated adult rat cardiomyocytes and the Langendorff-perfused rat hearts.

Thus, it appeared that sulfaphenazole not only upregulated autophagy but it also triggered PKC-δ translocation, a key component of the signaling pathways involved in IPC. Inhibition of PKC with chelerythrine, a potent selective PKC inhibitor, blocked the activation of autophagy. Therefore, the pharmacologic conditioning mediated by sulfaphenazole appears to be conferred in part by the up-regulation of autophagy and involves a PKC-dependent pathway not unlike IPC.

To better understand whether autophagy is necessary for cardioprotection conferred by IPC in mCherry-LC3 transgenic mice and Langendorff-perfused hearts, IPC was induced. The number of mCherry-LC3-labeled autophagosomes in the area of risk was measured in the transgenic mice. Cadaverine-dye binding and Western blot for p62 was used in the Langendorff hearts. The p62 protein, also known as sequestosome 1 (SQSTM1), is believed to link ubiquitinated proteins to autophagic machinery and enable their degradation within the lysosome[28]. In the transgenic mice, IPC resulted in increased levels of mCherry puncta consistent with upregulation of autophagy. IPC was associated with increased cadaverine dye-binding and an increase in p62 in the Langendorff model. When the specific autophagy inhibitor, Tat-Atg5^{K130R}, was infused prior to the IPC protocol, however, IPC was significantly less effective and the infarcts were larger. In addition to IPC, three structurally unrelated drugs known to be acutely cardioprotective and with different mechanisms of action were examined. All three cardioprotective agents, uridine-5′-triphosphate (UDP), diazoxide, and ranolazine were shown to upregulate autophagy, and their protective effects were significantly attenuated when autophagy was blocked[82]. These findings are consistent with others that the protection conferred by both IPC and pharmacologic preconditioning are associated with an upregulation in autophagy[83-85].

Recently, Huang *et al* reported that autophagy involving the selective degradation of mitochondria (mitophagy), may play a key role in IPC and is mediated by the proteins Parkin and p62[86]. Parkin is an E3 ubiquitin ligase which has been shown to be essential for the selective autophagic clearance of damaged mitochondria. The p62 protein recognizes toxic cellular waste which is scavenged by autophagy[67, 87]. While Parkin has been reported to mediate the targeting of proteins for degradation in the brain, very little is known about Parkin in the heart. In this study, the investigators found basal levels of Parkin present in the cytosol of HL-1 cells and neonatal rat cardiomyocytes. In response to the stress of simulated ischemia, however, endogenous Parkin translocated to mitochondria. A similar redistribution of Parkin to mitochondria was observed in preconditioned Langendorff hearts. When Parkin knockout mice were subjected to IPC followed by 20 minutes ischemia and 22 hours reperfusion, Parkin ablation abolished IPC-induced cardioprotection, thus revealing for the first time that Parkin translocation may play an important role in IPC. Based on previous observations that p62 translocation similarly occurs in response to simulated ischemia[2], p62 was also assessed in wild-type and Parkin knockout mice. After IPC, p62 translocation to the mitochondria increased in the wild type but not in the Parkin knockout mice. Taken together these findings suggest that Parkin is essential for mitochondrial p62 translocation in response to IPC and the selective quality control of mitochondria by way of Parkin and p62 play an important role in IPC.

In contrast to the acute studies in isolated cells and rodent models, much less is known about the role in autophagy in myocardial I/R in vivo chronic preparations. In 2005, Yan *et al* reported that repetitive myocardial ischemia in chronically instrumented pigs produced by one, two, or six episodes of 90 minutes coronary stenosis was associated with an increase in the expression of cathepsin D, cathepsin B, Hsc73, Beclin-1, and LC3, all markers of autophagy activation. They concluded that under these conditions ischemia-induced autophagy is an adaptive or protective mechanism.[9] Based on these findings, they later examined the role of autophagy in swine models of preconditioning that involved repetitive coronary occlusion (RCO) or repetitive coronary stenosis (RCS).[82,83] In the RCO model, IPC was induced by six episodes of 90 minutes coronary stenosis12 hours apart; in the RCS model, IPC was induced by six episodes of 2x10 minutes coronary occlusion 12 hours apart. Both models provided IPC protection as effective as the classical models of first and second window of preconditioning, i.e., a reduction in infarct size by 60%. The repetitive ischemia models, however, resulted in chronic myocardial stunning reflective of a hibernating myocardium.[83] In this study, when they compared their repetitive ischemia models with traditional first and second window IPC models, they found that autophagy was only activated in the two repetitive ischemia models. They concluded that autophagy is an important mechanism underlying preconditioning induced by repetitive episodes of ischemia/reperfusion in chronically ischemic hearts reflective of hibernating myocardium. Thus, it appears that in chronically ischemic hearts with hibernating myocardium, autophagy also plays a role in IPC.

CONCLUSION

There is increasing evidence that cardiac autophagy is an endogenous cellular process that protects the heart against acute and chronic ischemic injury. In response to an ischemic event, isolated cytosolic membranes or autophagosomes form, which in turn sequester aggregate-prone proteins and damaged organelles including mitochondria. The autophagosomes transport their cargo to lysosomes where they fuse to form autophagolysosomes and their contents are degraded. As a result, key nutrients such as amino acids and fatty acids are released and recycled as building blocks and metabolic fuel and the cardiomyocyte is repopulated with robust mitochondria. In this manner, cardiomyocytes at risk for lethal ischemia/reperfusion injury can be rescued.

The concept that cardiac autophagy is both a cardioprotective and adaptive process is supported by studies performed in isolated cells, isolated heart preparations, and intact animals subjected to various degrees of ischemia/reperfusion injury. There is compelling evidence that autophagy also plays an important role in the protection conferred by ischemic preconditioning and is involved in limiting adverse postinfarction remodeling. It is important to note, however, there are reports that under certain conditions cardiac autophagy may be maladaptive.

While there are several explanations for this ranging from duration of ischemia to experimental design, these observations need to be reconciled in order to realize the full therapeutic potential of autophagy. Ultimately this will require the development of more sophisticated tools for measuring autophagic formation and flux in vivo, the elucidation of the steps that are crucial (induction, formation, targeting, fusion, breakdown and recycling) and most importantly, the initiation of studies in humans that are designed to elucidate the role of cardiac autophagy in health and disease.

Acknowledgments

Mrs. Kristyn Hagood provided research support and was an active participant in the preparation of this manuscript.

References

1. Levine B, Klionsky DJ. Development by self-digestion: molecular mechanisms and biological functions of autophagy. *Dev Cell* 2004;**6**(4):463–77.
2. Gottlieb RA, Gustafsson AB. Mitochondrial turnover in the heart. *Biochim Biophys Acta* 2011;**1813**(7):1295–301.
3. Nishino I, Fu J, Tanji K, Yamada T, Shimojo S, Koori T, et al. Primary LAMP-2 deficiency causes X-linked vacuolar cardiomyopathy and myopathy (Danon disease). *Nature* 2000;**406**(6798):906–10.
4. Tanaka Y, Guhde G, Suter A, Eskelinen EL, Hartmann D, Lullmann-Rauch R, et al. Accumulation of autophagic vacuoles and cardiomyopathy in LAMP-2-deficient mice. *Nature* 2000;**406**(6798):902–6.
5. Nakai A, Yamaguchi O, Takeda T, Higuchi Y, Hikoso S, Taniike M, et al. The role of autophagy in cardiomyocytes in the basal state and in response to hemodynamic stress. *Nat Med* 2007;**13**(5):619–24.
6. Kuma A, Hatano M, Matsui M, Yamamoto A, Nakaya H, Yoshimori T, et al. The role of autophagy during the early neonatal starvation period. *Nature* 2004;**432**(7020):1032–6.
7. Zhu H, Tannous P, Johnstone JL, Kong Y, Shelton JM, Richardson JA, et al. Cardiac autophagy is a maladaptive response to hemodynamic stress. *J Clin Invest* 2007;**117**(7):1782–93.
8. Tannous P, Zhu H, Nemchenko A, Berry JM, Johnstone JL, Shelton JM, et al. Intracellular protein aggregation is a proximal trigger of cardiomyocyte autophagy. *Circulation* 2008;**117**(24):3070–8.
9. Yan L, Vatner DE, Kim SJ, Ge H, Masurekar M, Massover WH, et al. Autophagy in chronically ischemic myocardium. *Proc Natl Acad Sci USA* 2005;**102**(39):13807–12.
10. Hariharan N, Zhai P, Sadoshima J. Oxidative stress stimulates autophagic flux during ischemia/reperfusion. *Antioxid Redox Signal* 2011;**14**(11):2179–90.
11. Matsui Y, Takagi H, Qu X, Abdellatif M, Sakoda H, Asano T, et al. Distinct roles of autophagy in the heart during ischemia and reperfusion: roles of AMP-activated protein kinase and Beclin 1 in mediating autophagy. *Circ Res* 2007;**100**(6):914–22.
12. Zhai P, Sciarretta S, Galeotti J, Volpe M, Sadoshima J. Differential Roles of GSK-3β During Myocardial Ischemia and Ischemia/Reperfusion. *Circ Res* 2011.
13. Hariharan N, Maejima Y, Nakae J, Paik J, Depinho RA, Sadoshima J. Deacetylation of FoxO by Sirt1 Plays an Essential Role in Mediating Starvation-Induced Autophagy in Cardiac Myocytes. *Circ Res* 2010;**107**(12):1470–82.
14. Kanamori H, Takemura G, Maruyama R, Goto K, Tsujimoto A, Ogino A, et al. Functional significance and morphological characterization of starvation-induced autophagy in the adult heart. *Am J Pathol* 2009;**174**(5):1705–14.
15. Carreira RS, Lee Y, Ghochani M, Gustafsson AB, Gottlieb RA. Cyclophilin D is required for mitochondrial removal by autophagy in cardiac cells. *Autophagy* 2010;**6**(4):462–72.
16. Hamacher-Brady A, Brady NR, Gottlieb RA. Enhancing macroautophagy protects against ischemia/reperfusion injury in cardiac myocytes. *J Biol Chem* 2006;**281**(40):29776–87.
17. Hamacher-Brady A, Brady NR, Logue SE, Sayen MR, Jinno M, Kirshenbaum LA, et al. Response to myocardial ischemia/reperfusion injury involves Bnip3 and autophagy. *Cell Death Differ* 2007;**14**(1):146–57.
18. Sengupta A, Molkentin JD, Paik JH, DePinho RA, Yutzey KE. FoxO transcription factors promote cardiomyocyte survival upon induction of oxidative stress. *J Biol Chem* 2011;**286**(9):7468–78.
19. Yuan H, Perry CN, Huang C, Iwai-Kanai E, Carreira RS, Glembotski CC, et al. LPS-induced autophagy is mediated by oxidative signaling in cardiomyocytes and is associated with cytoprotection. *Am J Physiol Heart Circ Physiol* 2009;**296**(2):H470–9.
20. Dosenko VE, Nagibin VS, Tumanovskaya LV, Zagoriy VY, Moibenko AA, Vaage J. Proteasomal proteolysis in anoxia-reoxygenation, preconditioning and postconditioning of isolated cardiomyocytes. *Pathophysiology* 2006;**13**(2):119–25.
21. Dosenko VE, Nagibin VS, Tumanovska LV, Moibenko AA. Protective effect of autophagy in anoxia-reoxygenation of isolated cardiomyocyte? *Autophagy* 2006;**2**(4):305–6.

22. Gustafsson AB, Gottlieb RA. Bcl-2 Family Members and Apoptosis, Taken to Heart. *Am J Physiol Cell Physiol* 2006.
23. Pattingre S, Tassa A, Qu X, Garuti R, Liang XH, Mizushima N, et al. Bcl-2 antiapoptotic proteins inhibit Beclin 1-dependent autophagy. *Cell* 2005;**122**(6):927–39.
24. Maiuri MC, Criollo A, Tasdemir E, Vicencio JM, Tajeddine N, Hickman JA, et al. BH3-only proteins and BH3 mimetics induce autophagy by competitively disrupting the interaction between Beclin 1 and Bcl-2/Bcl-X(L). *Autophagy* 2007;**3**(4):374–6.
25. Maiuri MC, Le Toumelin G, Criollo A, Rain JC, Gautier F, Juin P, et al. Functional and physical interaction between Bcl-X(L) and a BH3-like domain in Beclin-1. *Embo J* 2007;**26**(10):2527–39.
26. Quinsay MN, Thomas RL, Lee Y, Gustafsson AB. Bnip3-mediated mitochondrial autophagy is independent of the mitochondrial permeability transition pore. *Autophagy* 2010;**6**(7):17–24.
27. Hamacher-Brady A, Brady NR, Gottlieb RA, Gustafsson AB. Autophagy as a Protective Response to Bnip3-Mediated Apoptotic Signaling in the Heart. *Autophagy* 2006;**2**(4):307–9.
28. Bjorkoy G, Lamark T, Pankiv S, Overvatn A, Brech A, Johansen T. Monitoring autophagic degradation of p62/SQSTM1. *Methods Enzymol* 2006;**452**:181–97.
29. Rikka S, Quinsay MN, Thomas RL, Kubli DA, Zhang X, Murphy AN, et al. Bnip3 impairs mitochondrial bioenergetics and stimulates mitochondrial turnover. *Cell Death Differ* 2011.
30. Schweers RL, Zhang J, Randall MS, Loyd MR, Li W, Dorsey FC, et al. NIX is required for programmed mitochondrial clearance during reticulocyte maturation. *Proc Natl Acad Sci USA* 2007;**104**(49):19500–5.
31. Sala-Mercado JA, Wider J, Undyala VV, Jahania S, Yoo W, Mentzer Jr RM, et al. Profound cardioprotection with chloramphenicol succinate in the swine model of myocardial ischemia-reperfusion injury. *Circulation* 2010;**122**(Suppl. 11):S179–84.
32. Berridge MJ. The endoplasmic reticulum: a multifunctional signaling organelle. *Cell Calcium* 2002;**32**(5-6):235–49.
33. Hoyer-Hansen M, Jaattela M. Connecting endoplasmic reticulum stress to autophagy by unfolded protein response and calcium. *Cell Death Differ* 2007;**14**(9):1576–82.
34. Petrovski G, Das S, Juhasz B, Kertesz A, Tosaki A, Das DK. Cardioprotection by endoplasmic reticulum stress-induced autophagy. *Antioxid Redox Signal* **14**(11): 2191–200.
35. Wullschleger S, Loewith R, Hall MN. TOR signaling in growth and metabolism. *Cell* 2006;**124**(3):471–84.
36. Ravikumar B, Vacher C, Berger Z, Davies JE, Luo S, Oroz LG, et al. Inhibition of mTOR induces autophagy and reduces toxicity of polyglutamine expansions in fly and mouse models of Huntington disease. *Nat Genet* 2004;**36**(6):585–95.
37. Mousavi SA, Brech A, Berg T, Kjeken R. Phosphoinositide 3-kinase regulates maturation of lysosomes in rat hepatocytes. *Biochem J* 2003;**372**(Pt 3):861–9.
38. Wouters BG, Koritzinsky M. Hypoxia signalling through mTOR and the unfolded protein response in cancer. *Nat Rev Cancer* 2008;**8**(11):851–64.
39. Khan S, Salloum F, Das A, Xi L, Vetrovec GW, Kukreja RC. Rapamycin confers preconditioning-like protection against ischemia-reperfusion injury in isolated mouse heart and cardiomyocytes. *J Mol Cell Cardiol* 2006;**41**(2):256–64.
40. Inoki K, Zhu T, Guan KL. TSC2 mediates cellular energy response to control cell growth and survival. *Cell* 2003;**115**(5):577–90.
41. Grimes CA, Jope RS. The multifaceted roles of glycogen synthase kinase 3beta in cellular signaling. *Prog Neurobiol* 2001;**65**(4):391–426.
42. Tong H, Imahashi K, Steenbergen C, Murphy E. Phosphorylation of glycogen synthase kinase-3beta during preconditioning through a phosphatidylinositol-3-kinase–dependent pathway is cardioprotective. *Circ Res* 2002;**90**(4):377–9.
43. Inoki K, Ouyang H, Zhu T, Lindvall C, Wang Y, Zhang X, et al. TSC2 integrates Wnt and energy signals via a coordinated phosphorylation by AMPK and GSK3 to regulate cell growth. *Cell* 2006;**126**(5):955–68.
44. Buller CL, Loberg RD, Fan MH, Zhu Q, Park JL, Vesely E, et al. A GSK-3/TSC2/mTOR pathway regulates glucose uptake and GLUT1 glucose transporter expression. *Am J Physiol Cell Physiol* 2008;**295**(3):C836–43.
45. Tong H, Chen W, Steenbergen C, Murphy E. Ischemic preconditioning activates phosphatidylinositol-3-kinase upstream of protein kinase C. *Circ Res* 2000;**87**(4):309–15.
46. Hirotani S, Zhai P, Tomita H, Galeotti J, Marquez JP, Gao S, et al. Inhibition of glycogen synthase kinase 3beta during heart failure is protective. *Circ Res* 2007;**101**(11):1164–74.

47. Greer EL, Brunet A. FOXO transcription factors at the interface between longevity and tumor suppression. *Oncogene* 2005;**24**(50):7410–25.
48. Hannenhalli S, Putt ME, Gilmore JM, Wang J, Parmacek MS, Epstein JA, et al. Transcriptional genomics associates FOX transcription factors with human heart failure. *Circulation* 2006;**114**(12):1269–76.
49. Sengupta A, Molkentin JD, Yutzey KE. FoxO transcription factors promote autophagy in cardiomyocytes. *J Biol Chem* 2009;**284**(41):28319–31.
50. Schips TG, Wietelmann A, Hohn K, Schimanski S, Walther P, Braun T, et al. FoxO3 induces reversible cardiac atrophy and autophagy in a transgenic mouse model. *Cardiovasc Res* **91**(4), 587-97
51. Ni YG, Berenji K, Wang N, Oh M, Sachan N, Dey A, et al. Foxo transcription factors blunt cardiac hypertrophy by inhibiting calcineurin signaling. *Circulation* 2006;**114**(11):1159–68.
52. Kanamori H, Takemura G, Goto K, Maruyama R, Tsujimoto A, Ogino A, et al. The role of autophagy emerging in postinfarction cardiac remodelling. *Cardiovasc Res* 2011;**91**(2):330–9.
53. Zhang JL, Lu JK, Chen D, Cai Q, Li TX, Wu LS, et al. Myocardial autophagy variation during acute myocardial infarction in rats: the effects of carvedilol. *Chin Med J (Engl)* 2009;**122**(19):2372–9.
54. Kanamori H, Takemura G, Goto K, Maruyama R, Ono K, Nagao K, et al. Autophagy limits acute myocardial infarction induced by permanent coronary artery occlusion. *Am J Physiol Heart Circ Physiol* 2011;**300**(6): H2261–71.
55. Gustafsson AB, Gottlieb RA. Heart mitochondria: gates of life and death. *Cardiovasc Res* 2008;**77**(2):334–43.
56. Doering CW, Jalil JE, Janicki JS, Pick R, Aghili S, Abrahams C, et al. Collagen network remodelling and diastolic stiffness of the rat left ventricle with pressure overload hypertrophy. *Cardiovasc Res* 1988;**22**(10):686–95.
57. Sabbah HN, Sharov VG, Lesch M, Goldstein S. Progression of heart failure: a role for interstitial fibrosis. *Mol Cell Biochem* 1995;**147**(1-2):29–34.
58. Weber KT, Brilla CG. Pathological hypertrophy and cardiac interstitium. Fibrosis and renin-angiotensin-aldosterone system. *Circulation* 1991;**83**(6):1849–65.
59. Buss SJ, Muenz S, Riffel JH, Malekar P, Hagenmueller M, Weiss CS, et al. Beneficial effects of Mammalian target of rapamycin inhibition on left ventricular remodeling after myocardial infarction. *J Am Coll Cardiol* 2009;**54**(25):2435–46.
60. Souders CA, Bowers SL, Baudino TA. Cardiac fibroblast: the renaissance cell. *Circ Res.* 2009;**105**(12):1164–76.
61. Dalla Costa AP, Clemente CF, Carvalho HF, Carvalheira JB, Nadruz W, Jr., Franchini KG. FAK mediates the activation of cardiac fibroblasts induced by mechanical stress through regulation of the mTOR complex. *Cardiovasc Res* 86(3):421-31.
62. Aranguiz-Urroz P, Canales J, Copaja M, Troncoso R, Vicencio JM, Carrillo C, et-al. Beta(2)-adrenergic receptor regulates cardiac fibroblast autophagy and collagen degradation: *Biochim Biophys Acta* **1812**(1):23–31.
63. Schwartz Longacre L, Kloner RA, Arai AE, Baines CP, Bolli R, Braunwald E, et al. New horizons in cardioprotection: recommendations from the 2010 national heart, lung, and blood institute workshop. *Circulation* 2011;**124**(10):1172–9.
64. Mentzer Jr RM. Myocardial protection in heart surgery. *J Cardiovasc Pharmacol Ther* 2011;**16**(3-4):290–7.
65. Ovize M, Baxter GF, Di Lisa F, Ferdinandy P, Garcia-Dorado D, Hausenloy DJ, et al. Postconditioning and protection from reperfusion injury: where do we stand? Position paper from the Working Group of Cellular Biology of the Heart of the European Society of Cardiology. *Cardiovasc Res* 2010;**87**(3):406–23.
66. Hausenloy DJ, Yellon DM. Preconditioning and postconditioning: underlying mechanisms and clinical application. *Atherosclerosis* 2009;**204**(2):334–41.
67. Zheng Q, Su H, Ranek MJ, Wang X. Autophagy and p62 in cardiac proteinopathy. *Circ Res* 2011;**109**(3):296–308.
68. Huffmyer J, Raphael J. Physiology and pharmacology of myocardial preconditioning and postconditioning. *Semin Cardiothorac Vasc Anesth* 2009;**13**(1):5–18.
69. Balakumar P, Rohilla A, Singh M. Pre-conditioning and postconditioning to limit ischemia-reperfusion-induced myocardial injury: what could be the next footstep? *Pharmacol Res* 2008;**57**(6):403–12.
70. Stanton LW, Garrard LJ, Damm D, Garrick BL, Lam A, Kapoun AM, et al. Altered patterns of gene expression in response to myocardial infarction. *Circ Res* 2000;**86**(9):939–45.
71. Mathiesen JM, Ulven T, Martini L, Gerlach LO, Heinemann A, Kostenis E. Identification of indole derivatives exclusively interfering with a G protein-independent signaling pathway of the prostaglandin D2 receptor CRTH2. *Mol Pharmacol* 2005;**68**(2):393–402.

72. Ferdinandy P, Schulz R, Baxter GF. Interaction of cardiovascular risk factors with myocardial ischemia/reperfusion injury, preconditioning, and postconditioning. *Pharmacol Rev* 2007;**59**(4):418–58.
73. Hausenloy DJ, Baxter G, Bell R, Botker HE, Davidson SM, Downey J, et al. Translating novel strategies for cardioprotection: the Hatter Workshop Recommendations. *Basic Res Cardiol* 2010.
74. Gurusamy N, Lekli I, Mukherjee S, Ray D, Ahsan MK, Gherghiceanu M, et al. Cardioprotection by resveratrol: a novel mechanism via autophagy involving the mTORC2 pathway. *Cardiovasc Res* 2010;**86**(1):103–12.
75. Qian J, Ren X, Wang X, Zhang P, Jones WK, Molkentin JD, et al. Blockade of Hsp20 phosphorylation exacerbates cardiac ischemia/reperfusion injury by suppressed autophagy and increased cell death. *Circ Res* 2009;**105**(12):1223–31.
76. Kanamori H, Takemura G, Goto K, Maruyama R, Tsujimoto A, Ogino A, et al. The role of autophagy emerging in postinfarction cardiac remodelling. *Cardiovasc Res* 2011.
77. Huang C, Liu W, Perry CN, Yitzhaki S, Lee Y, Yuan H, et al. Autophagy and protein kinase C are required for cardioprotection by sulfaphenazole. *Am J Physiol Heart Circ Physiol* 2010;**298**(2):H570–9.
78. Yitzhaki S, Huang C, Liu W, Lee Y, Gustafsson AB, Mentzer Jr RM, et al. Autophagy is required for preconditioning by the adenosine A1 receptor-selective agonist CCPA. *Basic Res Cardiol* 2009;**104**(2):157–67.
79. Nishida K, Kyoi S, Yamaguchi O, Sadoshima J, Otsu K. The role of autophagy in the heart. *Cell Death Differ* 2009;**16**(1):31–8.
80. Dosenko VE, Nagibin VS, Tumanovskaya LV, Zagoriy VY, Moibenko AA, Vaage J. Proteasome inhibitors eliminate protective effect of postconditioning in cultured neonatal cardiomyocytes. *Fiziol Zh* 2006;**52**(3):15–24.
81. Yellon DM, Downey JM. Preconditioning the myocardium: from cellular physiology to clinical cardiology. *Physiol Rev* 2003;**83**(4):1113–51.
82. Huang C, Yitzhaki S, Perry CN, Liu W, Giricz Z, Mentzer Jr RM, et al. Autophagy induced by ischemic preconditioning is essential for cardioprotection. *J Cardiovasc Transl Res* 2010;**3**(4):365–73.
83. Yan L, Sadoshima J, Vatner DE, Vatner SF. Autophagy in ischemic preconditioning and hibernating myocardium. *Autophagy* 2009;**5**(5):709–12.
84. Gurusamy N, Lekli I, Gorbunov NV, Gherghiceanu M, Popescu LM, Das DK. Cardioprotection by adaptation to ischaemia augments autophagy in association with BAG-1 protein. *J Cell Mol Med* 2009;**13**(2):373–87.
85. Park HK, Chu K, Jung KH, Lee ST, Bahn JJ, Kim M, et al. Autophagy is involved in the ischemic preconditioning. *Neurosci Lett* 2009;**451**(1):16–9.
86. Huang C, Andres AM, Ratliff EP, Hernandez G, Lee P, Gottlieb RA. Preconditioning involves selective mitophagy mediated by Parkin and p62/SQSTM1. *PLoS One* 2011;**6**(6):e20975.
87. Bjorkoy G, Lamark T, Brech A, Outzen H, Perander M, Overvatn A, et al. p62/SQSTM1 forms protein aggregates degraded by autophagy and has a protective effect on huntingtin-induced cell death. *J Cell Biol* 2005;**171**(4):603–14.
88. Sybers HD, Ingwall J, DeLuca M. Autophagy in cardiac myocytes. *Recent advances in studies on cardiac structure and metabolism* 1976;**12**:453–63.
89. Decker RS, Wildenthal K. Lysosomal alterations in hypoxic and reoxygenated hearts. I. Ultrastructural and cytochemical changes. *Am J Pathol* 1980;**98**(2):425–44.
90. Decker RS, Poole AR, Crie JS, Dingle JT, Wildenthal K. Lysosomal alterations in hypoxic and reoxygenated hearts. II. Immunohistochemical and biochemical changes in cathepsin D. *Am J Pathol* 1980;**98**(2):445–56.
91. Lekli I, Ray D, Mukherjee S, Gurusamy N, Ahsan MK, Juhasz B, et al. Co-ordinated autophagy with resveratrol and gamma-tocotrienol confers synergetic cardioprotection. *J Cell Mol Med* 2010;**14**(10):2506–18.
92. Petrovski G, Das S, Juhasz B, Kertesz A, Tosaki A, Das DK. Cardioprotection by endoplasmic reticulum stress-induced autophagy. *Antioxid Redox Signal* 2011;**14**(11):2191–200.
93. May D, Gilon D, Djonov V, Itin A, Lazarus A, Gordon O, et al. Transgenic system for conditional induction and rescue of chronic myocardial hibernation provides insights into genomic programs of hibernation. *Proc Natl Acad Sci USA* 2008;**105**(1):282–7.
94. Zhai P, Sciarretta S, Galeotti J, Volpe M, Sadoshima J. Differential roles of GSK-3beta during myocardial ischemia and ischemia/reperfusion. *Circ Res* 2011;**109**(5):502–11.
95. Dong Y, Undyala VV, Gottlieb RA, Mentzer Jr RM, Przyklenk K. Autophagy: definition, molecular machinery, and potential role in myocardial ischemia-reperfusion injury. *J Cardiovasc Pharmacol Ther* 2010;**15**(3):220–30.

SECTION IV

DIAGNOSIS AND THERAPEUTICS

CHAPTER

10

Drug Discovery in the Autophagy Pathways

Bo Liu[1], *Yan Cheng*[2], *and Jin-Ming Yang*[2]

[1]State Key Laboratory of Biotherapy and Cancer Center, West China Hospital, West China Medical School, Sichuan University, Chengdu 610064, China

[2]Department of Pharmacology and The Penn State Cancer Institute, The Pennsylvania State University College of Medicine, and Milton S. Hershey Medical Center, Hershey, PA, U.S.A.

INTRODUCTION: A BRIEF OVERVIEW OF AUTOPHAGIC PATHWAYS IN DRUG DISCOVERY

Autophagy, a term from Greek “auto” (self) and “phagy” (to eat), refers to an evolutionarily conserved, multi-step lysosomal degradation process in which a cell degrades long-lived proteins and damaged organelles[1]. Hitherto, three forms of autophagy have been identified, namely macroautophagy, microautophagy and chaperone-mediated autophagy. These three types of autophagy differ with respect to their modes of delivery to lysosome and physiological functions[2]. Macroautophagy (hereafter referred to as autophagy) is the major regulated catabolic mechanism that involves the delivery of cytoplasmic cargo sequestered inside double-membrane vesicles to the lysosome[3], and is highly regulated by a number of autophagy-related genes (ATGs) (Figure 10.1).

Originally discovered in yeast, ATGs were reported to play a crucial role in autophagosome formation and autophagy regulation. Thus far, the function and regulation of dozens of those ATGs have been elucidated, and their associations with physiologic or patho-physiologic processes including cancer, neuro-degeneration, aging, type-II diabetes, cardiac disease, muscular disorders, hepatic encephalopathy, infection, immunity and inflammatory diseases have been revealed[4, 5]. It appears that the links between autophagy and human diseases are multifaceted (Figure 10.2).

Autophagy is known as a physiological process that may serve as a means of temporary survival, and provide a means of recycling macromolecules as an alternative

Autophagy in Health and Disease
http://dx.doi.org/10.1016/B978-0-12-385101-7.00010-3

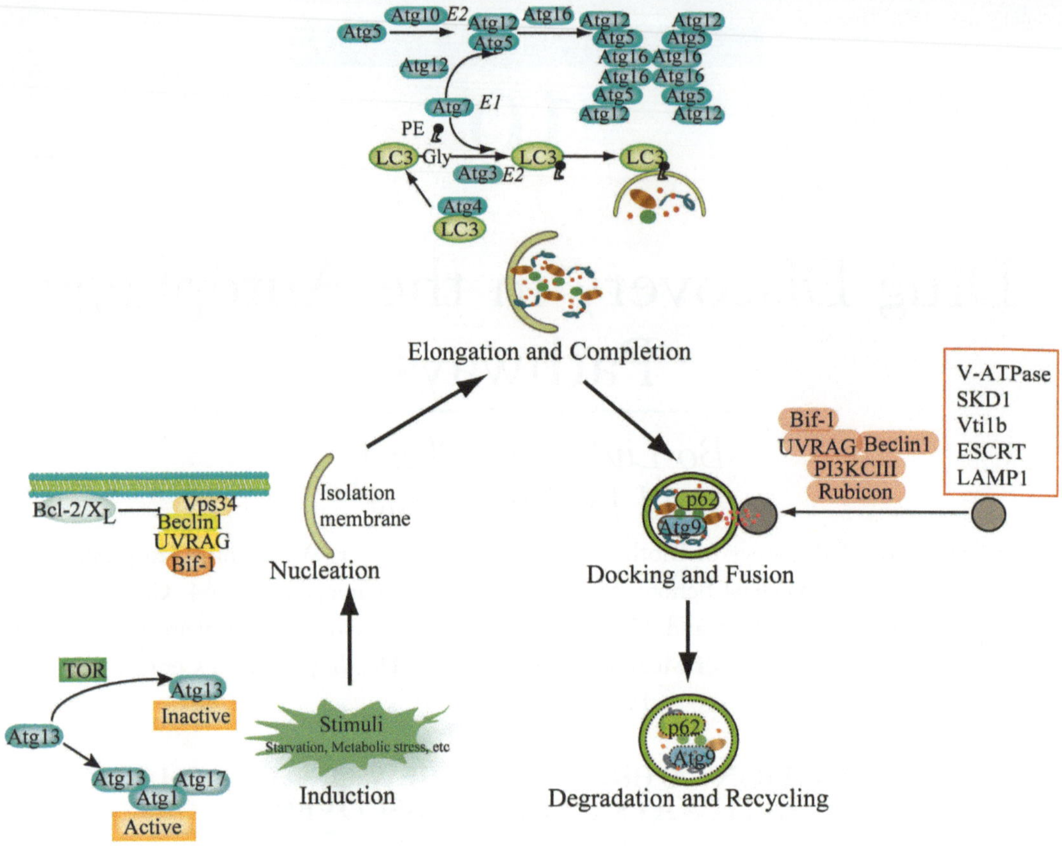

FIGURE 10.1 Multiple stages of autophagy and the molecular regulators involved.

source of energy supply. However, if the cellular stress leads to excessive autophagy, cell death will ensue (cannibalistic cell death)[6, 7]. These paradoxical studies are often viewed as confusing because, depending on different cell type and context, autophagy appears to play the opposite roles, which are regulated via several key mediators (e.g., Beclin-1, PI3Kinase, mTOR, Bcl-2 and p53). Therefore, these regulators can be manipulated to alter autophagic activity, and be utilized as therapeutic targets for treatment of human diseases[8, 9].

AUTOPHAGY IN HUMAN DISEASES: A GUARDIAN OR EXECUTIONER?

A growing number of studies of molecular mechanisms of autophagy are leading to a better understanding of the roles of autophagy in a variety of human diseases. However, when analyzing the intricate relationship between autophagy and certain diseases,

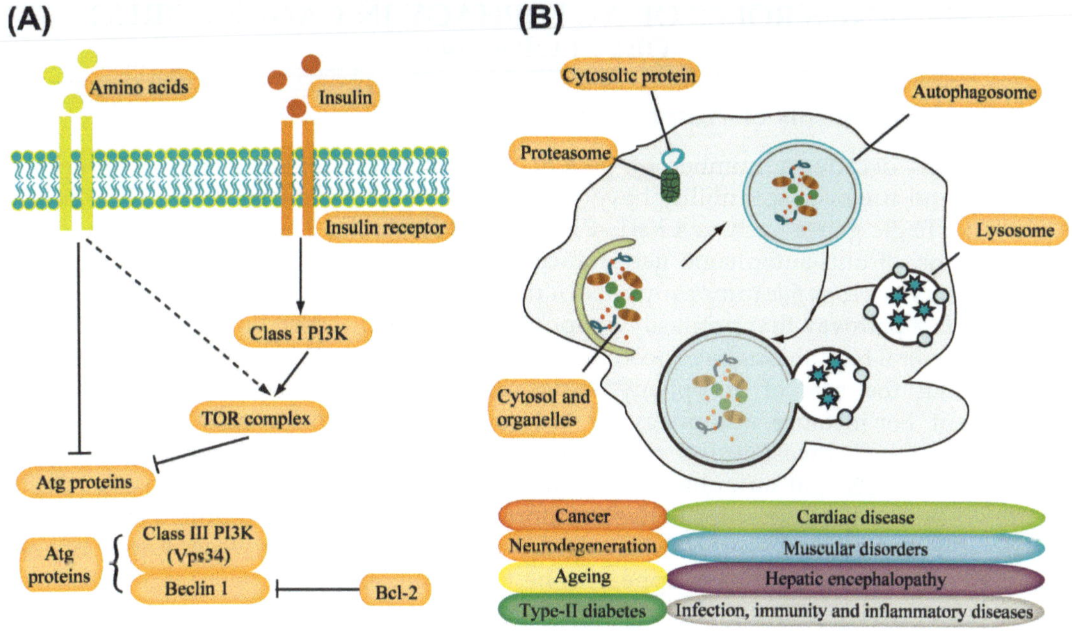

FIGURE 10.2 A simplified model linking autophagy regulation to human diseases.

a common challenge is to determine whether autophagy protects cell survival or contributes to cell death. Autophagy is well known to be crucial for cell survival under stress conditions, and degradation of intracellular macromolecules via autophagy provides the energy required for minimal cellular functioning when nutrients are deprived or scarce[10]. Autophagy-mediated elimination of altered cytosolic constituents, such as aggregated proteins or damaged organelles, also preserves cells from further damage[11]. Consequently, induction of autophagy plays a protective role in early stages of some disorders. On the other hand, many studies have demonstrated that autophagy plays a death-promoting role as type II programmed cell death (type II PCD), compared to apoptosis (type I PCD)[12]. Interestingly, although apoptosis and autophagy bear distinct morphological characteristics and biochemical processes, there exist intricate collaborations or cross-talks between these two important cellular events. Under certain circumstances, apoptosis and autophagy can exert synergetic effects. By contrast, sometimes autophagy would be triggered only when apoptosis is suppressed[13]. Recent studies have further pointed out that apoptosis and autophagy may be interconnected and even simultaneously regulated by the same trigger in cancer cells[14]. Nevertheless, a number of studies have reconciled these opposing effects of autophagy in diseases, and showed that autophagy can act as either guardian or executioner of the cell, depending on different stages of human disease, the surrounding cellular environment, or the exerted therapeutic interventions[15]. The dual roles of autophagy might complicate the targeting of this pathway as therapies for human diseases.

THE JANUS ROLES OF AUTOPHAGY IN CANCER DRUG DEVELOPMENT

Molecular Mechanisms of Autophagy in Cancer

In the past decade, a number of genetic links between deficiencies in autophagic machinery and tumor susceptibility have emerged. Those deficiencies involve inactivation of pro-autophagic genes such as LKB1, PTEN, TSC1/2, Beclin 1, UVRAG, Bif-1 and p53, and activation of anti-autophagic genes such as Ras, Bcl-2, PI3KCI and Akt, suggesting that autophagy is a *bona fide* tumor suppressor pathway[16, 17]. A seeming paradox is that while autophagy is a pathway that primarily supports cell survival, it also has tumor suppressive functions. Several hypotheses have been put forward to alleviate this puzzle. One hypothesis proposes that the role of autophagy varies depending on different stages of tumor development. For instance, autophagy limits tumor formation in the early stage but favors tumor cell survival, invasion and metastasis as soon as tumors have formed[18]. Another hypothesis suggests that autophagy can affect carcinogenesis in a cell- or tissue-specific manner[19]. The exact relationships between autophagy and cancer development and progression are still unclear; however, considering the close correlations between autophagy and cancer, the autophagic pathways appear to be promising and offer new targets in cancer drug development.

Of note, autophagy-related genes (ATGs) play a key role in the formation of the autophagosome and regulation of autophagic activity, which is closely linked to cancer initiation and progression. Silencing some essential autophagy-regulate genes such as Atg3, Atg4, Beclin-1/Atg6, Atg10 and Atg12 have been demonstrated to sensitize cancer cells to a wide spectrum of stressful conditions[20]. Thus, this strategy might be a useful approach to targeting protective autophagy in cancer therapy. In addition, targeting selected protein kinases involved in autophagy regulation by use of small molecule inhibitors may be another feasible approach in cancer treatment. A number of protein kinases have been known to regulate the induction of autophagy following nutrient deprivation or other cellular stresses; the following protein kinases have been reported to activate protective autophagy in cancer cells as a response to cytotoxic agents, including AMP-activated protein kinase (AMPK), glycogen synthase kinase 3 (GSK3) beta, extracellular signal-regulated kinases 1 and 2 (ERK1/2) and eukaryotic elongation factor-2 kinase (eEF-2K). On the other hand, some other protein kinases are involved in the promotion of autophagy but are not yet evaluated as potential therapeutic targets, including death-associated protein kinase (DAPK) and Unc-51-like kinase 1 (ULK-1)[21]. Thus, further studies are necessary to assess the precise functions of these protein kinases and the possibility of these kinases as drug targets for cancer treatment.

Autophagy-Modulating Agents for Cancer Treatment

In recent years, a number of autophagy-inducing drugs have been identified as potential cancer therapeutic agents[22]. For instances, tamoxifen, an antagonist of the estrogen receptor, interacts with high affinity with the microsomal antiestrogen binding site (AEBS) that is a hetero-oligomeric complex involved in cholesterol metabolism. Tamoxifen and other AEBS ligands induce breast cancer cell autophagy through the induction of sterol accumulation. These data provide a therapeutic implication for the selective AEBS ligands in breast cancer management and reveal a mechanism that may explain the induction of autophagy in MCF-7 cells by tamoxifen and other selective estrogen receptor modulators[23]. Also, chloroquine (CQ), an anti-malarial drug used world-wide, has recently been reported to inhibit therapy-induced autophagy and to increase cell death in established tumors, leading to better tumor regression and delayed tumor growth[24]. In addition, imatinib (gleevac), an inhibitor of tyrosine kinases, has been found to induce autophagy in multidrug resistant Kaposi's sarcoma cells as part of its mode of action[25]. Histone deacetylase inhibitors (HDACs) such as suberoylanilide hydroxamic acid (SAHA) have been reported to induce autophagy and cell death in HeLa cells, and the cell death is independent of caspase-dependent apoptosis. Thus, initiation of autophagic cell death by SAHA has clear therapeutic implications for apoptosis-defective tumors. Furthermore, it is well-known that mTOR is a major regulator of cell growth that has been implicated in tumorigenesis[26]. Rapamycin, which binds the 12 kDa immunophilin FK506-binding protein (FKBP12) and inhibits the mTORC1 complex, and its derivatives, CCI-779 and RAD001, has been used in clinical trials for the treatment of cancer. Recent evidence has indicated that tumor suppression following rapamycin treatment is linked to induction of autophagic cell death. mTOR inhibitors can also sensitize various tumor cells to radiation therapy[27]. For example, combined treatment of RAD001 with the caspase-3 inhibitor DEVD radiosensitized non-small cell lung cancer cells in mice models and leads to enhanced cytotoxicity through induction of autophagy and to delayed tumor growth. Thus, inhibition of mTOR by rapamycin or its derivatives may provide an effective therapeutic tool for the treatment of various malignancies. In addition to these agents, there are also other interesting examples of the autophagy-inducing agents from traditional Chinese medicine. Arsenic trioxide (As_2O_3), a classical toxin from traditional Chinese medicine, has been reported to induce apoptosis through cytochrome *c* release and caspase activation[28]. Interestingly, recent studies showed that treatment of human T-lymphocytic leukemia cells with arsenic trioxide caused cytotoxicity through induction of autophagy. The Bcl-2 family member, Bcl-2-adenovirus E1B 19-kDa-interacting protein 3 (BNIP3), was reported to play a pivotal role in arsenic trioxide-induced autophagic cell death in malignant glioma cells[29]. Additionally, *Polygonatum cyrtonema* lectin (PCL) has been shown to induce autophagic cell death via a mitochondria-mediated ROS-p38-p53 pathway in human melanoma A375 cells. PCL-induced autophagic cell death has been confirmed to block Ras-Raf and PI3K-Akt signaling pathways in murine fibrosarcoma L929 cells[30, 31]. NH125, a small molecule inhibitor of elongation factor-2 kinase[32], has the potential to become a clinically useful autophagy-modulating agent in the treatment of cancer, as elongation factor-2 kinase has been shown to be a positive regulator of autophagy when tumor cells are subjected to various stresses including therapeutic stress[33, 34, 53]. Based upon the aforementioned examples, autophagy may play an important role in the cytotoxic effects of these compounds that would spark off a new autophagy-targeted cancer therapeutic strategy. Nevertheless, much work would be needed to determine how the modulators of autophagy impact cancer initiation, progression and therapeutic response, and to elucidate why targeting autophagic pathways makes sense for cancer drug development (Figure 10.3).

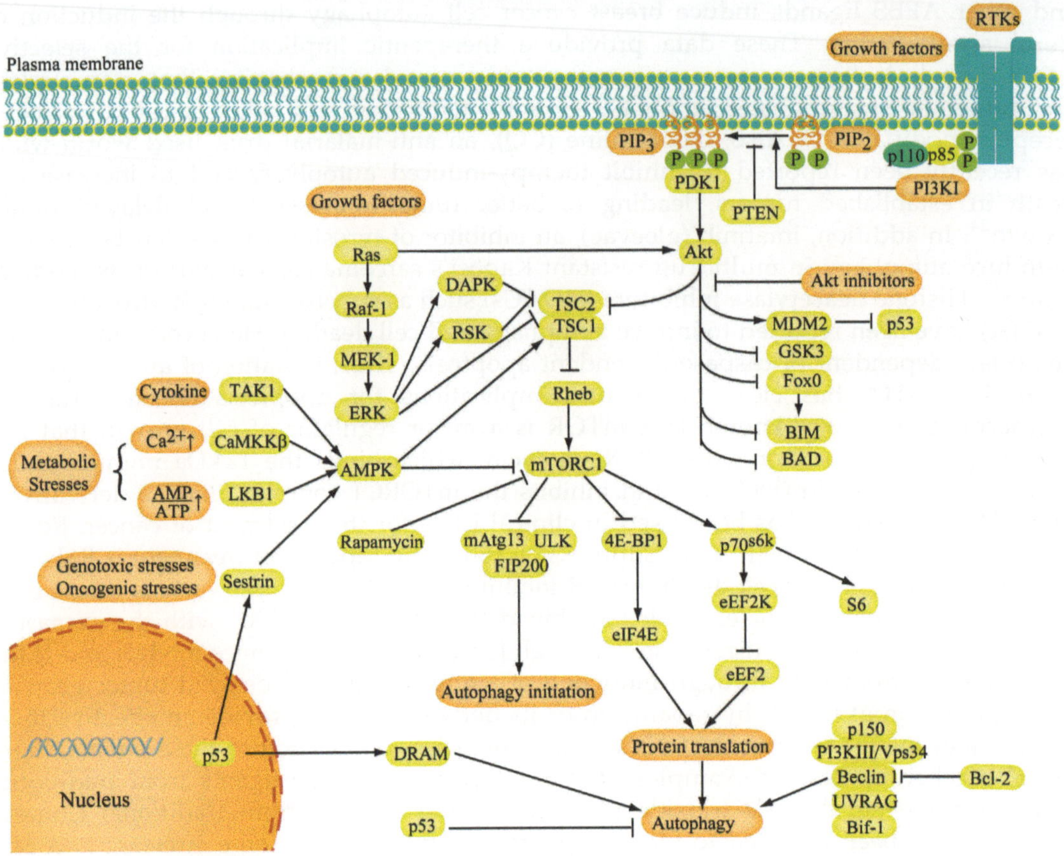

FIGURE 10.3 Autophagic pathways associated with cancer.

AUTOPHAGY AS DRUG TARGETS IN NEURODEGENERATION TREATMENT

Molecular Mechanisms of Autophagy in Neurodegeneration

In recent years, numerous studies have reported that autophagosomes accumulate in the brains of patients with diverse neurodegenerative diseases, including Alzheimer's disease, Parkinson's disease, Huntington's disease, and transmissible spongiform encephalopathies, suggesting that the alteration of autophagic pathways may contribute to neurodegenerative diseases[36]. Dopamine, a neurotransmitter with a well-defined role in neurodegeneration, can trigger autophagic cell death[37]. By contrast, a protective role for autophagy in neurodegeneration has also been reported. This study showed that activation of autophagy facilitated the removal of intracellular protein aggregates that are closely associated with different neurodegenerative diseases[38] (Figure 10.4).

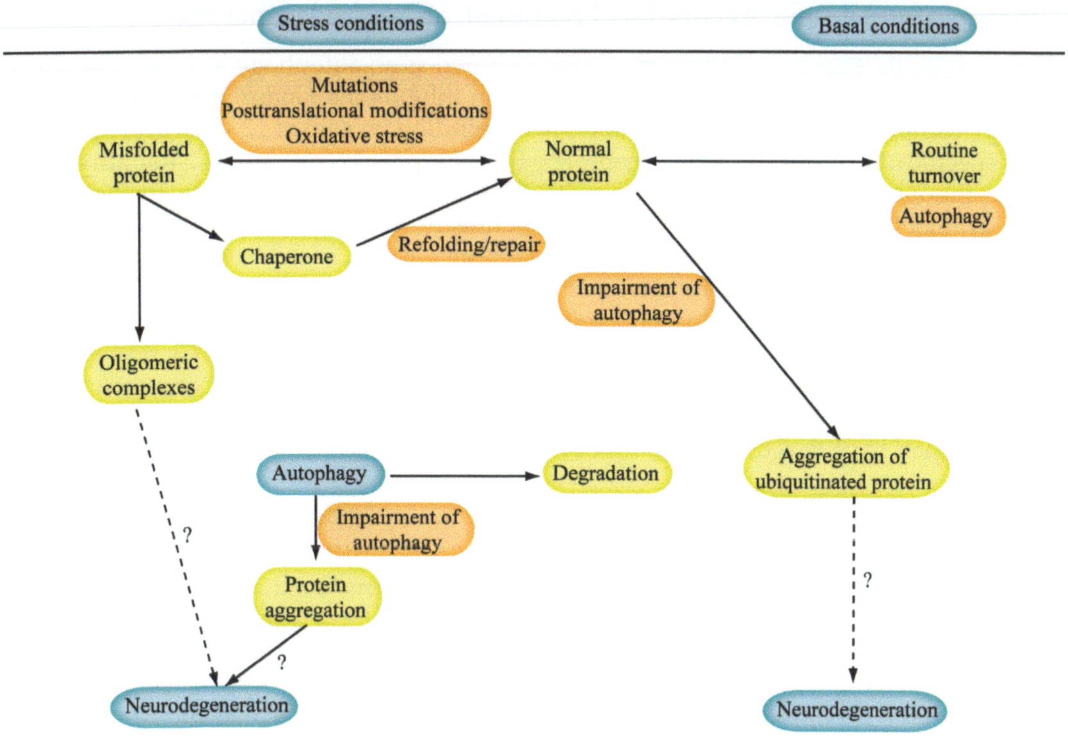

FIGURE 10.4 Involvement of autophagy in the progression of neurodegenerative diseases.

These proteins include polyglutamine (polyQ) expansion tracts that have been found in Huntington's disease and spinocerebellar ataxia, mutant α-synuclein that can cause familial Parkinson's disease, and different forms of tau including mutations causing frontotemporal dementia[36]. Because the substrates need to be unfolded to pass through the narrow pore of the proteasomal barrel, oligomeric and aggregated proteins are poor substrates for proteasomal degradation but are good targets for autophagic degradation. It is believed that those proteins are particularly toxic in oligomeric complexes and that higher-order protein aggregates may be formed as a last attempt to prevent toxicity in the absence of a properly functioning quality-control system[39]. This observation is in a good agreement with the model in which autophagy serves as a quality-control system that degrades oligomeric proteins; activation of autophagy reduces, whereas inhibition of autophagy increases, the formation of protein aggregates and the neurotoxicity of aggregate-prone proteins[40]. Because preclinical studies have demonstrated that autophagy plays an important role in protecting against neurodegenerative diseases, it would be interesting to test autophagy-inducing agents in clinical trials to determine whether or not modulating autophagy is beneficial in reducing neurotoxicity of mutant aggregate-prone proteins and can be used in the treatment of patients with neurodegenerative disease[41].

Autophagy-Modulating Agents for Treatment of Neurodegeneration

Rapamycin, the first known drug that was identified as an autophagy inducer, has already been used in clinic. In mammalian cells, rapamycin inhibits the kinase activity of mTOR by forming a complex with the immunophilin FK506-binding protein of 12 kDa (FKBP12)[42], and it acts specifically on the mTORC1 complex that suppresses autophagy. Rapamycin can enhance the clearance of mutant huntingtin fragments, reduce aggregate formation and protect against toxicity in cells, and in mouse models of Huntington's disease[43]. Rapamycin also promote the clearance of a wide range of aggregate-prone proteins with polyglutamine- or polyalanine-expansion, thereby attenuating the toxicity of those molecules[44]. CCI-779, an analogue of rapamycin, can reduce the levels of both mutant huntingtin and ataxin-3, thereby attenuating their toxicity in mouse models of Huntington's disease and SCA3, respectively[45]. In addition, rapamycin treatment and up-regulation of autophagy can protect cells against subsequent pro-apoptotic insults that are independent of the aggregates[46]. Therefore, autophagy may have two distinct beneficial effects in neurodegenerative diseases:

1. promoting clearance of aggregate-prone proteins
2. protecting cells against pro-apoptotic insults.

Although rapamycin is sometimes prescribed for long-term use, this agent has side effects that may make it unattractive to patients who need to take it for years. As far as we are aware at present, these side effects are unrelated to its autophagy-inducing effect. Thus, studies aiming to identify the autophagy up-regulating compounds that are independent of the targets of rapamycin have been reported. Inositol-lowering drugs such as lithium, valproate and carbamazepine can induce autophagy by inhibiting inositol monophosphatase (IMPase), leading to depletion of intracellular inositol levels and inhibition of the phospho-inositol cycle[47]. Indeed, these drugs have been shown to enhance the clearance of mutant huntingtin and to protect against the toxicity of this molecule in *in vitro* and *in vivo* models, and may be considered as potential therapeutic agents for neurodegenerative diseases. Another compound, trehalose, can inhibit aggregation of mutant huntingtin and reduces toxicity in cells, and alleviates disease pathology in a Huntington's disease mouse model[48]. Trehalose is a non-reducing disaccharide found in various non-mammalian species that protects cells against various environmental stresses, as it functions as a chemical chaperone for assisting in protein folding. Trehalose has been identified as an autophagy inducer that promotes the clearance of the aggregate-prone proteins (mutant huntingtin, α-synuclein, etc.) via an mTOR-independent pathway[49]. Trehalose also confers a protective mechanism against cell death. In view of the diverse protective effects of trehalose in different models of proteinopathies, the use of trehalose in the treatment of neurodegenerative diseases warrants further consideration for clinical use.

To identify and develop novel autophagy-modulating agents for treatment of neurodegeneration, high-throughput chemical screens have been conducted, and some small molecule inhibitors (SMIRs) and enhancers (SMERs) of the growth-suppressing properties of rapamycin were identified[50]. Three SMERs were identified to induce mTOR-independent autophagy in mammalian cells, increase clearance of autophagy substrates such as A53T α-synuclein and mutant huntingtin, and reduce huntingtin toxicity in the Drosophila Huntington's disease model[51]. A screening of a library of the FDA-approved drugs in search of alternative therapeutic candidates to test rapamycin revealed that the L-type Ca^{2+} channel antagonists such as verapamil, loperamide and amiodarone, the $K^{+}{}_{ATP}$ channel opener, minoxidil, and the Gi-signaling activator, clonidine, all induced autophagy[52]. It is notable that these drugs can enhance mutant huntingtin clearance via an mTOR-independent autophagy pathway, thus decreasing the levels of mutant huntingtin aggregate and conferring protection against toxicity in organism models[53].

Yet, the exploitation of up-regulating autophagy as a therapeutic strategy for neurodegenerative diseases still faces some challenges. In principle, earlier treatment shall be required to delay the onset of the diseases. Although rapamycin is often tailored to long-term use, this agent has certain side effects due to the inhibition of mTOR, which regulates many other cellular processes independent of autophagy. One possible approach could be to administer rapamycin intermittently. Another alternative approach is to use non-mTOR-targeted drugs or combine mTOR-targeted and non-mTOR-targeted therapeutics for a long-term period. In any event, a better understanding of pathways and mechanisms that regulate autophagy would be critically important in the context of treatment of neurodegenerative diseases.

THE ROLE OF AUTOPHAGY IN AGING: TO HELP LIVE LONGER?

A common feature of aging cells is the accumulation of damaged proteins and organelles, even in the absence of any mutation that predisposes cells to a pathogenic phenotype such as aggregate-prone mutant proteins[54]. Therefore, it is conceivable that a gradual decrease in autophagic activity with aging can play a critical role in functional deterioration of aging organisms (Figure 10.5).

Conversely, caloric restriction, an intervention known to slow down aging, seems to enhance autophagic activity, likely due to lower levels of insulin, an autophagy inhibitor. Thus, current efforts to prevent or restore the decline in autophagy with aging are mainly aimed at reproducing the observed beneficial effects of caloric restriction on the autophagic pathway by using anti-lipolytic drugs that mimic the starvation state induced by caloric restriction. Anti-lipolytic drugs such as 3,5-dimethylpyrazole or acipimox are known to decrease insulin secretion, suggesting that the treatment with anti-lipolytic drugs can delay the process of aging[55]. In addition, recent studies have reported that treatment with acipimox, a hypolipidemic agent with mild or severe caloric restriction, showed the age-related changes of accumulation of the membrane lipid and decline of the function of liver autophagy, which are known to be associated with life-expectancy[56].

Rapamycin, an inhibitor of mTOR and a well-characterized pharmacological inducer of autophagy, has been reported to prolong life in several types of organisms[57]. Resveratrol, a poly-phenol present in red wine and an indirect activator of sirtuin 1, was shown to prolong the lifespan of yeast, worms and flies[58, 59]. In mice, resveratrol extended lifespan only if the animals were kept on a high-calorie diet[60]. Resveratrol is also a potent inducer of canonical and non-canonical autophagy. It prolonged the lifespan of *C. elegans* only if the *C. elegans* expressed a functional sirtuin 1. Knockdown of ATG6/Beclin 1 abolished the beneficial effects of resveratrol on longevity[61]. These studies indicate that resveratrol increases lifespan of *C. elegans* through a sirtuin 1-dependent induction of autophagy. However, it is noteworthy that sirtuin 1 overexpression or resveratrol fails to extend the lifespan of mice on

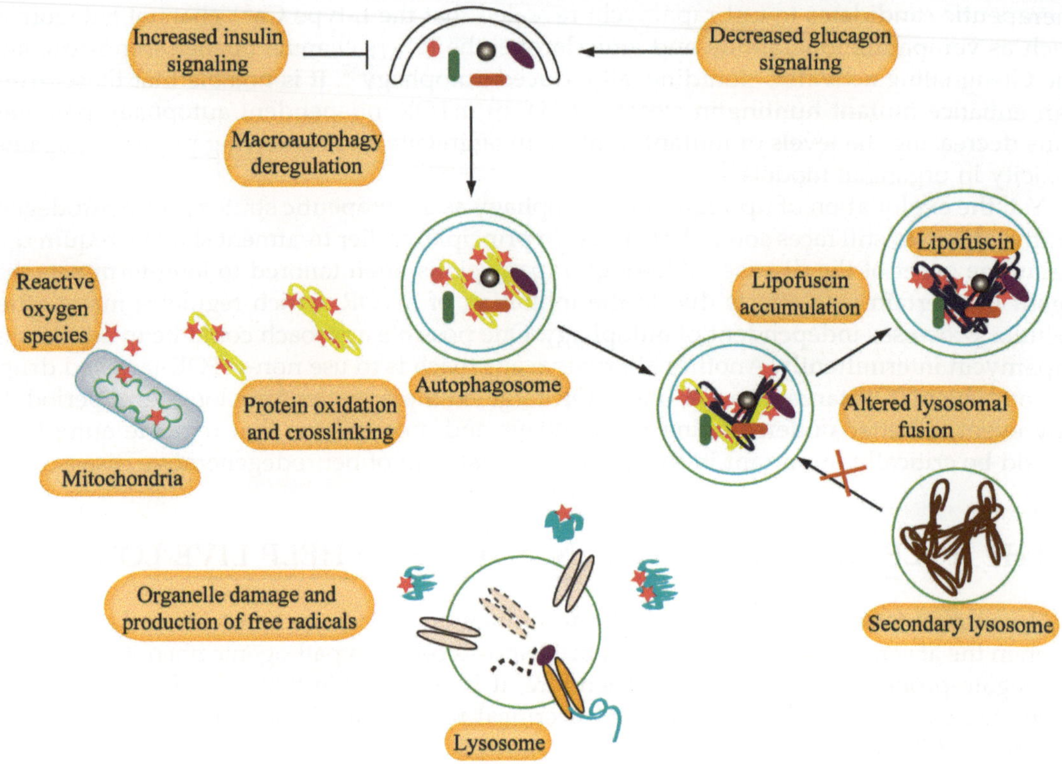

FIGURE 10.5 Autophagy and aging.

a normal diet. Also, the ability of sirtuins to increase lifespan of lower eukaryotes remains controversial, as yeast sirtuins can also shorten chronological lifespan. Moreover, the inhibition of sirtuin 1 actually prevents features of aging (decreased IGF-1 signaling and increased protection from oxidative stress in neurons) in mice, suggesting that the notion of sirtuins as solely anti-aging molecules constitutes an oversimplification[62]. Recent evidence has also indicated that resveratrol does not activate sirtuin 1 directly, but rather functions promiscuously on various targets including receptors, ion channels, enzymes and transporters[61]. Rapamycin-induced autophagy is independent of sirtuin 1, suggesting that rapamycin promotes autophagy through a distinct, non-overlapping mechanism[58] (Figure 10.6).

Polyamine spermidine has been recently discovered as a molecule that can promote autophagy and lead to longevity. Spermidine was demonstrated to prolong the lifespan of yeast, flies and worms in an autophagy-dependent fashion[59]. A diet enriched in physiologically relevant polyamines such as putrescine, spermidine and spermine also increases lifespan in mice. In yeast, spermidine caused a global hypo-acetylation of histone H3, but induced a selective histone acetylation at the promoter region of the ATG7 gene[63]. Thus, in the context of general gene silencing, transcription of ATG7 is enabled, resulting in a relative up-regulation of ATG7 transcription. Spermidine-mediated transcription of autophagy-relevant genes may account for the observed activation of autophagy. Spermidine was also

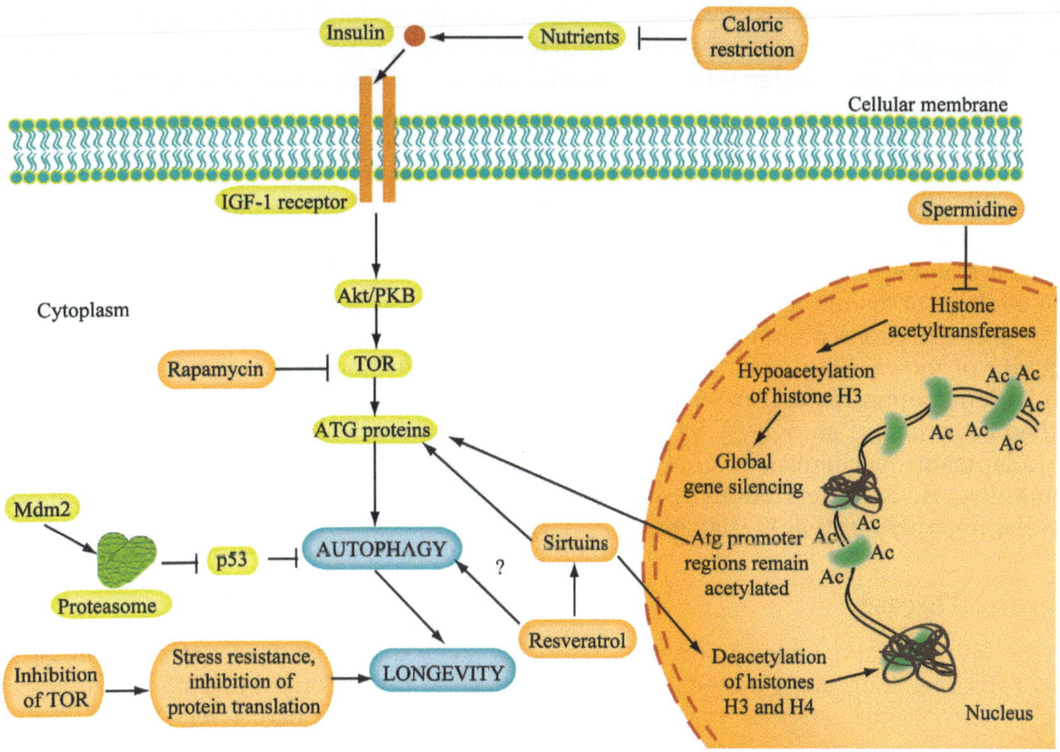

FIGURE 10.6 Autophagy and potential lifespan-prolonging treatments.

shown to inhibit histone acetyltransferases in cell-free assays. In support of a role for histone acetyltransferases in the lifespan of yeast, deletion of essential subunits in two histone acetyltransferases increased lifespan. Although treatment of this strain with polyamines further increased longevity, it was not to the same extent as that seen in wild-type cells[64]. Spermidine may induce autophagy through a combination of nuclear (transcriptional) and cytoplasmic (transcription-independent) mechanisms (Figure 10.6). A reasonable speculation is that inhibition of acetyltransferases (by spermidine) and activation of deacetylases (by resveratrol) might converge on the same autophagy regulators or effectors. However, a clear identification of the targets for these drugs remains elusive, and would need further investigation.

AUTOPHAGIC PATHWAYS AS POTENTIAL TARGETS FOR TREATMENT OF OTHER DISEASES

Type-II Diabetes

Type-II diabetes is characterized by insulin resistance and glucose intolerance, with decreased glucose consumption in muscle and fat cells and increased gluconeogenesis in the liver, which are caused by overfeeding when caloric intake exceeds energy expenditure[65]. Autophagy is increased in type-II diabetes because the insulin-sensitive cells are in a semi-starved state under these conditions. The major factor underlying this disease is the inability of the body to buffer the free fatty acid concentration, which is associated with an increased reduction of pressure on the mitochondrial respiratory chain, increased production of reactive oxygen species (ROS), reduced function of mitochondrial, and increased apoptosis, in β-cells[66]. Amongst the various fatty acids, palmitate is one of the most important members causing insulin resistance. Because palmitate is the precursor of ceramide and sphingolipid biosynthesis, the sphingolipid pathway has been implicated in the etiology of insulin resistance[67]. Ceramide stimulates autophagy and decreases intracellular concentrations of amino acids by inhibiting their transport, resulting in a decreased activity of the mTOR-dependent signaling pathways. Although it is not totally clear yet, inhibition of amino acid transport may also contribute to autophagy activated by ceramide. When activation of autophagy by ceramide occurs *in vivo*, this may provide the body with an adaptation mechanism of eliminating damaged mitochondria and avoiding apoptosis while fatty acid concentrations rise. These findings may help search for potential targets for the treatment of type-II diabetes.

Cardiac Disease

It is known that defective autophagy may play a key role in some rare forms of inherited heart disease. Autophagy may constitute an important physiological or pathophysiological response to cardiac stresses such as ischemia or pressure overload, which are frequently encountered in coronary artery disease, hypertension, aortic valvular disease, and congestive heart failure. The accumulation of autophagosomes has been observed in cardiac biopsy tissues from the patients with these disorders, rodent models of these cardiac diseases, and isolated stressed cardiomyocytes[68]. Prior to genetic studies, it was assumed that autophagy invariably contributed to myocyte degeneration; however, more recent data challenge this viewpoint, and support the viewpoint that the cyto-protective effects of autophagy may predominate in certain settings. Furthermore, the cardiomyocytes are post-mitotic cells in which basal autophagy may be important in protein and organelle quality control. Heart-specific knockout of ATG5 in adult mice results in cardiac hypertrophy and contractile dysfunction that is accompanied by increased levels of ubiquitinated proteins and sarcomere and mitochondrial structural abnormalities. Beside the need for basal level of autophagy, the heart may uniquely rely on stress-induced up-regulation of autophagy to ensure the availability of energy substrates and to promote cellular remodeling. The heart consumes more energy per gram than any other organ in the body, and common cardiac disorders (e.g., cardiac ischemia and heart failure) are characterized by a reduction in the availability of energy substrates, a factor that contributes to transient or sustained impairment of cardiac function[69]. Moreover, when the hearts are subjected to stress for a prolonged period of time, the myocytes will remodel cellular architecture to adapt to stressful conditions. The needs of the stressed heart for more energy substrates and for remodeling of cellular architecture may be met in part through the autophagy. The cardiac-specific deficiency of ATG5 in early cardiogenesis does not result in any phenotypic abnormalities under basal conditions but results in severe cardiac dysfunctions following treatment with pressure overload or β-adrenergic stress. These observations suggest that up-regulation of autophagy in failing hearts is an adaptive response that protects against hemodynamic or neuro-hormonal stresses. Although this is an attractive model, not all studies support an adaptive role for autophagy during cardiac stress. In a pressure overload model, heterozygous disruption of Beclin 1 led to preservation but not deterioration of contractile function. These results differ from those obtained in ATG5 knockout mice, which may reflect different contributions of ATG5 versus Beclin 1 or different effects of homozygous versus heterozygous disruption of an ATG gene[69]. Heterozygous disruption of Beclin 1 also decreases the size of the myocardial infarction after ischemia or reperfusion. However, in an *in vitro* model of cardiac ischemia or reperfusion, Beclin 1 over-expression decreased cell injury; expression of a dominant-negative Atg5 increased cell injury. These observations suggest a protective role for both ATG genes in ischemia/reperfusion[69]. Additional studies are needed to clarify whether activation of autophagy serves as an adaptive or maladaptive cardiac response to hemodynamic and ischemic stress. It is noteworthy, however, that in cardiac patients, interventions known to exacerbate ischemia or heart failure (e.g., β-adrenergic receptor agonists) reduce autophagy in the heart, whereas interventions known to ameliorate ischemia or heart failure tend to enhance autophagy in the heart[70]. These studies would provide hints for development and utilization of the autophagy-modulating agents in treatment of heart disease.

Muscular Disorders

Numerous studies demonstrated the presence of cytosolic vacuoles in the muscle cells of patients with altered muscular function, but only until recently, those vacuoles were linked to autophagy activity, due to the use of new autophagy markers[71]. Direct evidence for impaired autophagy in myopathy was first obtained in mice whose gene encoding a lysosomal membrane protein (lamp 2) was knocked out[72]. The predominant phenotype of these mice is massive accumulations of autophagic vacuoles in the cells of the liver, muscle and heart. Despite the increased number of autophagic vesicles, the rate of protein degradation in the lysosome is reduced because the autophagosome clearance through lysosomal fusion is impaired. The histological resemblance of the skeletal and heart muscles of the lamp2 knockout mice to those from patients with Danon disease led to the identification of mutations in lamp2 as the primary defect of this lysosomal storage disease, which is characterized clinically by cardiomyopathy, myopathy and variable mental retardation[73]. Recently, a second myopathy, characterized by muscle weaknesses at both birth and early death, has been linked to a defect in myotubularin, a phosphatase that modulates the levels of phosphatidylinositol (PtdIns); PtdIns has been implicated in vesicular trafficking and autophagy[74]. The identifications of these and other possible mutations in autophagy pathways and machinery are helping re-define and better classify these muscular pathologies. However, despite the ubiquitous distribution of the mutated genes, why vacuoles accumulate only in certain types of tissues but not in others remains a question. This question might be associated with alterations of autophagy pathways, as autophagy plays an important role in the clearance of such proteins. Thus, it will be logical to protect muscle cells against aggregate accumulation via developing drugs that target the autophagic pathways.

Hepatic Encephalopathy

Hepatic encephalopathy (HE) is characterized by neuropsychiatric manifestations, ranging from mild abnormalities in mental state to coma. Failure of liver function as a result of acute or chronic liver disease is the cause of this disease[75]. Under these conditions, ammonia, one of the major causative agents in HE, is not properly converted into either urea or glutamine by the liver. The mechanism of ammonia toxicity in the brain has been a question of debate for years and remains unclear. Astrocyte swelling, mitochondrial ROS production, and protein nitration have been proposed as the possible mechanisms[76]. Aberrant activation of the mTOR signaling pathway in astrocytes can be caused by a mutation in TSC1 and result in increased cell size, suggesting a potential new mechanism of the ammonia toxicity in hepatic encephalopathy. Increased intracellular glutamine (produced from ammonia) not only increases astrocyte volume during short term osmosis which may lead to brain swelling and increased intracranial pressure[77], but also during long term activation of the mTOR signaling pathway. This can then result in suppression of autophagy and the capacity of the astrocyte to remove aberrant cell structures such as mitochondria. Therefore, autophagy-modulating agents may represent a new class of drugs for treatment of HE.

Infection, Immunity and Inflammatory Diseases

The autophagic machinery is used in multipronged defenses against microbes, including the selective delivery of microorganisms to degradative lysosomes (xenophagy) and the delivery of microbial nucleic acids and antigens to endo/lysosomal compartments for activation of innate and adaptive immunity[78]. Numerous medically important pathogens are degraded by xenophagy, including bacteria, viruses and parasites. With the availability of tissue-specific ATG knockout mice, it should be possible to more thoroughly evaluate the role of xenophagy in microbial pathogenesis.

Recent studies have indicated that autophagy participates in trafficking events that activate innate and adaptive immunity[79]. With respect to innate immunity, ATG5 is required for the delivery of viral nucleic acids from Sendai virus and vesicular stomatitis virus to the endosomal toll-like receptor (TLR), TLR7, and subsequent activation of type I interferon signaling in plasmacytoid dendritic cells. With regard to adaptive immunity, autophagy is involved in the delivery of certain endogenously synthesized microbial antigens to MHC class II antigen-presenting molecules, leading to activation of CD4+ T lymphocytes. Not only does the autophagic machinery function in innate and adaptive immunity, but several innate and adaptive immune mediators involved in the control of intracellular pathogen stimulate autophagy, including the interferon-inducible antiviral molecule, PKR, CD40-CD40 ligand interactions, IFNγ, and its downstream immunity-related GTPases; TNFα; T-helper type 1 lymphocytes; and the cell-surface receptor, TLR4[78]. Successful intracellular microbes may antagonize both the signaling pathways that activate autophagy as well as the membrane trafficking events required for lysosomal delivery and degradation either directly or indirectly, by inhibiting the interactions with autophagy proteins[80]. Thus, selective disruption of the interactions between microbial virulence factors and the host autophagy proteins targeted by those factors may represent a new antimicrobial therapeutic strategy. Autophagy has additional effects on immunity that are not directly relevant to pathogen control. The role of autophagy in cellular homeostasis, life and death decisions may extend to the development and depletion of T lymphocyte during HIV infection. Additionally, autophagy may also be involved in the homeostasis, differentiation and function of other populations of immune cells. Another interesting question is whether autophagy plays a role in thymic selection and central tolerance. The role of autophagy in MHC class II presentation of self-antigens has not yet been well elucidated, but the high levels of autophagic activity in the thymic epithelial cells of newborn mice[81] suggest that autophagy may enable thymic epithelial cells to present self-antigens to lymphocytes during positive and negative selection. These findings should provide some new clues for development of novel approaches to therapeutic intervention of infection, immune and inflammatory diseases.

CONCLUDING REMARKS AND PERSPECTIVE

Autophagy has a Janus-face: while primarily being a survival mechanism, under certain circumstances it can also lead to autophagic cell death. Several key autophagic mediators, including ATGs, Bcl-2, PI3Kinase, mTOR and p53, have been known to play important roles in the complex autophagic network in the context of human diseases such as cancer, neurodegeneration, aging, type-II diabetes, cardiac diseases, muscular disorders, hepatic encephalopathy, as well as infection, immunity and inflammatory diseases. To a certain extent, our understanding of the links between autophagy and human diseases has benefited from the availability of rapamycin, an mTOR inhibitor that has been proved to be one of the powerful agents for discerning the major players in some human diseases associated with autophagy. However, there are also significant disadvantages inherent in the effects of rapamycin. Therefore, identifying and developing new and better small molecules that can modulate autophagy would be needed. In fact, some other drugs that can also modulate autophagy are already in clinical use for treatment of human diseases. These drugs include tamoxifen, chloroquine, imatinib, suberoylanilide hydroxamic acid (SAHA), lithium, valproate, carbamazepine, trehalose, 3, 5-dimethylpyrazole, acipimox, resveratrol, polyamine spermidine, etc. We believe that the recent biological insights, as discussed in this chapter, can provide foundations for launching the next round of small-molecule drug discovery. The drug discovery process can be driven by the availability of structural information on the potential targets. X-ray crystallography, nuclear magnetic resonance (NMR), and structural bioinformatics-docking techniques will be invaluable in the development and discovery of drugs that target key components and regulators of autophagic pathways. Hopefully, the emergence of new approaches such as accurate mathematics models (e.g., Naïve Bayesian framework), and the support vector machine (SVM) for the disruption of protein-protein interactions, would facilitate screening and identification of small molecules able to either inhibit or promote protein-protein interactions (PPIs), which affect the autophagic process, and those small molecules can be further considered, tested and developed to produce candidate drugs. Recent studies have implied that the best hope for targeting autophagy as therapeutic interventions may lie in the discovery of agents that are able to target the altered autophagy-regulating signaling pathways or even the autophagic network, rather than targeting the individual gene or protein molecule. Thus, a better understanding of the autophagic PPI network should provide useful insights into how these hub proteins and the

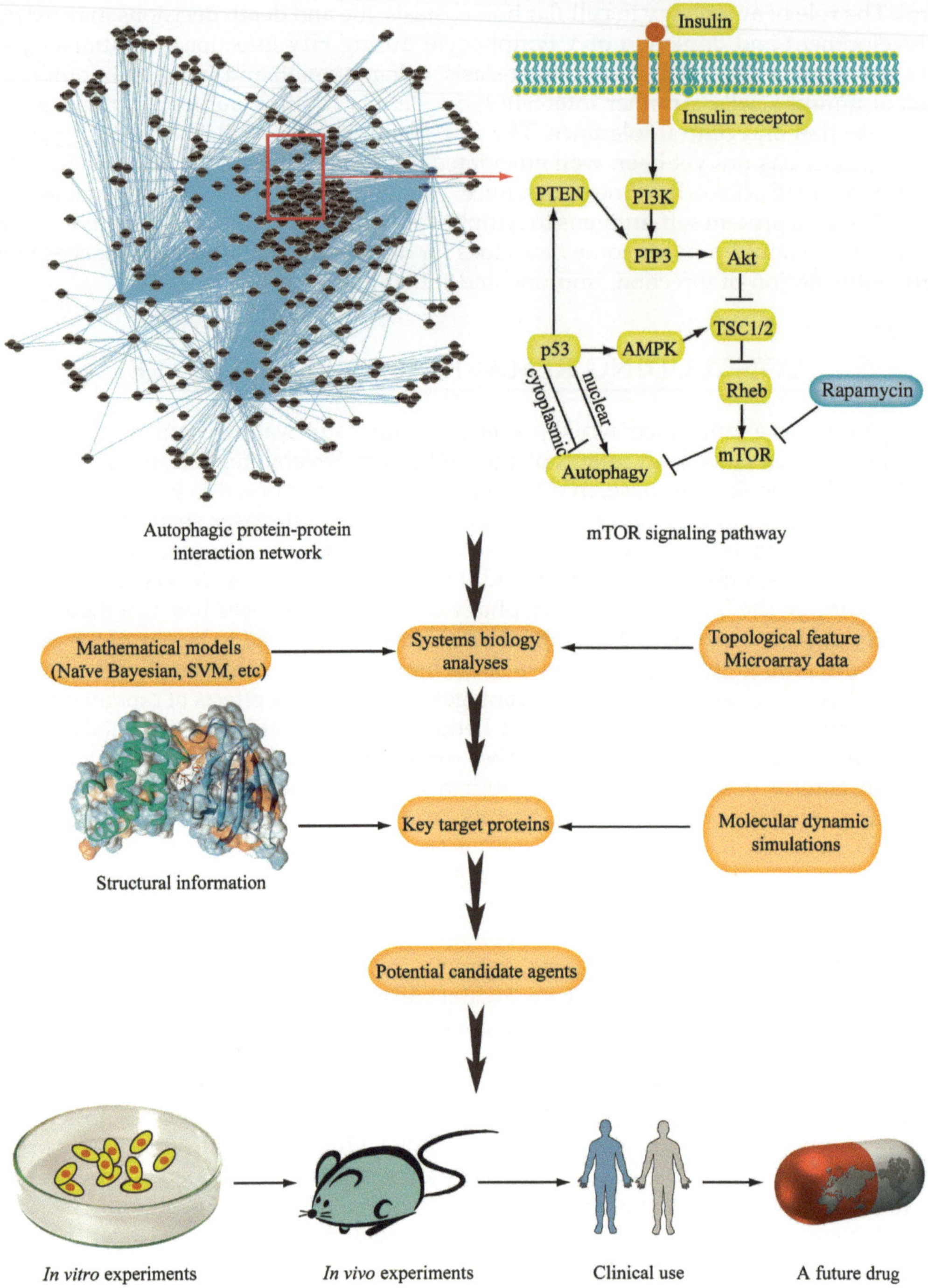

FIGURE 10.7 Autophagy network-based identification of novel targets for drug discovery.

autophagy-related signaling pathways can be exploited as potential therapeutic targets for treatment of human diseases. Another promising strategy is the screening of a chemical library to search for the compounds that have autophagy-modulating activity, and then assess whether and what kind of disorders those compounds can ameliorate.

References

1. Klionsky DJ, Emr SD. Autophagy as a regulated pathway of cellular degradation. *Science* 2000;**290**:1717–21.
2. Klionsky DJ. Autophagy: from phenomenology to molecular understanding in less than a decade. *Nat Rev Mol Cell Biol* 2007;**8**:931–7.
3. Yang Z, Klionsky DJ. Eaten alive: a history of macroautophagy. *Nat Cell Biol* 2010;**12**:814–22.
4. Mizushima N, Levine B, Cuervo AM, Klionsky DJ. Autophagy fights disease through cellular self-digestion. *Nature* 2008;**451**:1069–75.
5. Levine B, Kroemer G. Autophagy in the pathogenesis of disease. *Cell* 2008;**132**:27–42.
6. Yorimitsu T, Klionsky DJ. Autophagy: molecular machinery for self-eating. *Cell Death Differ* 2005;**12**:1542–52.
7. Huang J, Klionsky DJ. Autophagy and human disease. *Cell* 2007;**6**:1837–49.
8. Liu B, Cheng Y, Liu Q, Bao JK, Yang JM. Autophagic pathways as new targets for cancer drug development. *Acta Pharmacol Sin* 2010;**31**:1154–64.
9. Rubinsztein DC, Gestwicki JE, Murphy LO, Klionsky DJ. Potential therapeutic applications of autophagy. *Nat Rev Drug Discov* 2007;**6**:304–12.
10. Michael OH. The biochemistry of apoptosis. *Nature* 2000;**407**:770–7.
11. Levine B, Klionsky DJ. Development by self-digestion: molecular mechanisms and biological functions of autophagy. *Dev Cell* 2004;**6**:463–77.
12. Hannigan AM, Gorski SM. Macroautophagy: the key ingredient to a healthy diet? *Autophagy* 2009;**5**:140–51.
13. Andrew T. Apoptosis and autophagy: regulatory connections between two supposedly different processes. *Apoptosis* 2008;**13**:1–9.
14. Liu B, Cheng Y, Zhang B, Bian HJ, Bao JK. *Polygonatum cyrtonema* lectin induces apoptosis and autophagy in human melanoma A375 cells through a mitochondria-mediated ROS-p38-p53 pathway. *Cancer Lett* 2009;**275**:54–60.
15. Cheng Y, Qiu F, Ikejima T. Molecular mechanisms of oridonin-induced apoptosis and autophagy in murine fibrosarcoma L929 cells. *Autophagy* 2009;**5**:430–43.
16. Mathew R, Karantza-Wadsworth V, White E. Role of autophagy in cancer. *Nat Rev Cancer* 2007;**7**:961–7.
17. Kroemer G, Jaattela M. Lysosomes and autophagy in cell death control. *Nat Rev Cancer* 2005;**5**:886–97.
18. Gozuacik D, Kimchi A. Autophagy as a cell death and tumor suppressor mechanism. *Oncogene* 2004;**23**:2891–906.
19. Takeuchi H, Kondo Y, Fujiwara K, Kanzawa T, Aoki H, Mills GB, et al. Synergistic augmentation of rapamycin-induced autophagy in malignant glioma cells by phosphatidylinositol 3-kinase/protein kinase B inhibitors. *Cancer Res* 2005;**65**:3336–46.
20. Morselli E, Galluzzi L, Kepp O, Vicencio JM, Criollo A, Maiuri MC, et al. Anti- and pro-tumor functions of autophagy. *Biochim. Biophys Acta* 2009;**1793**:1524–32.
21. Dalby KN, Tekedereli I, Lopez-Berestein G, Ozpolat B. Targeting the prodeath and prosurvival functions of autophagy as novel therapeutic strategies in cancer. *Autophagy* 2010;**6**:322–9.
22. Elisabeth AC, Pietri P, Marja J. Apoptosis and autophagy: Targeting autophagy signaling in cancer cells – 'trick or treats'? *FEBS J* 2009;**276**:6084–96.
23. de Medina P, Payré B, Boubekeur N, Bertrand-Michel J, Tercé F, Silvente-Poirot S, et al. Ligands of the anti-estrogen-binding site induce active cell death and autophagy in human breast cancer cells through the modulation of cholesterol metabolism. *Cell Death Differ* 2009;**16**:1372–84.
24. Amaravadi RK, Yu D, Lum JJ, Bui T, Christophorou MA, Evan GI, et al. Autophagy inhibition enhances therapy-induced apoptosis in a Myc-induced model of lymphoma. *J Clin Invest* 2007;**117**:326–36.
25. Ertmer A, Huber V, Gilch S, Yoshimori T, Erfle V, Duyster J, et al. The anticancer drug matinib induces cellular autophagy. *Leukemia* 2007;**21**:936–42.
26. Marks PA. Discovery and development of SAHA as an anticancer agent. *Oncogene* 2007;**26**:1351–6.

27. Kim KW, Hwang M, Moretti L, Jaboin JJ, Cha YI, Lu B. Autophagy upregulation by inhibitors of caspase-3 and mTOR enhances radiotherapy in a mouse model of lung cancer. *Autophagy* 2008;**4**:659–68.
28. Miller WH, Schipper HM, Lee JS, Singer J, Waxman S. Mechanisms of action of arsenic trioxide. *Cancer Res* 2002;**62**:3893–903.
29. Kanzawa T, Zhang L, Xiao L, Germano IM, Kondo Y, Kondo S. Arsenic trioxide induces autophagic cell death in malignant glioma cells by upregulation of mitochondrial cell death protein BNIP3. *Oncogene* 2005;**24**:980–91.
30. Liu B, Cheng Y, Bian HJ, Bao JK. Molecular mechanisms of *Polygonatum cyrtonema* Lectin-induced apoptosis and autophagy in cancer cells. *Autophagy* 2009;**5**:253–5.
31. Liu B, Wu JM, Li J, Li WW, Liu JJ, Bao JK. *Polygonatum cyrtonema* lectin induces murine fibrosarcoma L929 cell apoptosis and autophagy via blocking Ras-Raf and PI3K-Akt signaling pathways. *Biochimie*; 2010. doi:10.1016/j.biochi.2010.08.009.
32. Arora S, Yang JM, Kinzy TG, Utsumi R, Okamoto T, Kitayama T, et al. Identification and Characterization of an Inhibitor of Eukaryotic Elongation Factor 2 Kinase against Human Cancer Cell Lines. *Cancer Res* 2003;**63**:6894–9.
33. Wu H, Yang JM, Jin S, Zhang H, Hait WN. Elongation Factor-2 Kinase Regulates Autophagy in Glioblastoma Cells. *Cancer Res* 2006;**66**:3015–23.
34. Wu H, Zhu H, Liu DX, Niu TK, Ren XC, Patel R, et al. Silencing of Elongation Factor-2 Kinase Potentiates the Effect of 2-Deoxy-D-Glucose against Human Glioma Cells through Blunting of Autophagy. *Cancer Res* 2009;**69**:2453–60.
35. Cheng Y, Li H, Ren XC, Niu TK, Hait WN, Yang JM. Cytoprotective Effect of the Elongation Factor-2 Kinase-mediated Autophagy in Breast Cancer Cells Subjected to Growth Factor Inhibition. *PLoS ONE* 2010;**3**:e9715.
36. Williams A, Jahreiss L, Sarkar S, Saiki S, Menzies FM, Ravikumar B, et al. Aggregate-prone proteins are cleared from the cytosol by autophagy: therapeutic implications. *Curr Top Dev Biol* 2006;**76**:89–101.
37. Berger Z, Davies JE, Luo S, Pasco MY, Majoul I, O'Kane CJ, et al. Deleterious and protective properties of an aggregate-prone protein with a polyalanine expansion. *Hum Mol Genet* 2006;**15**:453–65.
38. Hara T, Nakamura K, Matsui M, Yamamoto A, Nakahara Y, Suzuki-Migishima R, et al. Suppression of basal autophagy in neural cells causes neurodegenerative disease in mice. *Nature* 2006;**441**:885–9.
39. Martinez-Vicente M, Cuervo AM. Autophagy and neurodegeneration: when the cleaning crew goes on strike. *Lancet Neurol* 2007;**6**:352–61.
40. García-Arencibia M, Hochfeld WE, Toh PP, Rubinsztein DC. Autophagy, a guardian against neurodegeneration. *Semin. Cell Dev Biol* 2010;**21**:691–8.
41. Yue Z, Friedman L, Komatsu M, Tanaka K. The cellular pathways of neuronal autophagy and their implication in neurodegenerative diseases. *Biochim Biophys Acta* 2009;**1793**:1496–507.
42. Sarbassov DD, Ali SM, Sengupta S, Sheen JH, Hsu PP, Bagley AF, et al. Prolonged rapamycin treatment inhibits mTORC2 assembly and Akt/PKB. *Mol Cell* 2006;**22**:159–68.
43. Kim DH, Sarbassov DD, Ali SM, King JE, Latek RR, Erdjument-Bromage H, et al. mTOR interacts with raptor to form a nutrient-sensitive complex that signals to the cell growth machinery. *Cell* 2002;**110**:163–75.
44. Berger Z, Ravikumar B, Menzies FM, Oroz LG, Underwood BR, Pangalos MN, et al. Rapamycin alleviates toxicity of different aggregate-prone proteins. *Hum. Mol Genet* 2006;**15**:433–42.
45. Ravikumar B, Vacher C, Berger Z, Davies JE, Luo S, Oroz LG, et al. Inhibition of mTOR induces autophagy and reduces toxicity of polyglutamine expansions in fly and mouse models of Huntington disease. *Nat Genet* 2004;**36**:585–95.
46. Ravikumar B, Berger Z, Vacher C, O'Kane CJ, Rubinsztein DC. Rapamycin pretreatment protects against apoptosis. *Hum. Mol Genet* 2006;**15**:1209–16.
47. Sarkar S, Rubinsztein DC. Inositol and IP3 levels regulate autophagy: biology and therapeutic speculations. *Autophagy* 2006;**2**:132–4.
48. Sarkar S, Krishna G, Imarisio S, Saiki S, O'Kane CJ, Rubinsztein DC. A rational mechanism for combination treatment of Huntington's disease using lithium and rapamycin. *Hum. Mol Genet* 2008;**17**:170–8.
49. Tanaka M, Machida Y, Niu S, Ikeda T, Jana NR, Doi H, et al. Trehalose alleviates polyglutamine-mediated pathology in a mouse model of Huntington disease. *Nat Med* 2004;**10**:148–54.
50. Sarkar S, Perlstein EO, Imarisio S, Pineau S, Cordenier A, Maglathlin RL, et al. Small molecules enhance autophagy and reduce toxicity in Huntington's disease models. *Nat Chem Biol* 2007;**3**:331–8.

51. Imarisio S, Carmichael J, Korolchuk V, Chen CW, Saiki S, Rose C, et al. Huntington's disease: from pathology and genetics to potential therapies. *Biochem J* 2008;**412**:191–209.
52. Achilli F, Bros-Facer V, Williams HP, Banks GT, AlQatari M, Chia R, et al. An ENUinduced mutation in mouse glycyl-tRNA synthetase (GARS) causes peripheral sensory and motor phenotypes creating a model of Charcot-Marie-Tooth type 2D peripheral neuropathy. *Dis Model Mech* 2009;**2**:359–73.
53. Menzies FM, Rubinsztein DC. Broadening the therapeutic scope for rapamycin treatment. *Autophagy* 2010;**6**:286–7.
54. Cuervo AM, Bergamini E, Brunk UT, Droge W, French M, Terman A. Autophagy and aging: the importance of maintaining "clean" cells. *Autophagy* 2005;**1**:131–40.
55. Alessio D, Gabriella C, Cristiano C, Zina G, Ilaria P, Ettore B. Anti-aging effects of anti-lipolytic drugs. *Exp Gerontol* 2004;**39**:1061–7.
56. Cavallini G, Donati A, Gori Z, Pollera M, Bergamini E. The protection of rat liver autophagic proteolysis from the age-related decline co-varies with the duration of anti-ageing food restriction. *Exp Gerontol* 2001;**36**:497–506.
57. Harrison DE, Strong R, Sharp ZD, Nelson JF, Astle CM, Flurkery K, et al. Rapamycin fed late in life extends lifespan in genetically heterogeneous mice. *Nature* 2009;**460**:392–5.
58. Morselli E, Maiuri MC, Markaki M, Megalou E, Pasparaki A, Palikaras K, et al. Caloric restriction and resveratrol promote longevity through the Sirtuin-1-dependent induction of autophagy. *Cell Death Dis* 2010;**1**:10.
59. Wood JG, Rogina B, Lavu S, Howitz K, Helfand SL, Tatar M, et al. Sirtuin activators mimic caloric restriction and delay ageing in metazoans. *Nature* 2004;**430**:686 9.
60. Baur JA, Pearson KJ, Price NL, Jamieson HA, Lerin C, Kalra A, et al. Resveratrol improves health and survival of mice on a high-calorie diet. *Nature* 2006;**444**:337–42.
61. Kaeberlein M. The Ongoing saga of sirtuins and aging. *Cell Metab* 2008;**8**:4–5.
62. Li Y, Xu W, McBurney MW, Longo VD. SirT1 Inhibition Reduces IGF-I/IRS-2/Ras/ERK1/2 Signaling and Protects Neurons. *Cell Metab* 2008;**8**:38–48.
63. Eisenberg T, Knauer H, Schauer AI, Büttner S, Ruckenstuhl C, Carmona-Gutierrez D, et al. Induction of autophagy by spermidine promotes longevity. *Nat Cell Biol* 2009;**11**:1305–14.
64. Lee IH, Cao L, Mostoslavsky R, Lombard DB, Liu J, Bruns NE, et al. A role for the NAD-dependent deacetylase Sirt1 in the regulation of autophagy. *Proc Natl Acad Sci USA* 2008;**105**:3374–9.
65. Petersen KF, Shulman GI. Etiology of insulin resistance. *Am J Med* 2006;**119**:S10–6.
66. Evans JL, Goldfine ID, Maddux BA, Grodsky GM. Oxidative stress and stress-activated signaling pathways: a unifying hypothesis of type 2 diabetes. *Endocr Rev* 2002;**23**:599–622.
67. Yamashita T, Hashiramoto A, Haluzik M, Mizukami H, Beck S, Norton A, et al. Enhanced insulin sensitivity in mice lacking ganglioside GM3. *Proc Natl Acad Sci USA* 2003;**100**:3445–9.
68. Terman A, Brunk UT. Autophagy in cardiac myocyte homeostasis, aging, and pathology. *Cardiovasc Res* 2005;**68**:355–65.
69. Hamacher-Brady A, Brady NR, Gottlieb RA. Enhancing macroautophagy protects against ischemia/reperfusion injury in cardiac myocytes. *J Biol Chem* 2006;**281**:29776–87.
70. Bahro M, Pfeifer U. Short-term stimulation by propanolol and verapamil of cardiac cellular autophagy. *J Mol Cell Cardiol* 1987;**19**:1169–78.
71. Nishino I. Autophagic vacuolar myopathies. *Curr Neurol Neurosci Rep* 2003;**3**:64–9.
72. Suzuki T, Nakagawa M, Yoshikawa A, Sasagawa N, Yoshimori T, Ohsumi Y, et al. The first molecular evidence that autophagy relates rimmed vacuole formation in chloroquine myopathy. *J Biochem* 2002;**131**:647–51.
73. Yamamoto A, Morisawa Y, Verloes A, Murakami N, Hirano M, Nonaka I, et al. Infantile autophagic vacuolar myopathy is distinct from Danon disease. *Neurology* 2001;**57**:903–5.
74. Buj-Bello A, Laugel V, Messaddeq N, Zahreddine H, Laporte J, Pellissier JF, et al. The lipid phosphatase myotubularin is essential for skeletal muscle maintenance but not for myogenesis in mice. *Proc Natl Acad Sci USA* 2002;**99**:15060–5.
75. Jones EA, Meijer AJ, Chamuleau RA. *Encephalopathy and Nitrogen Metabolism in Liver Failure*. Dordrecht, Boston, London: Kluwer Academic Publishers; 2003.

76. Uhlmann EJ, Li W, Scheidenhelm DK, Gau CL, Tamanoi F, Gutmann DH. Loss of tuberous sclerosis complex 1 (Tsc1) expression results in increased Rheb/S6K pathway signaling important for astrocyte cell size regulation. *Glia* 2004;**47**:180–8.
77. Alfred JM, Patrice C. Signalling and autophagy regulation in health, aging and disease. *Molecular Aspects of Medicine* 2006;**27**:411–25.
78. Levine B, Deretic V. Unveiling the roles of autophagy in innate and adaptive immunity. *Nat Rev Immunol* 2007;**7**:767–77.
79. Schmid D, Munz C. Innate and adaptive immunity through autophagy. *Immunity* 2007;**27**:11–21.
80. Espert L, Denizot M, Grimaldi M, Robert-Hebmann V, Gay B, Varbanov M, et al. Autophagy is involved in T cell death after binding of HIV-1 envelope proteins to CXCR4. *J Clin Invest* 2006;**116**:2161–72.
81. Mizushima N, Yamamoto A, Matsui M, Yoshimori T, Ohsumi Y. In vivo analysis of autophagy in response to nutrient starvation using transgenic mice expressing a fluorescent autophagosome marker. *Mol Biol Cell* 2004;**15**:1101–11.

CHAPTER

11

Measuring Autophagy in Vivo

Roberta A. Gottlieb

Director, Donald P. Shiley BioScience Center, San Diego State University, 5500 Campanile Drive, San Diego, CA 92182-4650

PROTEIN DEGRADATION

The definitive method to monitor lysosomal degradation is measurement of turnover of long-lived radioactive proteins[2]. While this is regarded as the gold standard, it is not applicable to human studies, although the use of ^{15}N-labeled amino acids and mass spectrometry may eventually be feasible.

ELECTRON MICROSCOPY

Autophagosomes, originally termed "cytolysosomes," were first described in electron microscope images as vacuolar structures distinguished by the presence of a double membrane and often containing cytoplasmic contents (glycogen granules, mitochondria, or other structures) (Figure 11.1)[3]. This is still considered a definitive demonstration of autophagy, but is difficult to employ for quantitation of the process and does not allow determination of flux. However, when investigating the suspected occurrence of autophagy in a new setting, it provides conclusive evidence of the phenomenon.

WESTERN BLOT OF LC3-II

The field was stalled until many of the autophagy-related genes were cloned, allowing for the development of antibodies. Nowadays the most commonly used method to assess autophagy in cell culture and tissue is measurement of LC3-II content by Western blot. LC3-II is rather labile; therefore, care must be taken in tissue handling. Ideally, samples should be snap-frozen, then solubilized in RIPA buffer or ground to a powder at liquid nitrogen temperature and then solubilized. LC3-II should be normalized to a cytosolic protein that does not change during the intervention under study. Previously, many

Autophagy in Health and Disease
http://dx.doi.org/10.1016/B978-0-12-385101-7.00011-5

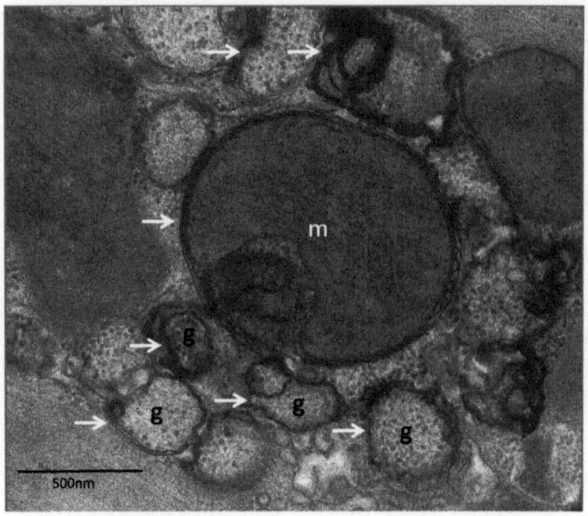

FIGURE 11.1 Electron microscope image of an autophagosome in a cardiomyocyte. The arrows indicate autophagosomes; m=mitochondria, g=glycogen granules.

investigators used the ratio of LC3-II to LC3-I as an indicator of autophagic activity. However, this has been discarded in favor of focusing on LC3-II normalized to a constant protein. In our experience, tubulin or actin are more stable than GAPDH for studies of cardiac autophagy, but this may not be the case when analyzing the nervous system (tubulin is variable) or skeletal muscle (actin levels change with exercise or atrophy). Because GAPDH is linked to energy metabolism, it may change in response to starvation, making it a potentially unreliable basis for normalization. Therefore, for the particular system under study, it is important to do comparisons and select the best marker to use for subsequent normalizations.

Acute changes in autophagy proteins can occur without a detectable change in mRNA; therefore mRNA can be a poor index of autophagic activity. However, in some settings, transcript levels can be informative. For instance, in mice, mRNA transcripts for many autophagy proteins diminish with age, and protein levels diminish in parallel. Changes in autophagy are not conserved across tissues, even in response to a generalized stimulus such as starvation, precluding generalization about responses in different tissues.

MEASURING AUTOPHAGIC FLUX

Autophagy is a dynamic process: autophagosome formation is followed by destruction in the lysosome. Thus the number of autophagosomes at any point in time is a reflection of their formation and their destruction in lysosomes. Autophagic flux can be inferred from comparison of parallel samples with and without lysosomal blockade (Figure 11.2). LC3-II abundance detected by Western blot parallels the autophagosome number assessed by fluorescence microscopy. If autophagosomes (or LC3-II levels) increase after blockade, then autophagic flux is intact. If, however, autophagosome (or LC3-II) abundance increases minimally after blockade, then flux is impaired. This should correlate with detergent-insoluble p62/SQSTM1 accumulation, which can be detected by immunofluorescence or Western blot.

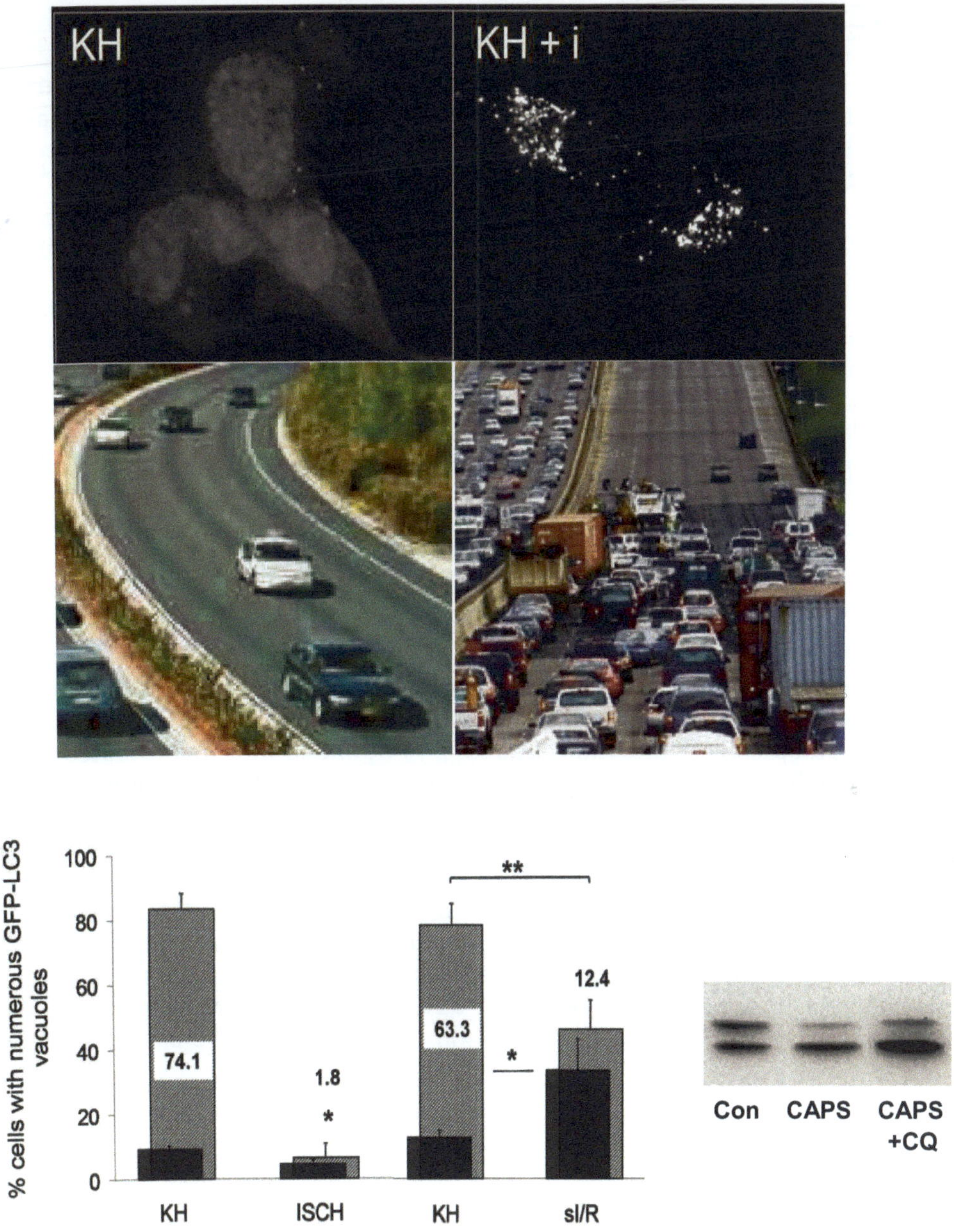

FIGURE 11.2 Diagram of autophagic flux. Image of HL1 cells in starvation buffer (KH) without and with (KH+i) flux inhibitos bafilomycin and pepstatin. Cells were transfected with LC3-GFP then subjected to 3.5h amino acid and serum starvation. Inset shows western blot of endogenous LC3. Traffic images convey the metaphor of lysosomal blockade equivalent to roadblock.

WESTERN BLOT OF OTHER AUTOPHAGY PROTEINS

Expression levels of most autophagy proteins do not change in response to starvation or other interventions; however, LC3-II, Beclin 1, and p62/SQSTM1 do change acutely in response to a stimulus. Notably, both p62 and Beclin 1 levels can rise within minutes after an autophagic stimulus, whereas levels of Atg3, Atg4, Atg5, and Atg7 do not[4]. Because p62/SQSTM1 can be incorporated into heavily ubiquitinated protein aggregates, levels of detergent-insoluble p62 reflect the inefficiency of protein aggregate removal by autophagic machinery[5].

Analysis of p62 coupled with LC3 can provide insights about autophagic flux in some tissues. If basal p62 levels are high and LC3-II is low, one can infer low levels of autophagic flux. Conversely, if LC3-II is high and p62 decreases after stimulation, this would be consistent with high flux.

Phosphorylation of ULK1 can be measured by Western blot; Since ULK1 is responsible for initiating autophagy, phosphorylation of Ser 317 and Ser 777 by AMPK parallels autophagic activity, whereas phosphorylation of Ser 757 by mTOR suppresses ULK1 and limits activation of autophagy (Figure 11.3).[6, 7]

IMAGING OF LC3-GFP

The introduction of LC3-GFP to image autophagosomes accelerated the study of autophagy[8]. Under resting conditions, LC3-GFP is diffusely distributed in the cytosol (as LC3-I). Under autophagic stimulation, LC3-GFP is lipidated and incorporated into autophagosomal membranes (as LC3-II)[9]. This can serve as a reliable and highly informative measure in which the number of autophagosomes, the total area or volume occupied by autophagosomes, or the percentage of cells with many autophagosomes, can be used as an index of autophagic activity. Some caveats exist: one must consider flux, as the number of autophagosomes at any point in time is a balance of formation and destruction[10]. Additionally, overexpression of LC3-GFP can result in protein aggregates that look like puncta[11]; therefore it is important to avoid massive overexpression and to confirm through other methods that the puncta actually represent autophagosomes. Transient transfection or adenoviral delivery of LC3-GFP can induce some autophagy above baseline, so again, it

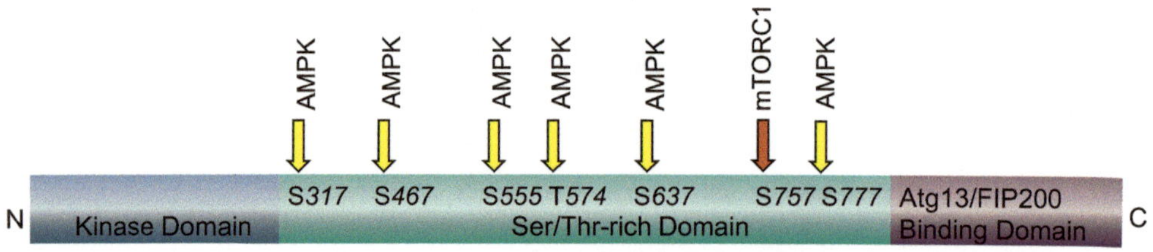

FIGURE 11.3 Sites of phosphorylation of ULK1. AMPK activates ULK1 by phosphorylating residues 317 and 777[6], and 467, 555, 574, and 637[7] (yellow arrows). Phosphorylation by mTORC1 on serine 757 (red arrow) prevents activation by AMPK[6].

is important to incorporate appropriate controls. Finally, many interventions (temperature shifts, DMSO used as a drug vehicle, anesthesia) can have their own effects on autophagy, again requiring attention to controls.

TANDEM RFP-GFP-LC3

Another approach to monitoring autophagy has been the use of tandem LC3-GFP-RFP[12] (or LC3-GFP and LC3-mCherry double-transgenic mice[13]). This is based on the happy circumstance that GFP is degraded in the lysosome more rapidly than mRFP or mCherry. Thus autophagosomes (before lysosomal fusion) would be both green and red, while autophagolysosomes would only have a red signal. In this way, autophagic flux can be inferred; impaired flux would be indicated by an abundance of green-and-red puncta and few or no red-only puncta.

DYES

Although the fluorescent LC3 constructs and transgenic mice have enabled great strides forward in our understanding of autophagy in vivo, there is a dearth of tools to measure autophagy in non-transgenic animals including humans. One reagent that has had limited acceptance is monodansylcadaverine (MDC)[14]. This probe accumulates in acidic compartments including autophagolysosomes, but also functions as a solvent polarity probe, thus yielding greater fluorescence in the hydrophobic environment of the autophagosome double membrane[15]. Unfortunately, MDC has weak fluorescence, which is further reduced upon fixation, and there is considerable nonspecific (diffuse cytosolic) staining. AlexaFluor488™-cadaverine has better quantum yield fluorescence but definitive evidence that it selectively labels autophagosomes is lacking. However, these dyes hold promise for use in tissues, if given systemically before tissue harvest[16]. In our experience they do not label autophagosomes if applied to fixed tissues or cryosections. There is a need to be able to monitor autophagy in vivo, including fluorescent dyes or agents suitable for medical imaging.

FUNCTIONAL INTERVENTION

The occurrence of autophagy in response to a stressor does not provide insight as to whether the response is adaptive or deleterious. To gain insight into the role of autophagy in a given setting, it is informative to either block autophagy or upregulate it. Pharmacologic agents to induce autophagy include rapamycin and a host of other small molecules. Drugs to inhibit autophagy initiation are rather limited: PI3-kinase inhibitors wortmannin and 3-methyladenine are commonly used but have off-target effects that can confound their interpretation. Drugs to inhibit lysosomal flux include bafilomycin A1 (inhibits the vacuolar proton ATPase that acidifies the lysosome), chloroquine (weak base that accumulates in the lysosome and neutralizes the pH, interfering with degradation), protease inhibitors such as leupeptin or pepstatin, or colchicine to prevent trafficking to the lysosome. A more specific inhibitor is the recombinant cell-permeable protein Tat-Atg5(K130R) which functions as a dominant negative; it has been used in isolated perfused hearts to demonstrate a role for autophagy in cardioprotection, but has not yet been applied in vivo[17, 18].

CIRCADIAN RHYTHMS AND AUTOPHAGY

Autophagy is highly regulated by nutritional and hormonal status, which are in turn tightly linked to circadian rhythm. Thus, it is not surprising that there are several reports that autophagy also varies markedly in accordance with diurnal cycle. In key studies it was shown that the abundance of autophagic vacuoles determined with EM changes according to time of day, not only in energy storage tissues such as liver but also in other organs including heart and kidney[19, 20] (Figure 11.4).

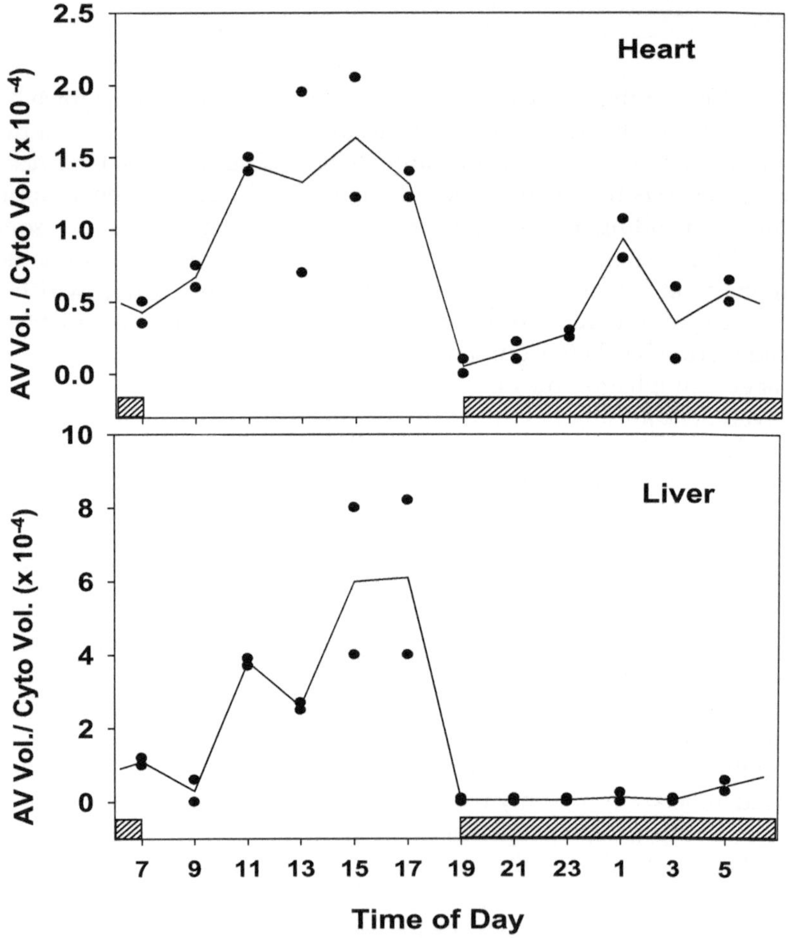

FIGURE 11.4 Circadian variation in autophagosome content in heart and liver. *Adapted from Pfeiffer[19].*

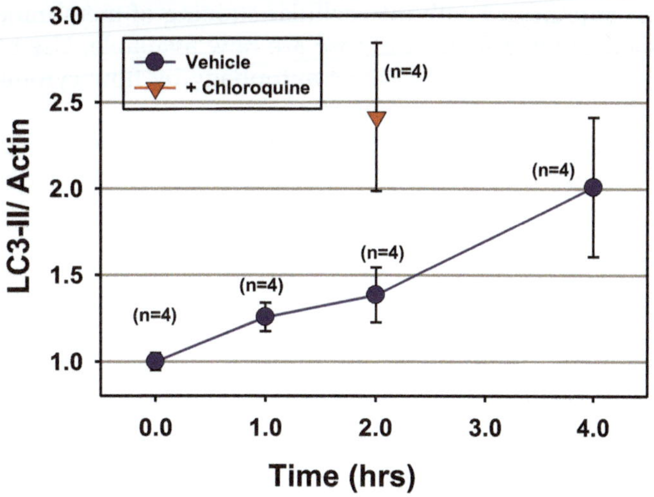

FIGURE 11.5 Changes in autophagy in mouse hearts over 4 hours, and flux determination with chloroquine.

The literature linking autophagy to circadian rhythm has recently been reviewed and potential implications to disease discussed[21]. More recently, it has been shown that the circadian variation in autophagy in mouse is also accompanied by a rhythmic expression of autophagy linked genes including ULK-1, GABARAP1, and Bnip3 which are in turn regulated via C/EBPβ[22]. This key transcription factor was suggested to link autophagy to circadian rhythm and is responsive to both nutritional as well as circadian pacemaker signals. The circadian variation in autophagy has important implications regarding the design, conduct, and translational interpretation of animal studies particularly when these studies are conducted in rodents which are nocturnal animals and are active and feeding during the dark phase. In our studies, we have found that the magnitude of change in cardiac autophagy as reflected by LC3-II can vary substantially by as much as 2-fold from morning to afternoon in ad-lib fed mice (Figure 11.5). Moreover, it is as yet unknown whether the ability to induce autophagy or the magnitude of the induced response will also vary relative to circadian cycle.

CLINICALLY FEASIBLE APPROACHES TO MONITORING AUTOPHAGY

It is possible to perform Western blot analysis of autophagy proteins from biopsy samples; analysis of LC3-II and detergent-insoluble p62 would be the most informative parameters. It is important to consider where the biopsy is obtained. For instance, slow-twitch vs. fast-twitch muscle vary widely in autophagic activity[23].

Lower levels of LC3-II (measured by Western blot) were observed in peripheral blood leukocytes in patients with coronary artery disease compared to age and sex-matched healthy controls[24]. Reagents are now available that would allow flow cytometric analysis of peripheral blood lymphocytes with intracellular staining of membrane-associated LC3[25]. A number of commercial kits for this purpose are now available. Use of a fluorescent dye such as AlexaFluor488™-cadaverine to detect autophagy by flow cytometry may be useful if the diffuse cytosolic staining can be overcome.

Agents suitable for detecting autophagy by magnetic resonance imaging, positron emission tomography, or computed tomography are needed but have not yet been developed.

CONCLUSION

Age, activity, metabolic status, infection, recent exercise, drug exposure, and disease processes can all influence autophagy in various tissues, as discussed in the other chapters in this book. These variables must be taken into account in experimental design and data interpretation. Given that autophagy is such a common response to cellular stress, it is likely to be encountered in many more disease contexts. The challenge will be to understand whether autophagy is functional (intact flux) and part of a protective response, or whether its activity contributes to the disease process. It is also important to be able to distinguish whether autophagy is impaired as a consequence of the disease pathology, or whether impaired autophagy is fundamentally responsible for the pathologic process.

References

1. Klionsky DJ, Abeliovich H, Agostinis P, Agrawal DK, Aliev G, Askew DS, et al. Guidelines for the use and interpretation of assays for monitoring autophagy in higher eukaryotes. *Autophagy* 2008;**4**(2):151–75.
2. Kisen GO, Tessitore L, Costelli P, Gordon PB, Schwarze PE, Baccino FM, et al. Reduced autophagic activity in primary rat hepatocellular carcinoma and ascites hepatoma cells. *Carcinogenesis* 1993;**14**(12):2501–5.
3. Novikoff AB. Biochemical and staining reactions of cytoplasmic constituents. In: Rudnick D, editor. *Developing Cell Systems and Their Control*. New York: Ronald Press Co; 1967. p. 167–203.
4. Barth S, Glick D, Macleod KF. Autophagy: assays and artifacts. *J Pathol* 2010;**221**(2):117–24.
5. Bartlett BJ, Isakson P, Lewerenz J, Sanchez H, Kotzebue RW, Cumming RC, et al. p62, Ref(2)P and ubiquitinated proteins are conserved markers of neuronal aging, aggregate formation and progressive autophagic defects. *Autophagy* 2011;**7**(6):572–83.
6. Kim J, Kundu M, Viollet B, Guan KL. AMPK and mTOR regulate autophagy through direct phosphorylation of Ulk1. *Nat Cell Biol* 2011;**13**(2):132–41.
7. Egan DF, Shackelford DB, Mihaylova MM, Gelino S, Kohnz RA, Mair W, et al. Phosphorylation of ULK1 (hATG1) by AMP-activated protein kinase connects energy sensing to mitophagy. *Science* 2011; **331**(6016):456–61.
8. Mizushima N, Yamamoto A, Matsui M, Yoshimori T, Ohsumi Y. In vivo analysis of autophagy in response to nutrient starvation using transgenic mice expressing a fluorescent autophagosome marker. *Mol Biol Cell* 2004;**15**(3):1101–11.
9. Ichimura Y, Kirisako T, Takao T, Satomi Y, Shimonishi Y, Ishihara N, et al. A ubiquitin-like system mediates protein lipidation. *Nature* 2000;**408**(6811):488–92.
10. Iwai-Kanai E, Yuan H, Huang C, Sayen MR, Perry-Garza CN, Kim L, et al. A method to measure cardiac autophagic flux in vivo. *Autophagy* 2008;**4**(3):322–9.
11. Kuma A, Matsui M, Mizushima N. LC3, an autophagosome marker, can be incorporated into protein aggregates independent of autophagy: caution in the interpretation of LC3 localization. *Autophagy* 2007;**3**(4):323–8.
12. Kimura S, Noda T, Yoshimori T. Dissection of the autophagosome maturation process by a novel reporter protein, tandem fluorescent-tagged LC3. *Autophagy* 2007;**3**(5):452–60.

13. Terada M, Nobori K, Munehisa Y, Kakizaki M, Ohba T, Takahashi Y, et al. Double transgenic mice crossed GFP-LC3 transgenic mice with alphaMyHC-mCherry-LC3 transgenic mice are a new and useful tool to examine the role of autophagy in the heart. *Circ J* 2010;**74**(1):203–6.
14. Munafo DB, Colombo MI. A novel assay to study autophagy: regulation of autophagosome vacuole size by amino acid deprivation. *J Cell Sci* 2001;**114**(Pt 20):3619–29.
15. Niemann A, Takatsuki A, Elsasser HP. The lysosomotropic agent monodansylcadaverine also acts as a solvent polarity probe. *J Histochem Cytochem* 2000;**48**(2):251–8.
16. Perry CN, Kyoi S, Hariharan N, Takagi H, Sadoshima J, Gottlieb RA. Novel methods for measuring cardiac autophagy in vivo. *Methods Enzymol* 2009;**453**:325–42.
17. Huang C, Liu W, Perry CN, Yitzhaki S, Lee Y, Yuan H, et al. Autophagy and Protein Kinase C Are Required for Cardioprotection by Sulfaphenazole. *Am J Physiol Heart Circ Physiol* 2010;**298**(2):H570–9.
18. Huang C, Yitzhaki S, Perry CN, Liu W, Giricz Z, Mentzer Jr RM, et al. Autophagy induced by ischemic preconditioning is essential for cardioprotection. *J Cardiovasc Transl Res* 2010;**3**(4):365–73.
19. Pfeifer U, Scheller H. A morphometric study of cellular autophagy including diurnal variations in kidney tubules of normal rats. *J Cell Biol* 1975;**64**(3):608–21.
20. Pfeifer U, Strauss P. Autophagic vacuoles in heart muscle and liver. A comparative morphometric study including circadian variations in meal-fed rats. *J Mol Cell Cardiol* 1981;**13**(1):37–49.
21. Sachdeva UM, Thompson CB. Diurnal rhythms of autophagy: implications for cell biology and human disease. *Autophagy* 2008;**4**(5):581–9.
22. Ma D, Panda S, Lin JD. Temporal orchestration of circadian autophagy rhythm by C/EBPbeta. *EMBO J* 2011;**30**(22):4642–51.
23. Mizushima N, Yamamoto A, Matsui M, Yoshimori T, Ohsumi Y9. Vivo Analysis of Autophagy in Response to Nutrient Starvation Using Transgenic Mice Expressing a Fluorescent Autophagosome Marker. *Mol Biol Cell* 2004;**15**(3):1101–11.
24. Wu G, Wei G, Huang J, Pang S, Liu L, Yan B. Decreased gene expression of LC3 in peripheral leucocytes of patients with coronary artery disease. *Eur J Clin Invest* 2011;**41**(9):958–63.
25. Eng KE, Panas MD, Hedestam GB, McInerney GM. A novel quantitative flow cytometry-based assay for autophagy. *Autophagy* 2010;**6**(5).

CHAPTER

12

Aging, Nutrition and Lifestyle

Stephanie Wohlgemuth

Ph.D. Assistant Professor, Department of Animal Sciences, University of Florida, FL, USA.

INTRODUCTION

Finding the cure for aging has been a quest for centuries, and the search for the fountain of youth has occupied poets and scientists alike. In recent years a plethora of theories of why we age has emerged. In his review "An attempt at a rational classification of theories of ageing." Medvedev[1] listed and classified 300 different theories of aging. Amongst the most prominent and most widely accepted theories, Denam Harman[2-4] first proposed the "Mitochondrial-Free Radical Theory of Aging". This theory postulates that an increased rate of mitochondrial free radical production and the concomitant oxidative damage to mitochondria themselves and other cellular constituents are the major driving forces for cellular aging (for review see[5]). While there is ample evidence for an age-related rise in cellular oxidative damage and mitochondrial oxygen radical production (reviewed in[6]), experimental genetic manipulation of the antioxidant defense system failed to modulate lifespan (summarized and reviewed by Perez *et al*[7]). Furthermore, Sanz and co-workers[8] have been able to dissociate lifespan from mitochondrial free radical production in Drosophila, diminishing support for Harman's aging theory. Evidently, there is considerable controversy in the literature about the role of oxidative stress for aging and lifespan. This important topic has recently been critically reviewed in more detail[6, 7, 9, 10], and its discussion is beyond the scope of this review.

Although the Mitochondrial-Free Radical Theory of Aging has been contested[7], an age-dependent accumulation of oxidative damage to intracellular macromolecules such as proteins, nucleic acids and lipids, as well as to cellular organelles such as the mitochondria, has repeatedly been demonstrated[9, 11-18]. While the antioxidant defense system appears highly effective, it apparently is not fully capable of preventing all oxidant-mediated protein damage[19]. Particularly, tissues that are composed of long-lived, postmitotic cells, such as the nervous system, heart, and skeletal muscle, tend to accrue relatively greater amounts of oxidative damage than those composed of short-lived cells[16]. The accumulation of damaged, oxidized proteins can result in organ dysfunction and disease, and several age-related diseases of the nervous system, including Alzheimer's disease, amyotrophic lateral

Autophagy in Health and Disease
http://dx.doi.org/10.1016/B978-0-12-385101-7.00012-7

sclerosis, and Parkinson's disease have been associated with oxidized proteins[20] (also reviewed in[9, 12, 14]).

It has subsequently been proposed that besides antioxidant defense systems, cellular quality control mechanisms, namely the ubiquitin-proteasome system and autophagy, play a major role in counteracting the age-related buildup of oxidatively damaged macromolecules and organelles within cells [21, 22]. Consequently, not only the age-related increase in oxidant production, oxidative damage, and the decrease in antioxidant defense, but also a decreasing efficiency of cellular quality control may exacerbate the accumulation of cellular "garbage", including protein aggregates, damaged mitochondria, and oxidatively damaged membranes. It seems therefore plausible that the quest for a cure of cellular aging includes cellular quality control mechanisms and possible ways to restore their function in order to clean up the accumulating cellular "garbage". This chapter will focus on autophagy, the cellular mechanism of "self-eating", and lifestyle interventions that have been described or that are proposed to attenuate its decline and increasing insufficiency with advanced age.

AUTOPHAGY AND AGING

Autophagy

Autophagy literally means "self-eating" and is a vital cellular process by which intracellular components are degraded within lysosomes[23, 24]. There are three classifications of autophagy:

1. microautophagy, in which lysosomes directly take up proteins for degradation[25, 26]
2. chaperone-mediated autophagy (subsequently referred to as CMA), in which soluble proteins with a particular pentapeptide motif are recognized and transported directly across the lysosomal membrane for degradation[27, 28]
3. macroautophagy, which is the sequestration of cytosol, containing misfolded proteins[29, 30], protein aggregates[31, 32], or organelles[33, 34], within a double membrane-bound compartment that ultimately fuses with a lysosome[35].

The following review of the literature will mainly focus on the process of macroautophagy, which will subsequently be referred to as autophagy. Autophagy has been shown to complement an impaired proteasomal system through degradation of ubiquitinated and aggregated proteins[36]. Importantly, it is the only mechanism in eukaryotic cells so far attributed to the degradation of dysfunctional and damaged mitochondria (namely "mitophagy")[34]. The decline of autophagic quality control can ultimately lead to loss of cellular function, cell death, and organ failure, as evidenced by pathologies in organ systems such as the nervous system[37] (for review see[38]), the heart[39, 40], the kidney[41], and pancreatic beta-cells[42]. However, despite autophagy's documented essential role in cellular health and survival, the balance between autophagic cell rescue and autophagic cell death can swing towards the latter (reviewed in[43]). Autophagic cell death has been described in diseases such as heart failure[44], traumatic brain injury[45], pancreatic beta-cell death[46], and muscle loss[47, 48]. It needs to be pointed out that it is and maybe cannot always be unambiguously tested whether autophagy is the primary cause of pathology and cell death, or merely constitutes a sinking ship in a lost battle.

Autophagy in Aging

The involvement of autophagy in life/death decisions of a cell has recently gained growing attention, and becomes even more important in aged cells. Autophagy has been implicated in cell death simultaneously or independently of apoptosis (reviewed in[49, 50]), but, on the other hand, autophagy as a housekeeping mechanism has been proven to be essential for the cell's survival. Therefore, a delicately balanced autophagic cellular quality control is crucial, especially when autophagy becomes induced by age-related increase in cellular damage. In addition, the autophagic machinery itself may be affected by age as well. While the effect of aging on microautophagy has not been studied so far, both CMA and macroautophagy have been shown to decline with increasing age[51-54]. The negative effect of aging on autophagy has been widely studied in liver cells[53, 55, 56], and postmitotic tissues such as the brain[14] and the peripheral nervous system[57]. However, we are only beginning to understand the importance of autophagy in the development of age-related functional impairment of post-mitotic tissues such as the skeletal muscle and the heart.

Consequential with the decline in autophagic cellular quality control, damaged cellular components accumulate, particularly in aged cells. Cellular waste matter such as lysosome-bound lipofuscin, protein aggregates, and damaged, dysfunctional mitochondria can subsequently cause further harm to the cell. For example, lipofuscin, an undegradable polymeric substance, also called the "age-pigment", is primarily composed of cross-linked protein residues and lipid degradation products[58]. Terman and Brunk[58] portrayed lipofuscin-loaded lysosomes as a sink for newly synthesized lysosomal enzymes, thereby interfering with efficient lysosomal autophagy and exacerbating the declining cellular quality control. These authors and colleagues[59, 60] furthermore suggested in their "Mitochondrial-Lysosomal Axis Theory of Aging" a link between dysfunctional mitochondria and lysosomal disturbance in the course of aging. Mitochondrial release of reactive oxygen species accelerates lipofuscin formation, which in turn impedes lysosomal autophagy and removal of dysfunctional mitochondria, initiating another round of the vicious cycle. The authors propose that increased oxidative stress ultimately causes lysosomal labilization and induction of apoptotic and necrotic cell death[60].

Autophagy and Longevity

It becomes apparent that autophagy is critical to overall cellular health, particularly in aged cells, which experience increasing cellular stress over time, and in which cellular defense mechanisms may become inherently defective. Correspondingly, the proper initiation and execution of autophagy has been associated with life-span extension in worms and flies[61-63]. For instance, in the fruit fly, *Drosophila sp.*, overall level of autophagy gene expression was reduced with age, and while mutations in the autophagy-gene Atg8a reduced lifespan, its overexpression in Drosophila brain extended lifespan by 56%[62]. Extensive genetic studies in the nematode *Caenorhabditis elegans* revealed a link between autophagy and life span. In the long-lived *C. elegans* mutant *daf-2*, deficient for the insulin-like receptor, and the dietary restriction mutant *eat-2*, the mutations' life-extending effects were abolished or reduced when autophagy genes were concurrently downregulated[61, 64, 65]. It was hypothesized that autophagy serves as a survival and cell-protective mechanism working in concert with other cellular programs to increase longevity. Hansen and co-workers[64] furthermore suggested that while autophagy is not sufficient to increase lifespan by itself, it plays an integral role in the lifespan extending mechanism regulated by environmental signals reflecting nutrient availability, such as the insulin/IGF-1 pathway and dietary restriction. Nonetheless, that the causality of autophagy and lifespan is more complex was revealed by studies in flies and worms, in which the suppression of autophagy genes extended lifespan in mature wild-type *C. elegans* and certain mutants (mutants deficient in daf-2, or daf-16, FOXO3 transcription factor, respectively[66]), and did not affect survival of fruit flies[67]. On the other hand, a recent multi-site longevity study, as part of the National Institute on Aging Intervention Testing Program, revealed that administration of rapamycin, a potent inhibitor of mammalian Target Of Rapamycin (mTOR) and, correspondingly, strong inducer of autophagy, extended lifespan in a population of genetically heterogeneous mice[68]. Whether autophagy was implicated in the longevity effect of rapamycin was not tested or determined, and therefore, a conclusion with regard to autophagy cannot be drawn. In a follow-up study, Miller and co-workers[69] compared the effects of rapamycin to the administration of Resveratrol. Resveratrol is a nutraceutical that has recently gained popularity based on reports of its numerous health effects. The polyphenol has been suggested to activate sirtuins, a class of protein deacetylases implicated in the health span in yeast, flies and worms, and in mice on a high fat diet (discussed in more detail below). Interestingly, sirtuins and Resveratrol have also been implicated in the regulation of autophagy, which will be highlighted below. However, despite its promise for lifespan extension in invertebrates and obese mice, Miller's study confirmed the notion that Resveratrol did not increase longevity in normal mice[69]. But again, the effect on autophagy regulation by Resveratrol was not investigated in this study and conclusions would be speculative.

Autophagy in Human Diseases

Several clinical investigations in humans have studied the role of autophagy in disease. Vanhorebeek and colleagues[70], for example, studied autophagy in critically ill patients, and found that autophagic vacuole formation was down-regulated and aberrant proteins and structurally abnormal mitochondria that are normally degraded and removed accumulated in liver and skeletal muscle from critically ill patients. Furthermore, defective autophagy was implicated in disease pathologies such as collagen VI muscular dystrophy[71], pancreatic beta cell death[72], cystic fibrosis[73], Crohn's disease[74], and Alzheimer disease[75], confirming the protective effect of autophagy in preclinical animal models. On the other hand, detrimental effects of up-regulated autophagy or autophagic cell death were reported for sporadic amyotrophic lateral sclerosis (ALS)[76], and heart failure[77]. Although studies in preclinical models strongly suggest a major role of autophagy in the progression of the aging phenotype, clinical trials on the association of autophagy and aging are scarce if not absent to date. Future studies in aged humans will need to be designed to unravel autophagy's role in human aging. This may provide potential therapeutic targets for the development of anti-aging and pro-healthy-aging interventions and treatments.

THE REGULATORY NETWORK OF AUTOPHAGY OFFERS POTENTIAL TARGETS FOR LIFESTYLE INTERVENTIONS

The attractive link between autophagy and lifespan revealed by investigations in worms and flies (and one could cautiously hypothesize a potential role for autophagy in the rapamycin-induced lifespan extension in rodents), and results from the preclinical aging studies indicate that the youthful balance of autophagic activity, sufficient to exert a protective role and yet preventing cellular self-digestion and death, becomes perturbed with advanced age, limiting life and health span. Madeo and colleagues[78] have summarized the results from longevity studies in yeast, worms, flies and mice that tested causality or association between autophagy and longevity. The longevity studies reviewed by the authors confirmed an autophagy-dependence of longevity interventions such as sirtuin activation, p53 depletion, TOR deficiency, rapamycin or spermidine administration, and genetic manipulation of the autophagy pathway itself, of mitochondrial metabolism, mTOR (mammalian target of rapamycin), and insulin/IGF-1 signaling, and finally dietary restriction. Consequently, in the quest for the fountain of youth it appears only rational to explore autophagy, the cellular quality control mechanism, and to identify targets in the autophagy regulatory system to design anti-aging interventions. Although attempts are being made to identify and therapeutically target the single most important factor responsible for the aging phenotype, to date more conservative or holistic approaches of anti-aging lifestyle interventions seem more attainable for the aging human population. Some of the lifestyle interventions that have attracted great attention in recent years with respect to their positive effects on aging and health span are caloric restriction, exercise or physical activity, and dietary nutraceuticals. Their association with autophagy will be discussed in this review.

Autophagy Regulation—Master Regulator mTOR

Mechanistic studies have unraveled an intricate system of signal transduction pathways and upstream regulators that control autophagy, which are reviewed in great detail elsewhere (for example[79, 80]). (Macro-)Autophagy is mainly controlled by nutrient availability[81]. Amino acids in particular suppress autophagy[82], with more pronounced effects of leucine, glutamine, tyrosine, phenylalanine, proline, methionine, histidine and tryptophan[81]. On the other hand, nutrient deprivation mediated through amino acid deprivation[83], serum deprivation[84, 85], and caloric restriction[86-89] are potent stimulators of autophagy. Autophagy may in fact play an important role in mediating the widely observed beneficial effect of calorie restriction (CR) for health and longevity. This suggestion is supported by the negative effect of autophagy-inhibition on dietary restriction-induced lifespan extension in *C. elegans*[65] (see above). In mammals, the major signaling pathways that regulate cellular energy metabolism converge on a master regulator, the serin/threonine kinase mTORC1 ("**m**ammalian **t**arget **o**f **r**apamycin" complex**1**) (reviewed in detail in[90-92]). Multiple upstream signals regulate its activity. For example, nutrient deprivation and calorie restriction (mainly through amino acid sensing and AMP-dependent kinase [AMPK] signaling), low cellular energy and hypoxia (through activation of AMPK) inhibit mTORC1 activity, while growth factor signaling (e.g., insulin, insulin-like growth factor 1 [IGF-1]) and amino acids activate the complex (through phosphatidylinositol 3-kinase [PI3K class I], and the AKT/PKB pathway).

Two of the well documented downstream effects of mTORC1 activation are stimulation of protein synthesis and inhibition of autophagy. It is interesting to note in the context of this review that genetically induced TOR deficiency extends lifespan in worms[93], and that this effect is dependent on autophagy[63]. The immunosuppressant drug rapamycin, on the other hand, is a well-studied inhibitor of mTORC1, and rapamycin administration leads to up-regulation of autophagic activity. Interestingly, the positive effect of mTOR inhibition through rapamycin on longevity in mice[68], summarized above, was observed even after late-in-life onset of treatment, at an approximate human-equivalent age of 60 years. However, whether this is, at least in part, due to increased autophagy has yet to be confirmed. A recent study, using a mouse model of Alzheimer's disease, showed that inhibition of mTOR with rapamycin decreased the accumulation of toxic amyloid-beta and Tau through induction of autophagy, and rescued cognitive deficits[94].

mTOR-Independent Regulation

Autophagy can also be regulated independently of TOR-signaling (for review see[95]). An alternative signaling pathway initiated by amino acid deprivation stimulates autophagy via GCN2 (general control nonderepressible-2, eIF2α kinase) and eIF2α (eukaryotic elongation factor-2)[96]. ER-stress and the unfolded-protein-response, elicited by expression of aggregate-prone proteins, glucose deprivation, hypoxia and oxidative stress, regulate autophagy via eIF2α, JNK1 (c-Jun N-terminal kinase), or through calcium-release (reviewed in[97]). Low cellular oxygen levels (hypoxia) induce autophagy through AMPK activity[98], or Hypoxia-Inducible-Factor 1 (HIF-1) and its downstream target BNIP3 (Bcl-2 adenovirus E1a nineteen kDa interacting protein 3)[99].

Reactive Oxygen Species

Reactive oxygen species (ROS) are important regulators of autophagy and can signal through mTOR-dependent (via AMPK activation) or mTOR-independent pathways (for review see[100, 101]). Scherz-Shouval and co-authors[102] suggested a direct interaction of ROS with one of the autophagy-regulators, the autophagy protein Atg4. The authors propose that during starvation-induced autophagy, increased production of ROS, specifically H_2O_2, inhibits Atg4 function. Atg4 oxidation, in turn, enables lipidation of the autophagy protein microtubule-associated protein 1A/1B-light chain 3 (LC3), thereby facilitating autophagosome formation. Particularly interesting is the ROS-induced removal of damaged mitochondria, which are believed to be the major source and immediate target of ROS. It is widely accepted that the process of mitochondrial clearance is accomplished by autophagy (mitophagy). While it had been commonly believed that autophagy is a random process, more and more evidence supporting selective autophagic removal has accumulated over the past decade. Specifically, the removal of damaged mitochondria, namely mitophagy, a term coined by Lemasters[34], has attracted attention. Dissipation of mitochondrial membrane potential has been identified as one factor initiating mitophagy[103]. Recently, some regulatory proteins associated with mitophagy have been identified. Several studies have shown that the E3 ubiquitin ligase Parkin, involved in the etiology of juvenile Parkinson's disease, is selectively recruited to damaged mitochondria and activated, a process that is dependent on PINK 1 (PTEN-induced kinase 1)[104, 105] (reviewed in[106]). Subsequently, mitochondrial outer membrane proteins are ubiquitinated, p62 may be recruited, which can then interact with the autophagy protein LC3, localized at the autophagosome[107]. A recent study furthermore identified the protein AMBRA to interact with Parkin, and to be essential for mitochondrial clearance[108]. The ROS-induced protein Sqstm1/p62[109] also interacts with ubiquitin-conjugated proteins destined for degradation and LC3, thereby mediating the autophagic removal of those proteins[110].

LIFESTYLE INTERVENTIONS THAT STIMULATE CELLULAR QUALITY CONTROL

Are any of these autophagy signaling pathways that regulate autophagic activity affected or can they be manipulated by lifestyle interventions especially geared to attenuate age-related pathologies and to preserve function in the elderly? A wealth of data exists that demonstrate beneficial effects of lifestyle choices such as exercise or nutritional interventions to functioning of older individuals[111-115]. Yet many older adults make poor dietary choices and engage in inadequate amounts of physical activity. The Federal Interagency Forum on Aging-Related Statistics (2010) reports that ~80% of individuals in the U.S. over the age of 65 years does not engage in regular physical activity. Furthermore, 88% do not perform muscle strengthening exercises, known to provide many beneficial effects to skeletal muscle. Meanwhile, the US Department of Agriculture's Healthy Eating Index indicates that approximately 80% of older individuals could benefit from improvements in their diet including increased protein and fiber intake and decreased consumption of saturated fats (Federal Interagency Forum on Aging-Related Statistics, 2010). The following part of this review will focus on the anti-aging lifestyle interventions:

1. dietary/calorie restriction
2. nutraceuticals
3. exercise, and their intersection with autophagy.

The studies reviewed here will emphasize data collected in preclinical studies, and translational clinical studies in human subjects with respect to the effect of intervention on autophagy. Skeletal muscle and the impact of age on muscle mass and function will be used as an example to discuss the effect of lifestyle interventions on autophagy aimed to attenuate age-related changes.

Sarcopenia

The loss of skeletal muscle mass, strength and function is a well described phenomenon occurring with age (for review see[116-122]). This condition, known as sarcopenia of aging, has significant effects on individual health and impacts the severity of frailty[119]. Moreover, poor muscular strength is highly predictive of disability[123] and mortality[124], and general weakness often results in the loss of independent living, thereby affecting individual quality of life and imposing a high burden on healthcare expenditure[125]. Discovering the underlying causes of age-related sarcopenia has therefore gained significant interest. The etiology of sarcopenia is complex and characterized by contribution of multiple factors (reviewed in[126, 127]). Among those, myonuclear apoptosis has emerged as an important underlying mechanism leading to muscle loss[128-130]. In particular, apoptosis induced by mitochondrial dysfunction and release of mitochondrial death effectors has been shown to promote muscle apoptosis and muscle loss[131-135]. Recently, increased age-related iron accumulation in skeletal muscle has been associated with sarcopenia in a rat model[136]. Seo and colleagues[137] found a strong correlation between age-related mitochondrial iron accumulation, mitochondrial dysfunction and oxidative damage in rat skeletal muscle, rendering the skeletal muscle cell more susceptible to (mitochondria-induced) apoptosis.

Autophagy and its Role in Aging Skeletal Muscle

From the collected evidence it becomes apparent that functional, healthy mitochondria are crucial for skeletal muscle homeostasis and maintenance, and their increasing damage and dysfunction with age is one important factor rendering myocytes susceptible to apoptotic cell death. The age-related decline of autophagy, as one of the major cellular housekeeping mechanisms, may be reflected in the increased accumulation of dysfunctional and damaged mitochondria, in addition to increase occurrence of damaged proteins and large protein aggregates. It could be hypothesized that defective cellular quality control and insufficient mitophagy in particular, are at the root of the vicious cycle of mitochondria-induced cell death[60] with subsequent loss of muscle cells. Consistent with this suggestion, we have recently reported that expression of autophagy-regulatory proteins and genes decline with age in skeletal muscle from Fischer 344 rats[89]. Interestingly, our results suggested that the regulation of induction, formation and elongation of autophagosomes was not negatively impacted by age, but instead lysosome-autophagosome fusion and final degradation were impaired. This decline in autophagic degradation correlated with increased oxidative damage and apoptosis[89]. In support of a suggested protective role of autophagy against muscle atrophy in skeletal muscle, Masiero and colleagues[138, 139] measured a significantly reduced muscle mass, muscle force production and aberrant morphology of skeletal muscle from autophagy-deficient mice. However, whether genetic upregulation of autophagy at an age at which autophagy becomes insufficient can attenuate these phenomena was not determined.

Interventions that sustain, promote or stimulate cellular quality control and ensure efficient cellular housekeeping during aging are sought-after to prevent accumulation of dysfunctional mitochondria and downstream detrimental effects such as muscle atrophy. We and others have tested whether nutritional and exercise interventions can stimulate autophagy and attenuate age-related adverse events.

CALORIE RESTRICTION

Calorie Restriction, Health and Lifespan

Calorie restriction (CR) without malnutrition is to date the most effective intervention for improving health, maintaining function and increasing mean and maximum lifespan in a variety of species [140].The beneficial effects of CR are manifold and extend from yeast to humans[141]. Its anti-aging properties reside in the prevention or retardation of several degenerative diseases, including cardiovascular and cerebrovascular disease[142-145], cancer[146], neurodegenerative disorders[57] and age-related damage to the nervous system[147, 148], diabetes and autoimmune diseases[149], and loss of muscle mass and function[150]. CR has also been demonstrated to modulate systemic processes such as inflammation[151], cellular oxidant production and oxidative damage[15, 152]. As a result, experimental rodents subjected to life-long CR display up to 60% maximum lifespan extension compared to *ad libitum* (AL) fed controls[153]. The magnitude of this effect suggests that dietary restriction affects global and fundamental biological processes underlying aging.

Calorie Restriction and Autophagy

Nutrient deprivation, or CR, is a potent inducer of autophagy in most species and in a variety of tissues[87] such as liver[83, 86], neuronal cells[154], and the peripheral nervous system[155]. We have recently demonstrated that CR stimulates autophagy in heart muscle of young rats that were life-long 40% calorie restricted, and, importantly, that the autophagic response to CR persists even at old age[88, 89]. The mechanisms by which nutrient availability impacts autophagy likely involves multiple pathways (see above), such as insulin signaling and mTOR pathways[156-158], or the activation of autophagy proteins through Sirt1 (mammalian ortholog of yeast Sir2, **S**ilent **I**nformation **R**egulator 2)[159]. In a recent study in Fischer 344 rats, we have compared the expression of autophagy proteins and genes in skeletal muscle from old rats that were either *ad libitum* fed or life-long mildly (8%) calorie restricted. We found that life-long mild CR stimulated autophagy in skeletal muscle from old rats, which correlated with decrease in oxidative damage and apoptosis[89]. That CR attenuates not only oxidative damage, but that it also slows the functional decline of skeletal muscle in old rats, as measured by grip strength, was demonstrated by Xu and co-workers[136]. Taken together, these results suggest that increased autophagy in the skeletal muscle of old rats may have removed harmful cellular constituents such as dysfunctional mitochondria, and thereby reduced the level of oxidative stress, mitochondria-induced apoptosis, and preserved physical function.

However, while the beneficial effects of both autophagy and CR on longevity and cellular homeostasis have found broad support, it is yet unclear whether an upregulation of autophagy in skeletal muscle is always beneficial. Recent studies in mouse skeletal muscle[160] and in cultured muscle cells[161] revealed a pivotal role for the forkhead transcription factor FoxO3 and its downstream autophagy-regulating targets BNIP3 and LC3 in skeletal muscle atrophy. Another study associated the upregulation of autophagy-related proteins Beclin-1 and LC3-II with muscle atrophy in an animal model of muscle denervation[48]. However, these studies did not assess the effect of aging and calorie restriction on autophagy and ensuing consequences for muscle mass, and an unambiguous correlation between autophagy stimulation and age-related muscle atrophy cannot be concluded. A recent study by Willcox and colleagues[162] identified FoxO3a genotype as associated with longevity in humans. It may be speculated that expression of FoxO3a and its downstream targets assist cellular health at old age when cellular garbage builds up. Future research needs to identify whether a shift in the activity of autophagy-regulatory proteins and autophagy-dependent degradation of cellular components might contribute to disruption of myocyte function and muscle atrophy, or whether autophagic removal of cellular waste overall promotes myocyte health.

Calorie Restriction in Humans

Human studies on the effect of calorie restriction on lifespan are naturally limited. However, there are several populations with proportionally high numbers of centenarians across the world which have drawn the attention of the scientific community. The most intriguing example of beneficial effects of dietary lifestyle for longevity comes from the population of Okinawans. The inhabitants of Okinawa, a Japanese prefecture consisting of hundreds of islands, are characterized by a higher number of centenarians and increased mean and maximum lifespan compared to other industrialized countries. Interestingly, Okinawans have on average a 10-20% lower caloric intake compared to mainland Japan[163, 164], and have lower mortality from age-associated diseases compared to mainland Japan and the US[164]. Other examples of dietary restriction studies in humans, unintentionally or planned, and its consequences are the unexpected food shortage in the *Biosphere 2* project[165], and results from the phase I trials preceding the clinical randomized controlled trial *CALERIE* (Comprehensive Assessment of the Long-term Effect of Reducing Intake of Energy), reviewed in[114]. Preliminary results from this phase I trial testing a 25% calorie restriction regimen for a maximum of six months include decreased fasting insulin concentrations, reduced systolic blood pressure, improvement in insulin sensitivity, and reduced oxidative DNA damage. In another trial, Civitarese and co-workers[166] assessed in healthy, overweight young subjects the effects of a 25% calorie restricted diet on mitochondrial function. They found an increased expression of genes involved in mitochondrial biogenesis (e.g., PGC1α, TFAM), elevated content of mitochondrial DNA, and reduced DNA damage. The biological effects of CR in older persons have not yet been thoroughly investigated. However, in a large-scale trial conducted in obese, older adults ($\geq$ 60 years, n = 316), 18-month diet-induced weight loss reduced markers of systemic inflammation (i.e., CRP, IL-6 and soluble TNF-α receptor-1)[167]. For further readings about the outcomes of calorie restriction in humans, the reader may be referred to several review articles in the recent literature[122, 168-170].

A beneficial effect of CR on autophagy in human skeletal muscle from older, overweight women was recently reported[171]. The objective of this trial was to assess the effects of a six-month weight loss program combined with moderate-intensity exercise on the cellular quality control mechanisms autophagy and the ubiquitin-proteasome, as well as on inflammation, apoptosis, and mitochondrial function in the skeletal muscle. Since the intervention combined dietary restriction with exercise the results will be discussed below.

CR MIMETICS AND THEIR EFFECTS ON AUTOPHAGY

Despite the obvious health benefits of CR outlined above, the discussion about the applicability and effectiveness of CR in humans, particularly regarding extension of lifespan, is ongoing and controversial[172-174]. Findings from the obesity literature indicate that most persons are reluctant to engage in long-term CR. In addition, many individuals are unable to sustain CR-induced weight loss, possibly due to internal feedback systems that signal the body to increase food intake or decrease energy expenditure in response to weight loss. Moreover, weight loss may not be advisable in older persons, as it can accelerate age-related muscle loss[175]. Importantly, low body mass index has been associated with increased risk of disability and mortality in older populations[176, 177]. Furthermore, people practicing long-term severe CR may experience several adverse events, including undesired changes in physical appearance, loss of strength and stamina, menstrual irregularities, infertility, loss of libido, osteoporosis, cold sensitivity, slower wound healing, and psychological conditions such as food obsession, depression and irritability[178]. Thus, a critical question that deserves to be thoroughly investigated is what degree of CR is tolerable in humans in order to obtain beneficial physiological changes without incurring adverse events. Animal studies have shown that even mild CR (i.e., 8% calorie intake reduction) may elicit cardioprotective, anti-inflammatory, and anti-apoptotic effects in skeletal muscle[89, 179, 180], thus obviating the need for substantial food intake reductions.

Calorie Restriction Mimetics

Given the questionable feasibility of long-term dietary restriction, especially in the elderly population, the field of CR mimetics has become a topic of increasing scientific focus. As a general definition, CR mimetics are agents or interventions that are capable of reproducing the effects of CR without requiring food intake reduction [181]. Since the identification of the first agent (2-deoxy-D-glucose) by Lane and colleagues in 1998 (cited in[182]), the list of putative CR mimetics has increasingly grown, including compounds such as Quercetin, Rapamycin and Resveratrol. For many of these agents, however, there is little, if any, scientific evidence supporting their efficacy and/or safety.

Resveratrol

Among CR mimetics, Resveratrol has received the greatest attention. Resveratrol is a naturally occurring polyphenol found in red wine, the notorious cardioprotective effects of which are invoked to explain the so-called "French paradox"[183]. One salient feature of Resveratrol resides in its ability to activate sirtuins, phylogenetically conserved deacetylases that senses the NAD^+/NADH ratio[184], which in turn are prominent mediators of lifespan extension by CR [185]. In fact, Resveratrol was found to extend the lifespan and delay the onset of aging phenotypes in short-lived organisms by modulating sirtuin signaling[186-188]. Among CR mimetics, Resveratrol and its analogue pterostilbene have received the greatest attention (for review see[184, 189, 190]). Amongst a plethora of health benefits, Resveratrol has been shown in animal models to reduce inflammation in cardiac tissue[191], inhibit cardiomyocyte apoptosis[192], protect the myocardium against ischemia-reperfusion injury[193], to improve in skeletal muscle muscular endurance, mitochondrial biogenesis, expression of genes for oxidative phosphorylation[194], and endogenous antioxidant enzymes[195], and to reduce inflammation[196], and oxidative stress[197]. In a recent study, Resveratrol improved survival and reduced the prevalence of cardiac pathology in mice fed a high-calorie diet[198]. Moreover, Resveratrol-supplemented mice displayed better insulin sensitivity and enhanced liver mitochondrial biogenesis compared to animals fed either a standard or high-calorie diet.

Several recent studies provided some evidence for the indirect activation of SIRT1 by Resveratrol through AMP-activated protein kinase (AMPK)[199-201], and/or by phosphodiesterase 4 (PDE4), an enzyme further upstream of AMPK activation[202]. AMPK is a major energy-sensing enzyme and its activation turns on cellular catabolic processes[203]. Autophagic proteolysis has been identified as one of the cellular processes that are regulated by AMPK[204-206]. It therefore seems consequential to suggest that autophagy can be stimulated by Resveratrol.

Resveratrol and Autophagy

Morselli and colleagues[207] recently described that CR and Resveratrol activate autophagy through sirtuin-1. Importantly, the authors found that autophagy is required for the life extending effect of CR in a human cancer cell line and in the nematode *C. elegans*. As was discussed above, nutrient deprivation acts (mostly) through the mTOR signaling pathway. However, Morselli and colleagues[207] found in their study that sirtuins can operate independently of mTOR, and can induce autophagy directly through deacetylation of the autophagy proteins Atg5 and Atg7. Most recently, Wu and colleagues[208] demonstrated that AMPK and/or SIRT1 are required in Resveratrol-mediated autophagy induction in SH-SY5Y neuronal cells. However, to illustrate the complexity of the network that regulates and fine tunes autophagic activity, others have reported that Resveratrol inhibits autophagy induced by rapamycin or nutrient-deprivation independently of sirtuins in several human tumor and non-tumor cell lines[209]. Interestingly, in this study the p70S6 kinase (S6K), a kinase downstream of mTOR, was identified as a target of Resveratrol. Resveratrol-induced inactivation of S6K was associated with the disruption of starvation-induced autophagy. The authors explain the apparent discrepancy to previous reports on autophagy activation by Resveratrol as well as by mTOR inhibition with a cellular "dual-switch" regulatory mechanism that enables a tighter control of autophagic activity[209]. Further studies will have to identify the downstream targets of S6K in charge of autophagy-regulation, and, importantly, the conditions and cellular pathways by which Resveratrol affects autophagy one way or another.

To this date, there are only a few upcoming, ongoing or completed Phase I/II trials in human participants in the US investigating the effects of Resveratrol supplementation, and one Phase II trial using pterostilbene supplementation (clinicaltrials.gov). These studies are set to examine a number of conditions, including metabolic syndrome, insulin resistance, diabetes, hyperlipidemia and blood pressure. None of these trials, however, has investigated or has explicitly planned to investigate the effect of Resveratrol or pterostilbene administration on autophagy in human subjects, and the concomitant consequences for health and disease conditions. For more information on clinical Resveratrol trials the reader is referred to a comprehensive review[210] of studies on Resveratrol bioavailability, cellular effects such as oxidative stress, inflammation, and on human disease conditions such as diabetes, obesity and cardiovascular disease in general. Resveratrol and other (indirect or direct) Sirt1 activators are presently under intense investigation and it will be exciting to see the regulatory network unravel that may underlie Resveratrol's well-documented health benefits.

Interestingly, Resveratrol has not only been described as CR-mimetic, but also as an exercise-mimetic. Lagouge and colleagues[194] tested the endurance capacity of young mice receiving Resveratrol supplementation over a period of 15 weeks. Endurance capacity was significantly enhanced in the treatment group compared to controls. The Resveratrol-induced improvement in fatigue resistance was also evident in mice on a high fat diet, accompanied by increases in mitochondrial enzyme activity and mitochondrial number[194]. In support of the protective effect of Resveratrol for muscle, administration of the drug during 15 days of hind limb suspension in rats (age was not disclosed), a model of muscle wasting induced by unloading, maintained muscle mass, maximal force contraction, mitochondrial capacity to oxidize fatty acids, and protected insulin sensitivity[211]. The results led the authors to the conclusion that Resveratrol likely acts as an exercise-mimetic. Although Resveratrol did not prevent the age-related loss of muscle mass in old rodents[212], fast-twitch muscle fiber contractile function in the aged animals was maintained and age-related[212] and exercise-induced oxidative stress[213] attenuated. Whether or not this effect of Resveratrol was accompanied or supported by an increase in myocellular housekeeping through autophagy was not determined, but could be speculated.

EXERCISE

Like CR, exercise and physical activity have a plethora of health benefits. A discussion of the wealth of data on beneficial effects of exercise in health and disease of various organ systems and tissues, in rodent models as well as humans, is beyond the scope of this review, and the reader is referred to more specific literature focused on exercise and its health effects. Interestingly, the role of autophagy has recently been investigated more intensively in:

1. obesity and Type 2 Diabetes (T2D)
2. sarcopenia, pathologies which have collectively shown to improve in response to physical activity or exercise training, and which will be discussed here.

The pathophysiological metabolic cluster termed "metabolic syndrome" is defined as the combination of obesity and abnormal glucose metabolism (insulin resistance, impaired glucose tolerance), and/or dyslipidemia, and/or hypertension, and is associated with increased risk for cardiovascular disease, diabetes, and all-cause mortality[214, 215]. It has been recognized that exercise is an essential part of necessary lifestyle changes that can help to improve the cluster of metabolic syndrome in general (for review see[216-218]), and insulin resistance in particular (reviewed in[219]). Similarly, T2D, often associated with obesity, can be improved by exercise, as two recent meta-analyses of randomized controlled clinical trials summarized[220, 221]. Interestingly, data provided by several studies suggest that autophagic function is required for functional glucose metabolism. Defective or impaired autophagy diminished structure, mass and function of pancreatic β-cells[42], promoted hepatic ER stress cells and caused insulin resistance in lean mice while restoration of autophagy in obese mice improved insulin sensitivity[222]. A link between autophagy and obesity has recently been provided by Singh and colleagues[223], who have uncovered a novel function of autophagy in regulation of intracellular lipid stores. The authors demonstrated that lipid droplets and autophagic components associate, and that loss of autophagy decreases hepatic triglyceride breakdown and increases hepatic triglycerides and lipid droplets. Furthermore, they observed that intracellular lipid accumulation had an inhibitory effect on autophagic clearance, which in turn increased lipid accumulation even further. This vicious cycle may be one mechanism by which a decrease of autophagy (for example in the course of aging or self-perpetuating in obese individuals) may contribute to abnormal and progressive lipid accumulation associated with obesity and the cluster of metabolic syndrome.

Sarcopenia, the age-related loss of skeletal muscle mass, is a significant public health concern and associated with development of functional disability and loss of independence[125, 224]. Exercise and physical activity have been added to the successful recipe for healthy aging and the prevention of sarcopenia (for review see[122, 225]). In particular, preclinical studies have demonstrated that exercise interventions can combat age-related muscle loss through attenuation of (muscle cell) apoptosis[136, 226-229]. Although autophagy has been shown to protect skeletal muscle from atrophy[138], other studies have proposed that muscle wasting is associated with increased autophagy[47, 230]. These discrepancies could be explained by the diverse muscle wasting conditions that were investigated such as denervation, fasting, cachexia and doxorubicin-induced muscle atrophy. Interestingly, there is a lack of systematic studies, clinical and preclinical, that address the question whether exercise modulates skeletal muscle autophagy, especially in aged animals and humans, and whether exercise acts inhibiting or stimulating (for review on the double function of autophagy in skeletal muscle see[231]). The question whether autophagy is directly affected by physical activity/exercise has just recently gained more attention.

Exercise and Autophagy

Exercise stimulates a number of metabolic regulatory pathways, and a few have been mentioned above in the context of CR and CR mimetics. AMPK is one of the energy-sensing metabolic master regulators, situated at the interface between exercise stimulus and metabolic response[232, 233]. Based on evidence that link AMPK activation with autophagy regulation[205], it seems reasonable to hypothesize that autophagy could play a role in the effects induced by exercise. As outlined above, the effects of autophagy on skeletal muscle homeostasis have been regarded from different perspectives: stimulation of cellular quality control providing protection from damage and damage accumulation on the one hand, and autophagy-mediated atrophy, which can be activated by exercise-induced myocellular damage on the other. These two sides of the autophagy-coin will be reflected in the following brief review of the most recent literature on the effects of exercise on autophagy.

Most recently, He and colleagues[234] elegantly unraveled the physiological function of exercise-induced autophagy *in vivo*. Both acute and chronic treadmill exercise induced autophagy in a variety of tissues from young-adult (8-12 weeks of age) mice, including skeletal muscle. When exercise-inducible autophagy, but not basal autophagy, was genetically disabled, the positive effects of exercise on insulin sensitivity, translocation of the glucose transporter GLUT4 to the plasma membrane, glucose uptake, and AMPK activation were abolished. Importantly, these mice had a lower maximal endurance capacity. Similarly, diet-induced obese mice with defective exercise-induced autophagy could not benefit from exercise with regard to glucose tolerance. The results of this study overall demonstrated, at least in young-adult mice, that autophagy mediates the beneficial effects of exercise on glucose tolerance and metabolism. Another study using myostatin null mice ($MSTN^{-/-}$) has shed an interesting light on autophagy function. These mice, in which myostatin, a regulator of muscle size, had been genetically disabled, are characterized by a hyper-muscular phenotype with muscle hypertrophy and hyperplasia, but compromised muscle force-generating capacity. Aerobic exercise training in $MSTN^{-/-}$ mice led to an increase in aerobic muscle phenotype, a decrease in muscle fiber cross-sectional area, and, importantly, an increase in muscle force[235]. Interestingly, BNIP3, a protein associated with autophagy regulation was upregulated in muscle from exercised $MSTN^{-/-}$ mice. It was speculated that the reduction in muscle fiber cross-sectional area was mediated through autophagy. Maybe more importantly, the authors proposed that stimulation of autophagy by exercise could have favored the beneficial removal of damaged cellular constituents, which is in contrast to sedentary $MSTN^{-/-}$ mice, in which accumulation of cytoplasmic inclusions, termed as tubular aggregates, has previously been observed[236].

The effect of strenuous exercise may not be beneficial for muscle homeostasis, as a study by Feng and colleagues [237] suggests. Young-mature rats that underwent a daily strenuous endurance exercise regimen over a period of eight weeks displayed elevated mRNA levels of the muscle atrophy markers atrogin-1 and MuRF1. This increase was accompanied by upregulated autophagy as indicated by Atg7, Beclin-1 and LC3 protein expression, increased markers of mitochondrial fission, and a non-significant tendency for reduced endurance capacity. These effects were blunted by administration of hydroxytyrosol, a polyphenol with antioxidant properties. The results of this study suggest that exercise-induced oxidant production exceeded endogenous antioxidant capacity, and ensuing intracellular damage, indicated by increased mitochondrial fission and stimulated autophagy. Whether the increased level of autophagy actually led to degradation of contractile proteins and reduction of fiber cross-sectional area, and could have contributed to the slight decrease in exercise capacity was not definitively explored. It remains to be determined whether autophagy in muscle during and after strenuous exercise serves as a cell survival mechanism and/or whether the level of autophagic degradation causes muscle atrophy associated with functional impairment. Another study by Smuder and colleagues[230] investigated Doxorubicin-induced muscle wasting. The chemotherapeutic drug is used in cancer treatment, and its administration is associated with generation of reactive oxygen species[238], and can result in severe muscle wasting. The authors of this study reported that Doxorubicin in young-adult rats caused an increase in autophagy markers that was significantly blunted when the animals were exercised before drug treatment[230]. The authors propose that exercise protects the muscle from doxorubicin-induced autophagy and concomitant muscle atrophy. They furthermore suggest investigating a prophylactic treatment with autophagy inhibitors prior to Doxorubicin treatment. However, if autophagy is acting as a cell survival and housekeeping mechanism that protects the cells from increased (Doxorubicin-induced) oxidative stress, then its inhibition may prove detrimental. One could argue that the exercise pre-treatment could have stimulated the endogenous antioxidant capacity of the muscle cells, thereby rendering an increase in autophagic housekeeping expendable. MacKenzie and colleagues[239] observed that an acute bout of resistance exercise stimulated both protein synthesis and degradation, and activated mammalian Vsp34, a class III PI3K (Phosphatidylinositol 3-kinase) implicated in induction of autophagy. The authors propose that resistance exercise increased protein synthesis, and that the consequential demand for amino acids would then drive mVsp34-stimulated autophagic protein degradation to release amino acids for further protein synthesis. From these studies it becomes apparent that autophagy might play multiple roles in muscle homeostasis, depending on demand for cellular building blocks and generation of cellular damage. Future therapeutic or prophylactic treatment strategies may need to take this into account and balance the double-edged sword as autophagy is often described.

We have recently tested the effect of life-long, voluntary wheel running in aging Fischer 344 rats[89]. In our study, voluntary wheel running was combined with a mild reduction of caloric intake (8% CR), necessary to stimulate the running activity (rats fed an *ad libitum* diet tend to decrease their running activity abruptly, but slight food restriction (8-10%) has been shown to prevent this decline). Life-long voluntary wheel running exercise stimulated the expression of autophagy markers in muscle from old rats, but it did not result in an additive effect on autophagic activity when compared to CR alone. Both treatments, CR alone and CR combined with exercise, were associated with decreased muscle oxidative damage and apoptosis. Further studies using an animal model in which consistent exercise can be achieved without confounding dietary intervention will have to isolate the effects of exercise on muscle cell autophagy and homeostasis in old animals. However, despite this limitation the design of this preclinical study proved somewhat translatable to clinical intervention studies designed to support weight loss in elderly obese humans while at the same time improving physical performance and muscle strength through an exercise regimen.

COMBINING EXERCISE AND DIETARY RESTRICTION INTERVENTIONS

The beneficial effects of a reduced caloric intake and of physical activity for health span and longevity, particularly in light of the current obesity epidemic, are undisputed. However, as discussed above, a calorie restriction diet for elderly humans may not be advisable, as it can accelerate age-related muscle loss[175]. Diet-induced weight loss typically leads to a decrease of fat-free mass, a significant concern for older adults. In contrast, exercise interventions may increase muscle mass, improve muscle quality[166, 240] and increase muscle protein synthesis [241], but rarely produce significant weight loss unless individuals engage in very high amounts of moderate or vigorous activity[242]. In a recent study, we conducted an intervention program in which moderate reduction of caloric intake was combined with an exercise regimen consisting of strength/resistance and aerobic exercises[171]. The target subjects were overweight, older women, and the treatment was performed over a period of six months. The intervention targeted a weight loss of around 6% or greater, and was coupled with exercise sessions. The treatment group was contrasted with an educational "healthy living" control group that attended monthly health education lectures. As functional outcome, we determined before and after treatment the time to walk 400m, a predictor of mortality, mobility limitation, disability and cardiovascular disease[243]. Muscle biopsies from the *vastus lateralis*, a thigh muscle, were taken before and after the six-months intervention to assess pre and post values of biological parameters of cellular quality control (autophagy-regulators, FoxO3a and atrogenes), inflammation and apoptosis (TNF-α, caspases), and mitochondrial function (PGC1α, TFAm, mitochondrial respiratory complexes). Finally, these parameters were compared between the treatment and the control group. We found that compared to the control group, six months of weight loss and exercise treatment, which produced significant weight loss, improved physical performance, and stimulated pathways of cellular quality control and markers of mitochondrial function[171]. Although these data are compelling, more work needs to be done to unravel the mechanistic link(s) between calorie restriction, exercise and autophagy. Furthermore, this study informs only about the effects of a short-term restriction of caloric intake combined with exercise on cellular quality control. Future studies will have to reveal whether consistent, long-term change of lifestyle will improve muscular health and physical function (even more). Nevertheless, the results are encouraging, as this study demonstrates that a late-in-life-onset of lifestyle changes can battle age-related adverse phenomena.

CONCLUDING REMARKS

Autophagy has gained increasing attention over the past two decades in the context of its role in pathophysiology and aging. Under physiological conditions its impact on cellular quality control seems to be tightly regulated and fine-tuned to balance between degradation, housekeeping and anabolic pathways in order to avoid cellular waste accumulation and self-digestion. However, once out-of-balance cellular homeostasis is disturbed, and lack of cellular quality control on the one hand as much as excessive autophagic degradation on the other hand can result in cellular dysfunction and cell death. In the process of aging, it appears that autophagic housekeeping becomes less efficient, which provides some guidance for lifestyle strategies in support of healthy aging. This review has summarized current research with regard to some of the most prominent healthy-aging lifestyle interventions that have or may potentially have a positive effect on autophagic cellular quality control. However, it needs to be emphasized that further research will have to distinguish the multiple roles autophagy plays in the balance of cellular quality control and catabolism. This will be particularly important for the development and evaluation of potential healthy-aging therapeutic or prophylactic treatments and pharmaceuticals or nutraceuticals that target autophagy.

References

1. Medvedev ZA. An attempt at a rational classification of theories of ageing. *Biol Rev Camb Philos Soc* 1990;**65**(3):375–98.
2. Harman D. Aging: a theory based on free radical and radiation chemistry. *J Gerontol* 1956;**11**(3):298–300.
3. Harman D. Free radical theory of aging: dietary implications. *Am J Clin Nutr* 1972;**25**(8):839–43.
4. Harman D. Free radical theory of aging. *Mutat Res* 1992;**275**(3-6):257–66.
5. Barja G. Free radicals and aging. *Trends Neurosci* 2004;**27**(10):595–600.
6. Sanz A, Pamplona R, Barja G. Is the mitochondrial free radical theory of aging intact? *Antioxid Redox Signal* 2006;**8**(3-4):582–99.
7. Perez VI, Bokov A, Van Remmen H, Mele J, Ran Q, Ikeno Y, et al. Is the oxidative stress theory of aging dead? *Biochim Biophys Acta* 2009;**1790**(10):1005–14.
8. Sanz A, Fernandez-Ayala DJ, Stefanatos RK, Jacobs HT. Mitochondrial ROS production correlates with, but does not directly regulate lifespan in Drosophila. *Ageing (Albany NY)* 2010;**2**(4):200–23.
9. Bokov A, Chaudhuri A, Richardson A. The role of oxidative damage and stress in aging. *Mech Ageing Dev* 2004;**125**(10-11):811–26.
10. Sanz A, Stefanatos RK. The mitochondrial free radical theory of aging: a critical view. *Curr Aging Sci* 2008;**1**(1):10–21.

11. Cristofalo VJ, Lorenzini A, Allen RG, Torres C, Tresini M. Replicative senescence: a critical review. *Mech Ageing Dev* 2004;**125**(10-11):827–48.
12. Grune T, Jung T, Merker K, Davies KJ. Decreased proteolysis caused by protein aggregates, inclusion bodies, plaques, lipofuscin, ceroid, and "aggresomes" during oxidative stress, aging, and disease. *Int J Biochem Cell Biol* 2004;**36**(12):2519–30.
13. Judge S, Leeuwenburgh C. Cardiac mitochondrial bioenergetics, oxidative stress, and aging. *Am J Physiol Cell Physiol* 2007;**292**(6):C1983–92.
14. Keller JN, Dimayuga E, Chen Q, Thorpe J, Gee J, Ding Q. Autophagy, proteasomes, lipofuscin, and oxidative stress in the aging brain. *Int J Biochem Cell Biol* 2004;**36**(12):2376–91.
15. Leeuwenburgh C, Wagner P, Holloszy JO, Sohal RS, Heinecke JW. Caloric restriction attenuates dityrosine cross-linking of cardiac and skeletal muscle proteins in aging mice. *Arch Biochem Biophys* 1997;**346**(1):74–80.
16. Sohal RS, Weindruch R. Oxidative stress, caloric restriction, and aging. *Science* 1996;**273**(5271):59–63.
17. Stadtman ER. Protein oxidation and aging. *Science* 1992;**257**(5074):1220–4.
18. Yan LJ, Levine RL, Sohal RS. Oxidative damage during aging targets mitochondrial aconitase. *Proc Natl Acad Sci USA* 1997;**94**(21):11168–72.
19. Merker K, Stolzing A, Grune T. Proteolysis, caloric restriction and aging. *Mech Ageing Dev* 2001;**122**(7):595–615.
20. Li S, Zheng J, Carmichael ST. Increased oxidative protein and DNA damage but decreased stress response in the aged brain following experimental stroke. *Neurobiol Dis* 2005;**18**(3):432–40.
21. Koga H, Kaushik S, Cuervo AM. Protein homeostasis and aging: The importance of exquisite quality control. *Ageing Res Rev* 2011;**10**(2):205–15.
22. Yen WL, Klionsky DJ. How to live long and prosper: autophagy, mitochondria, and aging. *Physiology (Bethesda)* 2008;**23**:248–62.
23. Meijer AJ, Codogno P. Regulation and role of autophagy in mammalian cells. *Int J Biochem Cell Biol* 2004;**36**(12):2445–62.
24. Xie Z, Klionsky DJ. Autophagosome formation: core machinery and adaptations. *Nat Cell Biol* 2007;**9**(10):1102–9.
25. Kunz JB, Schwarz H, Mayer A. Determination of four sequential stages during microautophagy in vitro. *J Biol Chem* 2004;**279**(11):9987–96.
26. Sahu R, Kaushik S, Clement CC, Cannizzo ES, Scharf B, Follenzi A, et al. Microautophagy of cytosolic proteins by late endosomes. *Dev Cell* 2011;**20**(1):131–9.
27. Chiang HL, Dice JF. Peptide sequences that target proteins for enhanced degradation during serum withdrawal. *J Biol Chem* 1988;**263**(14):6797–805.
28. Dice JF. Chaperone-mediated autophagy. *Autophagy* 2007;**3**(4):295–9.
29. Kubota H. Quality control against misfolded proteins in the cytosol: a network for cell survival. *J Biochem* 2009;**146**(5):609–16.
30. Zheng Q, Su H, Tian Z, Wang X. Proteasome malfunction activates macroautophagy in the heart. *Am J Cardiovasc Dis* 2011;**1**(3):214–26.
31. Fan W, Tang Z, Chen D, Moughon D, Ding X, Chen S, et al. Keap1 facilitates p62-mediated ubiquitin aggregate clearance via autophagy. *Autophagy* 2010;**6**(5).
32. Rubinsztein DC. The roles of intracellular protein-degradation pathways in neurodegeneration. *Nature* 2006;**443**(7113):780–6.
33. Gottlieb RA, Carreira RS. Autophagy in health and disease 5. Mitophagy as a way of life. *Am J Physiol Cell Physiol* 2010;**299**(2):C203–10.
34. Lemasters JJ. Selective mitochondrial autophagy, or mitophagy, as a targeted defense against oxidative stress, mitochondrial dysfunction, and aging. *Rejuvenation Res* 2005;**8**(1):3–5.
35. Klionsky DJ, editor. *Autophagy.* Georgetown, Texas: Eurekah.com/Landes Bioscience; 2004.
36. Pandey UB, Nie Z, Batlevi Y, McCray BA, Ritson GP, Nedelsky NB, et al. HDAC6 rescues neurodegeneration and provides an essential link between autophagy and the UPS. *Nature* 2007;**447**(7146):859–63.
37. Fortun J, Verrier JD, Go JC, Madorsky I, Dunn Jr WA, Notterpek L. The formation of peripheral myelin protein 22 aggregates is hindered by the enhancement of autophagy and expression of cytoplasmic chaperones. *Neurobiology of Disease*; 2006. In press.
38. Cheung ZH, Ip NY. Autophagy deregulation in neurodegenerative diseases—recent advances and future perspectives. *J Neurochem* 2011;**118**(3):317–25.

39. Pattison JS, Osinska H, Robbins J. Atg7 induces basal autophagy and rescues autophagic deficiency in CryABR120G cardiomyocytes. *Circ Res* 2011;**109**(2):151–60.
40. Taneike M, Yamaguchi O, Nakai A, Hikoso S, Takeda T, Mizote I, et al. Inhibition of autophagy in the heart induces age-related cardiomyopathy. *Autophagy* 2010;**6**(5).
41. Jiang M, Liu K, Luo J, Dong Z. Autophagy is a renoprotective mechanism during in vitro hypoxia and in vivo ischemia-reperfusion injury. *Am J Pathol* 2010;**176**(3):1181–92.
42. Jung HS, Chung KW, Won Kim J, Kim J, Komatsu M, Tanaka K, et al. Loss of autophagy diminishes pancreatic beta cell mass and function with resultant hyperglycemia. *Cell Metab* 2008;**8**(4):318–24.
43. Hill JA. Autophagy in cardiac plasticity and disease. *Pediatr Cardiol* 2011;**32**(3):282–9.
44. Akazawa H, Komazaki S, Shimomura H, Terasaki F, Zou Y, Takano H, et al. Diphtheria toxin-induced autophagic cardiomyocyte death plays a pathogenic role in mouse model of heart failure. *J Biol Chem* 2004;**279**(39):41095–103.
45. Luo CL, Li BX, Li QQ, Chen XP, Sun YX, Bao HJ, et al. Autophagy is involved in traumatic brain injury-induced cell death and contributes to functional outcome deficits in mice. *Neuroscience* 2011;**184**:54–63.
46. Fujimoto K, Hanson PT, Tran H, Ford EL, Han Z, Johnson JD, et al. Autophagy regulates pancreatic beta cell death in response to Pdx1 deficiency and nutrient deprivation. *J Biol Chem* 2009;**284**(40):27664–73.
47. McClung JM, Judge AR, Powers SK, Yan Z. p38 MAPK links oxidative stress to autophagy-related gene expression in cachectic muscle wasting. *Am J Physiol Cell Physiol* 2010;**298**(3):C542–9.
48. O'Leary MF, Hood DA. Denervation-induced oxidative stress and autophagy signaling in muscle. *Autophagy* 2009;**5**(2).
49. Eisenberg-Lerner A, Bialik S, Simon HU, Kimchi A. Life and death partners: apoptosis, autophagy and the cross-talk between them. *Cell Death Differ* 2009;**16**(7):966–75.
50. Maiuri MC, Zalckvar E, Kimchi A, Kroemer G. Self-eating and self-killing: crosstalk between autophagy and apoptosis. *Nat Rev Mol Cell Biol* 2007;**8**(9):741–52.
51. Cuervo AM, Bergamini E, Brunk UT, Droge W, French M, Terman A. Autophagy and aging: the importance of maintaining "clean" cells. *Autophagy* 2005;**1**(3):131–40.
52. Donati A, Cavallini G, Paradiso C, Vittorini S, Pollera M, Gori Z, et al. Age-related changes in the regulation of autophagic proteolysis in rat isolated hepatocytes. *J Gerontol A Biol Sci Med Sci* 2001;**56**(7):B288–93.
53. Cuervo AM, Dice JF. Age-related decline in chaperone-mediated autophagy. *J Biol Chem* 2000;**275**(40): 31505–13.
54. Terman A. The effect of age on formation and elimination of autophagic vacuoles in mouse hepatocytes. *Gerontology* 1995;**41**(Suppl. 2):319–26.
55. Bergamini E, Cavallini G, Donati A, Gori Z. The role of macroautophagy in the ageing process, anti-ageing intervention and age-associated diseases. *Int J Biochem Cell Biol* 2004;**36**(12):2392–404.
56. Del Roso A, Vittorini S, Cavallini G, Donati A, Gori Z, Masini M, et al. Ageing-related changes in the in vivo function of rat liver macroautophagy and proteolysis. *Exp Gerontol* 2003;**38**(5):519–27.
57. Rangaraju S, Hankins D, Madorsky I, Madorsky E, Lee WH, Carter CS, et al. Molecular architecture of myelinated peripheral nerves is supported by calorie restriction with aging. *Aging Cell* 2009;**8**(2):178–91.
58. Terman A, Brunk UT. Lipofuscin. *Int J Biochem Cell Biol* 2004;**36**(8):1400–4.
59. Terman A, Brunk UT. Oxidative stress, accumulation of biological "garbage", and aging. *Antioxid Redox Signal* 2006;**8**(1-2):197–204.
60. Terman A, Kurz T, Navratil M, Arriaga EA, Brunk UT. Mitochondrial turnover and aging of long-lived postmitotic cells: the mitochondrial-lysosomal axis theory of aging. *Antioxid Redox Signal* 2010;**12**(4): 503–35.
61. Melendez A, Talloczy Z, Seaman M, Eskelinen EL, Hall DH, Levine B. Autophagy genes are essential for dauer development and life-span extension in C. elegans. *Science* 2003;**301**(5638):1387–91.
62. Simonsen A, Cumming RC, Brech A, Isakson P, Schubert DR, Finley KD. Promoting basal levels of autophagy in the nervous system enhances longevity and oxidant resistance in adult Drosophila. *Autophagy* 2008;**4**(2):176–84.
63. Toth ML, Sigmond T, Borsos E, Barna J, Erdelyi P, Takacs-Vellai K, et al. Longevity pathways converge on autophagy genes to regulate life span in Caenorhabditis elegans. *Autophagy* 2008;**4**(3):330–8.
64. Hansen M, Chandra A, Mitic LL, Onken B, Driscoll M, Kenyon C. A role for autophagy in the extension of lifespan by dietary restriction in C. elegans. *PLoS Genet* 2008;**4**(2):e24.

65. Jia K, Levine B. Autophagy is required for dietary restriction-mediated life span extension in C. elegans. *Autophagy* 2007;**3**(6):597–9.
66. Hashimoto Y, Ookuma S, Nishida E. Lifespan extension by suppression of autophagy genes in Caenorhabditis elegans. *Genes Cells* 2009;**14**(6):717–26.
67. Ren C, Finkel SE, Tower J. Conditional inhibition of autophagy genes in adult Drosophila impairs immunity without compromising longevity. *Exp Gerontol* 2009;**44**(3):228–35.
68. Harrison DE, Strong R, Sharp ZD, Nelson JF, Astle CM, Flurkey K, et al. Rapamycin fed late in life extends lifespan in genetically heterogeneous mice. *Nature* 2009;**460**(7253):392–5.
69. Miller RA, Harrison DE, Astle CM, Baur JA, Boyd AR, de Cabo R, et al. Rapamycin, but not resveratrol or simvastatin, extends life span of genetically heterogeneous mice. *J Gerontol A Biol Sci Med Sci* 2011;**66**(2):191–201.
70. Vanhorebeek I, Gunst J, Derde S, Derese I, Boussemaere M, Guiza F, et al. Insufficient activation of autophagy allows cellular damage to accumulate in critically ill patients. *J Clin Endocrinol Metab* 2011;**96**(4):E633–45.
71. Grumati P, Coletto L, Sabatelli P, Cescon M, Angelin A, Bertaggia E, et al. Autophagy is defective in collagen VI muscular dystrophies, and its reactivation rescues myofiber degeneration. *Nat Med* 2010;**16**(11):1313–20.
72. Masini M, Bugliani M, Lupi R, del Guerra S, Boggi U, Filipponi F, et al. Autophagy in human type 2 diabetes pancreatic beta cells. *Diabetologia* 2009;**52**(6):1083–6.
73. Luciani A, Villella VR, Esposito S, Brunetti-Pierri N, Medina D, Settembre C, et al. Defective CFTR induces aggresome formation and lung inflammation in cystic fibrosis through ROS-mediated autophagy inhibition. *Nat Cell Biol* 2010;**12**(9):863–75.
74. Cadwell K, Liu JY, Brown SL, Miyoshi H, Loh J, Lennerz JK, et al. A key role for autophagy and the autophagy gene Atg16l1 in mouse and human intestinal Paneth cells. *Nature* 2008;**456**(7219):259–63.
75. Pickford F, Masliah E, Britschgi M, Lucin K, Narasimhan R, Jaeger PA, et al. The autophagy-related protein beclin 1 shows reduced expression in early Alzheimer disease and regulates amyloid beta accumulation in mice. *J Clin Invest* 2008;**118**(6):2190–9.
76. Sasaki S. Autophagy in spinal cord motor neurons in sporadic amyotrophic lateral sclerosis. *J Neuropathol Exp Neurol* 2011;**70**(5):349–59.
77. Hein S, Arnon E, Kostin S, Schonburg M, Elsasser A, Polyakova V, et al. Progression from compensated hypertrophy to failure in the pressure-overloaded human heart: structural deterioration and compensatory mechanisms. *Circulation* 2003;**107**(7):984–91.
78. Madeo F, Tavernarakis N, Kroemer G. Can autophagy promote longevity? *Nat Cell Biol* 2010;**12**(9):842–6.
79. Yorimitsu T, Klionsky DJ. Autophagy: molecular machinery for self-eating. *Cell Death Differ* 2005; **12**(Suppl. 2):1542–52.
80. Yang YP, Liang ZQ, Gu ZL, Qin ZH. Molecular mechanism and regulation of autophagy. *Acta Pharmacol Sin* 2005;**26**(12):1421–34.
81. Kadowaki M, Karim MR, Carpi A, Miotto G. Nutrient control of macroautophagy in mammalian cells. *Mol Aspects Med* 2006;**27**(5-6):426–43.
82. Kovacs AL, Grinde B, Seglen PO. Inhibition of autophagic vacuole formation and protein degradation by amino acids in isolated hepatocytes. *Exp Cell Res* 1981;**133**(2):431–6.
83. Mortimore GE, Schworer CM. Induction of autophagy by amino-acid deprivation in perfused rat liver. *Nature* 1977;**270**(5633):174–6.
84. Mitchener JS, Shelburne JD, Bradford WD, Hawkins HK. Cellular autophagocytosis induced by deprivation of serum and amino acids in HeLa cells. *Am J Pathol* 1976;**83**(3):485–92.
85. Kim I, Lemasters JJ. Mitochondrial degradation by autophagy (mitophagy) in GFP-LC3 transgenic hepatocytes during nutrient deprivation. *Am J Physiol Cell Physiol* 2011;**300**(2):C308–17.
86. Bergamini E, Cavallini G, Donati A, Gori Z. The anti-ageing effects of caloric restriction may involve stimulation of macroautophagy and lysosomal degradation, and can be intensified pharmacologically. *Biomed Pharmacother* 2003;**57**(5-6):203–8.
87. Mizushima N, Yamamoto A, Matsui M, Yoshimori T, Ohsumi Y. In vivo analysis of autophagy in response to nutrient starvation using transgenic mice expressing a fluorescent autophagosome marker. *Mol Biol Cell* 2004;**15**(3):1101–11.
88. Wohlgemuth SE, Julian D, Akin DE, Fried J, Toscano K, Leeuwenburgh C, et al. Autophagy in the heart and liver during normal aging and calorie restriction. *Rejuvenation Res* 2007;**10**(3):281–92.

89. Wohlgemuth SE, Seo AY, Marzetti E, Lees HA, Leeuwenburgh C. Skeletal muscle autophagy and apoptosis during aging: effects of calorie restriction and life-long exercise. *Exp Gerontol* 2010;**45**(2):138–48.
90. Corradetti MN, Guan KL. Upstream of the mammalian target of rapamycin: do all roads pass through mTOR? *Oncogene* 2006;**25**(48):6347–60.
91. Evans DS, Kapahi P, Hsueh WC, Kockel L. TOR signaling never gets old: aging, longevity and TORC1 activity. *Ageing Res Rev* 2011;**10**(2):225–37.
92. Kim J, Guan KL. Amino acid signaling in TOR activation. *Annu Rev Biochem* 2011;**80**:1001–32.
93. Vellai T, Takacs-Vellai K, Zhang Y, Kovacs AL, Orosz L, Muller F. Genetics: influence of TOR kinase on lifespan in C. elegans. *Nature* 2003;**426**(6967):620.
94. Caccamo A, Majumder S, Richardson A, Strong R, Oddo S. Molecular interplay between mammalian target of rapamycin (mTOR), amyloid-beta, and Tau: effects on cognitive impairments. *J Biol Chem* 2010;**285**(17):13107–20.
95. He C, Klionsky DJ. Regulation mechanisms and signaling pathways of autophagy. *Annu Rev Genet* 2009;**43**: 67–93.
96. Talloczy Z, Jiang W, Virgin HWt, Leib DA, Scheuner D, Kaufman RJ, et al. Regulation of starvation- and virus-induced autophagy by the eIF2alpha kinase signaling pathway. *Proc Natl Acad Sci USA* 2002;**99**(1):190–5.
97. Kroemer G, Marino G, Levine B. Autophagy and the integrated stress response. *Mol Cell* 2010;**40**(2):280–93.
98. Papandreou I, Lim AL, Laderoute K, Denko NC. Hypoxia signals autophagy in tumor cells via AMPK activity, independent of HIF-1, BNIP3, and BNIP3L. *Cell Death Differ* 2008;**15**(10):1572–81.
99. Zhang H, Bosch-Marce M, Shimoda LA, Tan YS, Baek JH, Wesley JB, et al. Mitochondrial autophagy is an HIF-1-dependent adaptive metabolic response to hypoxia. *J Biol Chem* 2008;**283**(16):10892–903.
100. Huang J, Lam GY, Brumell JH. Autophagy signaling through reactive oxygen species. *Antioxid Redox Signal* 2011;**14**(11):2215–31.
101. Scherz-Shouval R, Elazar Z. Regulation of autophagy by ROS: physiology and pathology. *Trends Biochem Sci* 2011;**36**(1):30–8.
102. Scherz-Shouval R, Shvets E, Fass E, Shorer H, Gil L, Elazar Z. Reactive oxygen species are essential for autophagy and specifically regulate the activity of Atg4. *EMBO J* 2007;**26**(7):1749–60.
103. Narendra D, Tanaka A, Suen DF, Youle RJ. Parkin is recruited selectively to impaired mitochondria and promotes their autophagy. *J Cell Biol* 2008;**183**(5):795–803.
104. Vives-Bauza C, Zhou C, Huang Y, Cui M, de Vries RL, Kim J, et al. PINK1-dependent recruitment of Parkin to mitochondria in mitophagy. *Proc Natl Acad Sci USA* 2010;**107**(1):378–83.
105. Narendra DP, Jin SM, Tanaka A, Suen DF, Gautier CA, Shen J, et al. PINK1 is selectively stabilized on impaired mitochondria to activate Parkin. *PLoS Biol* 2010;**8**(1):e1000298.
106. Springer W, Kahle PJ. Regulation of PINK1-Parkin-mediated mitophagy. *Autophagy* 2011;**7**(3).
107. Geisler S, Holmstrom KM, Skujat D, Fiesel FC, Rothfuss OC, Kahle PJ, et al. PINK1/Parkin-mediated mitophagy is dependent on VDAC1 and p62/SQSTM1. *Nat Cell Biol* 2010;**12**(2):119–31.
108. Van Humbeeck C, Cornelissen T, Hofkens H, Mandemakers W, Gevaert K, De Strooper B, et al. Parkin interacts with Ambra1 to induce mitophagy. *J Neurosci* 2011;**31**(28):10249–61.
109. Jain A, Lamark T, Sjottem E, Larsen KB, Awuh JA, Overvatn A, et al. p62/SQSTM1 is a target gene for transcription factor NRF2 and creates a positive feedback loop by inducing antioxidant response element-driven gene transcription. *J Biol Chem* 2010;**285**(29):22576–91.
110. Pankiv S, Clausen TH, Lamark T, Brech A, Bruun JA, Outzen H, et al. p62/SQSTM1 binds directly to Atg8/LC3 to facilitate degradation of ubiquitinated protein aggregates by autophagy. *J Biol Chem* 2007; **282**(33):24131–45.
111. Goodpaster BH, Chomentowski P, Ward BK, Rossi A, Glynn NW, Delmonico MJ, et al. Effects of physical activity on strength and skeletal muscle fat infiltration in older adults: a randomized controlled trial. *J Appl Physiol* 2008;**105**(5):1498–503.
112. Kalapotharakos V, Smilios I, Parlavatzas A, Tokmakidis SP. The effect of moderate resistance strength training and detraining on muscle strength and power in older men. *J Geriatr Phys Ther* 2007;**30**(3):109–13.
113. Manini TM, Pahor M. Physical activity and maintaining physical function in older adults. *Br J Sports Med* 2009;**43**(1):28–31.
114. Redman LM, Ravussin E. Caloric restriction in humans: impact on physiological, psychological, and behavioral outcomes. *Antioxid Redox Signal* 2011;**14**(2):275–87.

115. Sugawara J, Miyachi M, Moreau KL, Dinenno FA, DeSouza CA, Tanaka H. Age-related reductions in appendicular skeletal muscle mass: association with habitual aerobic exercise status. *Clin Physiol Funct Imaging* 2002;**22**(3):169–72.
116. Buford TW, Anton SD, Judge AR, Marzetti E, Wohlgemuth SE, Carter CS, et al. Models of accelerated sarcopenia: critical pieces for solving the puzzle of age-related muscle atrophy. *Ageing Res Rev* 2010;**9**(4):369–83.
117. Karakelides H, Nair KS. Sarcopenia of aging and its metabolic impact. *Curr Top Dev Biol* 2005;**68**:123–48.
118. Marzetti E, Anne Lees H, Eva Wohlgemuth S, Leeuwenburgh C. Sarcopenia of aging: underlying cellular mechanisms and protection by calorie restriction. *Biofactors* 2009;**35**(1):28–35.
119. Roubenoff R. Sarcopenia and its implications for the elderly. *Eur J Clin Nutr* 2000;**54**(Suppl. 3):S40–7.
120. Roubenoff R. Physical activity, inflammation, and muscle loss. *Nutr Rev* 2007;**65**(12 Pt 2):S208–12.
121. Visser M, Schaap LA. Consequences of sarcopenia. *Clin Geriatr Med* 2011;**27**(3):387–99.
122. Waters DL, Baumgartner RN, Garry PJ, Vellas B. Advantages of dietary, exercise-related, and therapeutic interventions to prevent and treat sarcopenia in adult patients: an update. *Clin Interv Aging* 2010;**5**:259–70.
123. Rantanen T, Guralnik JM, Foley D, Masaki K, Leveille S, Curb JD, et al. Midlife hand grip strength as a predictor of old age disability. *Jama* 1999;**281**(6):558–60.
124. Metter EJ, Talbot LA, Schrager M, Conwit R. Skeletal muscle strength as a predictor of all-cause mortality in healthy men. *J Gerontol A Biol Sci Med Sci* 2002;**57**(10):B359–65.
125. Janssen I, Shepard DS, Katzmarzyk PT, Roubenoff R. The healthcare costs of sarcopenia in the United States. *J Am Geriatr Soc* 2004;**52**(1):80–5.
126. Ryall JG, Schertzer JD, Lynch GS. Cellular and molecular mechanisms underlying age-related skeletal muscle wasting and weakness. *Biogerontology* 2008;**9**(4):213–28.
127. Lang T, Streeper T, Cawthon P, Baldwin K, Taaffe DR, Harris TB. Sarcopenia: etiology, clinical consequences, intervention, and assessment. *Osteoporos Int* 2010;**21**(4):543–59.
128. Alway SE, Siu PM. Nuclear apoptosis contributes to sarcopenia. *Exerc Sport Sci Rev* 2008;**36**(2):51–7.
129. Leeuwenburgh C. Role of apoptosis in sarcopenia. *J Gerontol A Biol Sci Med Sci* 2003;**58**(11):999–1001.
130. Marzetti E, Privitera G, Simili V, Wohlgemuth SE, Aulisa L, Pahor M, et al. Multiple pathways to the same end: mechanisms of myonuclear apoptosis in sarcopenia of aging. *ScientificWorldJournal* 2010;**10**:340–9.
131. Chabi B, Ljubicic V, Menzies KJ, Huang JH, Saleem A, Hood DA. Mitochondrial function and apoptotic susceptibility in aging skeletal muscle. *Aging Cell* 2008;**7**(1):2–12.
132. Hiona A, Sanz A, Kujoth GC, Pamplona R, Seo AY, Hofer T, et al. Mitochondrial DNA mutations induce mitochondrial dysfunction, apoptosis and sarcopenia in skeletal muscle of mitochondrial DNA mutator mice. *PLoS One* 2010;**5**(7):e11468.
133. Kujoth GC, Hiona A, Pugh TD, Someya S, Panzer K, Wohlgemuth SE, et al. Mitochondrial DNA mutations, oxidative stress, and apoptosis in mammalian aging. *Science* 2005;**309**(5733):481–4.
134. Marzetti E, Hwang JC, Lees HA, Wohlgemuth SE, Dupont-Versteegden EE, Carter CS, et al. Mitochondrial death effectors: relevance to sarcopenia and disuse muscle atrophy. *Biochim Biophys Acta* 2010;**1800**(3):235–44.
135. Marzetti E, Wohlgemuth SE, Lees HA, Chung HY, Giovannini S, Leeuwenburgh C. Age-related activation of mitochondrial caspase-independent apoptotic signaling in rat gastrocnemius muscle. *Mech Ageing Dev* 2008;**129**(9):542–9.
136. Xu J, Knutson MD, Carter CS, Leeuwenburgh C. Iron accumulation with age, oxidative stress and functional decline. *PLoS ONE* 2008;**3**(8):e2865.
137. Seo AY, Xu J, Servais S, Hofer T, Marzetti E, Wohlgemuth SE, et al. Mitochondrial iron accumulation with age and functional consequences. *Aging Cell* 2008;**7**(5):706–16.
138. Masiero E, Agatea L, Mammucari C, Blaauw B, Loro E, Komatsu M, et al. Autophagy is required to maintain muscle mass. *Cell Metab* 2009;**10**(6):507–15.
139. Masiero E, Sandri M. Autophagy inhibition induces atrophy and myopathy in adult skeletal muscles. *Autophagy* 2010;**6**(2):307–9.
140. Weindruch R, Walford RL. *The retardation of aging and disease by dietary restriction*. Springfield: Charles C. Thomas; 1988.
141. Fontana L, Partridge L, Longo VD. Extending healthy life span–from yeast to humans. *Science* 2010; **328**(5976):321–6.

142. Kemi M, Keenan KP, McCoy C, Hoe CM, Soper KA, Ballam GC, et al. The relative protective effects of moderate dietary restriction versus dietary modification on spontaneous cardiomyopathy in male Sprague-Dawley rats. *Toxicol Pathol* 2000;**28**(2):285–96.
143. Fontana L, Meyer TE, Klein S, Holloszy JO. Long-term calorie restriction is highly effective in reducing the risk for atherosclerosis in humans. *Proc Natl Acad Sci USA* 2004;**101**(17):6659–63.
144. Mattson MP, Wan R. Beneficial effects of intermittent fasting and caloric restriction on the cardiovascular and cerebrovascular systems. *J Nutr Biochem* 2005;**16**(3):129–37.
145. Marzetti E, Wohlgemuth SE, Anton SD, Bernabei R, Carter CS, Leeuwenburgh C. Cellular mechanisms of cardioprotection by calorie restriction: state of the science and future perspectives. *Clin Geriatr Med* 2009;**25**(4):715–32. ix.
146. Lee C, Longo VD. Fasting vs dietary restriction in cellular protection and cancer treatment: from model organisms to patients. *Oncogene* 2011;**30**(30):3305–16.
147. Ingram DK, Young J, Mattison JA. Calorie restriction in nonhuman primates: assessing effects on brain and behavioral aging. *Neuroscience* 2007;**145**(4):1359–64.
148. Whidden MA, Kirichenko N, Halici Z, Erdos B, Foster TC, Tumer N. Lifelong caloric restriction prevents age-induced oxidative stress in the sympathoadrenal system of Fischer 344 x Brown Norway rats. *Biochem Biophys Res Commun* 2011;**408**(3):454–8.
149. Gonzalez OA, Tobia C, Ebersole JL, Novak MJ. Caloric restriction and chronic inflammatory diseases. *Oral Dis* 2012;**18**(1):16–31.
150. Marzetti E, Lees HA, Wohlgemuth SE, Leeuwenburgh C. Sarcopenia of aging: underlying cellular mechanisms and protection by calorie restriction. *Biofactors* 2009;**35**(1):28–35.
151. Chung HY, Kim HJ, Kim JW, Yu BP. The inflammation hypothesis of aging: molecular modulation by calorie restriction. *Ann NY Acad Sci* 2001;**928**:327–35.
152. Drew B, Phaneuf S, Dirks A, Selman C, Gredilla R, Lezza A, et al. Effects of aging and caloric restriction on mitochondrial energy production in gastrocnemius muscle and heart. *Am J Physiol Regul Integr Comp Physiol* 2003;**284**(2):R474–80.
153. Weindruch R, Walford RL, Fligiel S, Guthrie D. The retardation of aging in mice by dietary restriction: longevity, cancer, immunity and lifetime energy intake. *J Nutr* 1986;**116**(4):641–54.
154. Alirezaei M, Kemball CC, Flynn CT, Wood MR, Whitton JL, Kiosses WB. Short-term fasting induces profound neuronal autophagy. *Autophagy* 2010;**6**(6):702–10.
155. Madorsky I, Opalach K, Waber A, Verrier JD, Solmo C, Foster T, et al. Intermittent fasting alleviates the neuropathic phenotype in a mouse model of Charcot-Marie-Tooth disease. *Neurobiol Dis* 2009;**34**(1):146–54.
156. Jia G, Cheng G, Gangahar DM, Agrawal DK. Insulin-like growth factor-1 and TNF-alpha regulate autophagy through c-jun N-terminal kinase and Akt pathways in human atherosclerotic vascular smooth cells. *Immunol Cell Biol* 2006;**84**(5):448–54.
157. Yamamoto A, Cremona ML, Rothman JE. Autophagy-mediated clearance of huntingtin aggregates triggered by the insulin-signaling pathway. *J Cell Biol* 2006;**172**(5):719–31.
158. Young JE, Martinez RA, La Spada AR. Nutrient Deprivation Induces Neuronal Autophagy and Implicates Reduced Insulin Signaling in Neuroprotective Autophagy Activation. *Journal of Biological Chemistry* 2009;**284**(4):2363–73.
159. Lee IH, Cao L, Mostoslavsky R, Lombard DB, Liu J, Bruns NE, et al. A role for the NAD-dependent deacetylase Sirt1 in the regulation of autophagy. *Proceedings of the National Academy of Sciences* 2008;**105**(9):3374–9.
160. Mammucari C, Milan G, Romanello V, Masiero E, Rudolf R, Del Piccolo P, et al. FoxO3 Controls Autophagy in Skeletal Muscle *In Vivo*. *Cell Metab* 2007;**6**(6):458–71.
161. Zhao J, Brault JJ, Schild A, Cao P, Sandri M, Schiaffino S, et al. FoxO3 Coordinately Activates Protein Degradation by the Autophagic/Lysosomal and Proteasomal Pathways in Atrophying Muscle Cells. *Cell Metab* 2007;**6**(6):472–83.
162. Willcox BJ, Donlon TA, He Q, Chen R, Grove JS, Yano K, et al. FOXO3A genotype is strongly associated with human longevity. *Proc Natl Acad Sci USA* 2008;**105**(37):13987–92.
163. Suzuki M, Wilcox BJ, Wilcox CD. Implications from and for food cultures for cardiovascular disease: longevity. *Asia Pac J Clin Nutr* 2001;**10**(2):165–71.

164. Willcox BJ, Willcox DC, Todoriki H, Fujiyoshi A, Yano K, He Q, et al. Caloric restriction, the traditional Okinawan diet, and healthy aging: the diet of the world's longest-lived people and its potential impact on morbidity and life span. *Ann NY Acad Sci* 2007;**1114**:434–55.
165. Walford RL, Mock D, Verdery R, MacCallum T. Calorie restriction in biosphere 2: alterations in physiologic, hematologic, hormonal, and biochemical parameters in humans restricted for a 2-year period. *J Gerontol A Biol Sci Med Sci* 2002;**57**(6):B211–24.
166. Civitarese AE, Carling S, Heilbronn LK, Hulver MH, Ukropcova B, Deutsch WA, et al. Calorie restriction increases muscle mitochondrial biogenesis in healthy humans. *PLoS Med* 2007;**4**(3):e76.
167. Nicklas BJ, Ambrosius W, Messier SP, Miller GD, Penninx BW, Loeser RF, et al. Diet-induced weight loss, exercise, and chronic inflammation in older, obese adults: a randomized controlled clinical trial. *Am J Clin Nutr* 2004;**79**(4):544–51.
168. Blagosklonny MV. Calorie restriction: decelerating mTOR-driven aging from cells to organisms (including humans). *Cell Cycle* 2010;**9**(4):683–8.
169. Omodei D, Fontana L. Calorie restriction and prevention of age-associated chronic disease. *FEBS Lett* 2011;**585**(11):1537–42.
170. Varady KA, Hellerstein MK. Alternate-day fasting and chronic disease prevention: a review of human and animal trials. *Am J Clin Nutr* 2007;**86**(1):7–13.
171. Wohlgemuth SE, Lees HA, Marzetti E, Manini TM, Aranda JM, Daniels MJ, et al. An exploratory analysis of the effects of a weight loss plus exercise program on cellular quality control mechanisms in older overweight women. *Rejuvenation Res* 2011;**14**(3):315–24.
172. Viidik A. The biological aging is our inescapable fate—but can we modify it? *Z Gerontol Geriatr* 1999;**32**(6):384–9.
173. Phelan JP, Rose MR. Why dietary restriction substantially increases longevity in animal models but won't in humans. *Ageing Res Rev* 2005;**4**(3):339–50.
174. Yu BP. Why calorie restriction would work for human longevity. *Biogerontology* 2006.
175. Miller SL, Wolfe RR. The danger of weight loss in the elderly. *J Nutr Health Aging* 2008;**12**(7):487–91.
176. Landi F, Zuccala G, Gambassi G, Incalzi RA, Manigrasso L, Pagano F, et al. Body mass index and mortality among older people living in the community. *J Am Geriatr Soc* 1999;**47**(9):1072–6.
177. Landi F, Onder G, Gambassi G, Pedone C, Carbonin P, Bernabei R. Body mass index and mortality among hospitalized patients. *Arch Intern Med* 2000;**160**(17):2641–4.
178. Dirks AJ, Leeuwenburgh C. Caloric restriction in humans: potential pitfalls and health concerns. *Mech Ageing Dev* 2006;**127**(1):1–7.
179. Judge S, Jang YM, Smith A, Selman C, Phillips T, Speakman JR, et al. Exercise by lifelong voluntary wheel running reduces subsarcolemmal and interfibrillar mitochondrial hydrogen peroxide production in the heart. *Am J Physiol Regul Integr Comp Physiol* 2005;**289**(6):R1564–72.
180. Kalani R, Judge S, Carter C, Pahor M, Leeuwenburgh C. Effects of caloric restriction and exercise on age-related, chronic inflammation assessed by C-reactive protein and interleukin-6. *J Gerontol A Biol Sci Med Sci* 2006;**61**(3):211–7.
181. Roth GS, Lane MA, Ingram DK. Caloric restriction mimetics: the next phase. *Ann N Y Acad Sci* 2005;**1057**:365–71.
182. Lane MA, Roth GS, Ingram DK. Caloric restriction mimetics: a novel approach for biogerontology. *Methods Mol Biol* 2007;**371**:143–9.
183. Kopp P. Resveratrol, a phytoestrogen found in red wine. A possible explanation for the conundrum of the "French paradox"? *Eur J Endocrinol* 1998;**138**(6):619–20.
184. Baur JA, Sinclair DA. Therapeutic potential of resveratrol: the in vivo evidence. *Nat Rev Drug Discov* 2006;**5**(6):493–506.
185. Guarente L, Picard F. Calorie restriction—the SIR2 connection. *Cell* 2005;**120**(4):473–82.
186. Howitz KT, Bitterman KJ, Cohen HY, Lamming DW, Lavu S, Wood JG, et al. Small molecule activators of sirtuins extend Saccharomyces cerevisiae lifespan. *Nature* 2003;**425**(6954):191–6.
187. Viswanathan M, Kim SK, Berdichevsky A, Guarente L. A role for SIR-2.1 regulation of ER stress response genes in determining C. elegans life span. *Dev Cell* 2005;**9**(5):605–15.
188. Valenzano DR, Terzibasi E, Genade T, Cattaneo A, Domenici L, Cellerino A. Resveratrol prolongs lifespan and retards the onset of age-related markers in a short-lived vertebrate. *Curr Biol* 2006;**16**(3):296–300.
189. Pterostilbene. Monograph. *Altern Med Rev* 2010;**15**(2):159–63.
190. Baur JA. Resveratrol, sirtuins, and the promise of a DR mimetic. *Mech Ageing Dev* 2010;**131**(4):261–9.

191. Csiszar A, Labinskyy N, Podlutsky A, Kaminski PM, Wolin MS, Zhang C, et al. Vasoprotective effects of resveratrol and SIRT1: attenuation of cigarette smoke-induced oxidative stress and proinflammatory phenotypic alterations. *Am J Physiol Heart Circ Physiol* 2008;**294**(6):H2721–35.
192. Seya K, Kanemaru K, Sugimoto C, Suzuki M, Takeo T, Motomura S, et al. Opposite effects of two resveratrol (trans-3,5,4′-trihydroxystilbene) tetramers, vitisin A and hopeaphenol, on apoptosis of myocytes isolated from adult rat heart. *J Pharmacol Exp Ther* 2009;**328**(1):90–8.
193. Ray PS, Maulik G, Cordis GA, Bertelli AA, Bertelli A, Das DK. The red wine antioxidant resveratrol protects isolated rat hearts from ischemia reperfusion injury. *Free Radic Biol Med* 1999;**27**(1-2):160–9.
194. Lagouge M, Argmann C, Gerhart-Hines Z, Meziane H, Lerin C, Daussin F, et al. Resveratrol improves mitochondrial function and protects against metabolic disease by activating SIRT1 and PGC-1alpha. *Cell* 2006;**127**(6):1109–22.
195. Hasegawa K, Wakino S, Yoshioka K, Tatematsu S, Hara Y, Minakuchi H, et al. Sirt1 protects against oxidative stress-induced renal tubular cell apoptosis by the bidirectional regulation of catalase expression. *Biochem Biophys Res Commun* 2008;**372**(1):51–6.
196. Ghanim H, Sia CL, Abuaysheh S, Korzeniewski K, Patnaik P, Marumganti A, et al. An antiinflammatory and reactive oxygen species suppressive effects of an extract of Polygonum cuspidatum containing resveratrol. *J Clin Endocrinol Metab* 2010;**95**(9):E1–8.
197. Ungvari Z, Labinskyy N, Mukhopadhyay P, Pinto JT, Bagi Z, Ballabh P, et al. Resveratrol attenuates mitochondrial oxidative stress in coronary arterial endothelial cells. *Am J Physiol Heart Circ Physiol* 2009;**297**(5):H1876–81.
198. Baur JA, Pearson KJ, Price NL, Jamieson HA, Lerin C, Kalra A, et al. Resveratrol improves health and survival of mice on a high-calorie diet. *Nature* 2006;**444**(7117):337–42.
199. Canto C, Auwerx J. AMP-activated protein kinase and its downstream transcriptional pathways. *Cell Mol Life Sci* 2010;**67**(20):3407–23.
200. Dasgupta B, Milbrandt J. Resveratrol stimulates AMP kinase activity in neurons. *Proc Natl Acad Sci USA* 2007;**104**(17):7217–22.
201. Um JH, Park SJ, Kang H, Yang S, Foretz M, McBurney MW, et al. AMP-activated protein kinase-deficient mice are resistant to the metabolic effects of resveratrol. *Diabetes* 2010;**59**(3):554–63.
202. Park SJ, Ahmad F, Philp A, Baar K, Williams T, Luo H, et al. Resveratrol ameliorates aging-related metabolic phenotypes by inhibiting cAMP phosphodiesterases. *Cell* 2012;**148**(3):421–33.
203. Hardie DG. AMP-activated/SNF1 protein kinases: conserved guardians of cellular energy. *Nat Rev Mol Cell Biol* 2007;**8**(10):774–85.
204. Alers S, Loffler AS, Wesselborg S, Stork B. Role of AMPK-mTOR-Ulk1/2 in the regulation of autophagy: cross talk, shortcuts, and feedbacks. *Mol Cell Biol* 2012;**32**(1):2–11.
205. Meley D, Bauvy C, Houben-Weerts JH, Dubbelhuis PF, Helmond MT, Codogno P, et al. AMP-activated protein kinase and the regulation of autophagic proteolysis. *J Biol Chem* 2006;**281**(46):34870–9.
206. Sanchez AM, Csibi A, Raibon A, Cornille K, Gay S, Bernardi H, et al. AMPK promotes skeletal muscle autophagy through activation of forkhead FoxO3a and interaction with Ulk1. *J Cell Biochem* 2012;**113**(2):695–710.
207. Morselli E, Maiuri MC, Markaki M, Megalou E, Pasparaki A, Palikaras K, et al. Caloric restriction and resveratrol promote longevity through the Sirtuin-1-dependent induction of autophagy. *Cell Death Dis* 2010;**1**:e10.
208. Wu Y, Li X, Zhu JX, Xie W, Le W, Fan Z, et al. Resveratrol-activated AMPK/SIRT1/autophagy in cellular models of Parkinson's disease. *Neurosignals* 2011;**19**(3):163–74.
209. Armour SM, Baur JA, Hsieh SN, Land-Bracha A, Thomas SM, Sinclair DA. Inhibition of mammalian S6 kinase by resveratrol suppresses autophagy. *Aging (Albany NY)* 2009;**1**(6):515–28.
210. Smoliga JM, Baur JA, Hausenblas HA. Resveratrol and health—a comprehensive review of human clinical trials. *Mol Nutr Food Res* 2011;**55**(8):1129–41.
211. Momken I, Stevens L, Bergouignan A, Desplanches D, Rudwill F, Chery I, et al. Resveratrol prevents the wasting disorders of mechanical unloading by acting as a physical exercise mimetic in the rat. *FASEB J* 2011;**25**(10):3646–60.
212. Jackson JR, Ryan MJ, Alway SE. Long-term supplementation with resveratrol alleviates oxidative stress but does not attenuate sarcopenia in aged mice. *J Gerontol A Biol Sci Med Sci* 2011;**66**(7):751–64.

213. Ryan MJ, Jackson JR, Hao Y, Williamson CL, Dabkowski ER, Hollander JM, et al. Suppression of oxidative stress by resveratrol after isometric contractions in gastrocnemius muscles of aged mice. *J Gerontol A Biol Sci Med Sci* 2010;**65**(8):815–31.
214. Reisin E, Alpert MA. Definition of the metabolic syndrome: current proposals and controversies. *Am J Med Sci* 2005;**330**(6):269–72.
215. Alberti KG, Zimmet P, Shaw J. Metabolic syndrome—a new world-wide definition. A Consensus Statement from the International Diabetes Federation. *Diabet Med* 2006;**23**(5):469–80.
216. Carroll S, Dudfield M. What is the relationship between exercise and metabolic abnormalities? A review of the metabolic syndrome. *Sports Med* 2004;**34**(6):371–418.
217. Lakka TA, Laaksonen DE. Physical activity in prevention and treatment of the metabolic syndrome. *Appl Physiol Nutr Metab* 2007;**32**(1):76–88.
218. Slentz CA, Houmard JA, Kraus WE. Exercise, abdominal obesity, skeletal muscle, and metabolic risk: evidence for a dose response. *Obesity (Silver Spring)* 2009;**17**(Suppl. 3):S27–33.
219. Hawley JA. Exercise as a therapeutic intervention for the prevention and treatment of insulin resistance. *Diabetes Metab Res Rev* 2004;**20**(5):383–93.
220. Boule NG, Haddad E, Kenny GP, Wells GA, Sigal RJ. Effects of exercise on glycemic control and body mass in type 2 diabetes mellitus: a meta-analysis of controlled clinical trials. *JAMA* 2001;**286**(10):1218–27.
221. Umpierre D, Ribeiro PA, Kramer CK, Leitao CB, Zucatti AT, Azevedo MJ, et al. Physical activity advice only or structured exercise training and association with HbA1c levels in type 2 diabetes: a systematic review and meta-analysis. *JAMA* 2011;**305**(17):1790–9.
222. Yang L, Li P, Fu S, Calay ES, Hotamisligil GS. Defective hepatic autophagy in obesity promotes ER stress and causes insulin resistance. *Cell Metab* 2010;**11**(6):467–78.
223. Singh R, Kaushik S, Wang Y, Xiang Y, Novak I, Komatsu M, et al. Autophagy regulates lipid metabolism. *Nature* 2009;**458**(7242):1131–5.
224. Visser M, Goodpaster BH, Kritchevsky SB, Newman AB, Nevitt M, Rubin SM, et al. Muscle mass, muscle strength, and muscle fat infiltration as predictors of incident mobility limitations in well-functioning older persons. *J Gerontol A Biol Sci Med Sci* 2005;**60**(3):324–33.
225. Marzetti E, Lawler JM, Hiona A, Manini T, Seo AY, Leeuwenburgh C. Modulation of age-induced apoptotic signaling and cellular remodeling by exercise and calorie restriction in skeletal muscle. *Free Radic Biol Med* 2008;**44**(2):160–8.
226. Dirks AJ, Leeuwenburgh C. Aging and lifelong calorie restriction result in adaptations of skeletal muscle apoptosis repressor, apoptosis-inducing factor, X-linked inhibitor of apoptosis, caspase-3, and caspase-12. *Free Radic Biol Med* 2004;**36**(1):27–39.
227. Phillips T, Leeuwenburgh C. Muscle fiber specific apoptosis and TNF-alpha signaling in sarcopenia are attenuated by life-long calorie restriction. *Faseb J* 2005;**19**(6):668–70.
228. Siu PM, Bryner RW, Martyn JK, Alway SE. Apoptotic adaptations from exercise training in skeletal and cardiac muscles. *Faseb J* 2004;**18**(10):1150–2.
229. Song W, Kwak HB, Lawler JM. Exercise training attenuates age-induced changes in apoptotic signaling in rat skeletal muscle. *Antioxid Redox Signal* 2006;**8**(3-4):517–28.
230. Smuder AJ, Kavazis AN, Min K, Powers SK. Exercise protects against doxorubicin-induced markers of autophagy signaling in skeletal muscle. *J Appl Physiol* 2011.
231. Sandri M. Autophagy in skeletal muscle. *FEBS Lett* 2010;**584**(7):1411–6.
232. Jensen TE, Wojtaszewski JF, Richter EA. AMP-activated protein kinase in contraction regulation of skeletal muscle metabolism: necessary and/or sufficient? *Acta Physiol (Oxf)* 2009;**196**(1):155–74.
233. Ruderman NB, Park H, Kaushik VK, Dean D, Constant S, Prentki M, et al. AMPK as a metabolic switch in rat muscle, liver and adipose tissue after exercise. *Acta Physiol Scand* 2003;**178**(4):435–42.
234. He C, Bassik MC, Moresi V, Sun K, Wei Y, Zou Z, et al. Exercise-induced BCL2-regulated autophagy is required for muscle glucose homeostasis. *Nature* 2012;**481**(7382):511–5.
235. Matsakas A, Macharia R, Otto A, Elashry MI, Mouisel E, Romanello V, et al. Exercise training attenuates the hypermuscular phenotype and restores skeletal muscle function in the myostatin null mouse. *Exp Physiol* 2012;**97**(1):125–40.
236. Amthor H, Macharia R, Navarrete R, Schuelke M, Brown SC, Otto A, et al. Lack of myostatin results in excessive muscle growth but impaired force generation. *Proc Natl Acad Sci USA* 2007;**104**(6):1835–40.

237. Feng Z, Bai L, Yan J, Li Y, Shen W, Wang Y, et al. Mitochondrial dynamic remodeling in strenuous exercise-induced muscle and mitochondrial dysfunction: regulatory effects of hydroxytyrosol. *Free Radic Biol Med* 2011;**50**(10):1437–46.

238. Doroshow JH. Effect of anthracycline antibiotics on oxygen radical formation in rat heart. *Cancer Res* 1983;**43**(2):460–72.

239. Mackenzie MG, Hamilton DL, Murray JT, Baar K. mVps34 is activated by an acute bout of resistance exercise. *Biochem Soc Trans* 2007;**35**(Pt 5):1314–6.

240. Adhihetty PJ, Irrcher I, Joseph AM, Ljubicic V, Hood DA. Plasticity of skeletal muscle mitochondria in response to contractile activity. *Exp Physiol* 2003;**88**(1):99–107.

241. Dreyer HC, Fujita S, Glynn EL, Drummond MJ, Volpi E, Rasmussen BB. Resistance exercise increases leg muscle protein synthesis and mTOR signalling independent of sex. *Acta Physiol (Oxf)* 2010;**199**(1):71–81.

242. Menshikova EV, Ritov VB, Toledo FG, Ferrell RE, Goodpaster BH, Kelley DE. Effects of weight loss and physical activity on skeletal muscle mitochondrial function in obesity. *Am J Physiol Endocrinol Metab* 2005;**288**(4):E818–25.

243. Newman AB, Simonsick EM, Naydeck BL, Boudreau RM, Kritchevsky SB, Nevitt MC, et al. Association of long-distance corridor walk performance with mortality, cardiovascular disease, mobility limitation, and disability. *JAMA* 2006;**295**(17):2018–26.

Sea of Knowledge

Index

Note: Page numbers with "f" denote figures; "t" tables.

E

F

G

H

I

N

O

P

Kolibri 360 GmbH
Ferdinand-Jühlke-Straße 7
99095 Erfurt, Deutschland
produktsicherheit@kolibri360.de